Blowout and Well Control Handbook

Blowout and Well Control Handbook

By

Robert D. Grace

With Contributions By
Bob Cudd, Richard S. Carden,
and Jerald L. Shursen

G|P **Gulf Professional Publishing**
P| an imprint of Butterworth-Heinemann

An imprint of Elsevier Science
Amsterdam Boston London New York Oxford Paris San Diego
San Francisco Singapore Sydney Tokyo

Gulf Professional Publishing is an imprint of Elsevier Science

Library of Congress Cataloging-in-Publication Data

Blowout and well control handbook / by Robert D. Grace with
   contributions by Bob Cudd ... [et al.].
     p. cm.
   Includes bibliographical references and index.
   ISBN 0-7506-7708-2 (hardcover : alk. paper)
    1. Oil wells–Blowouts–Prevention. 2. Oil wells–Safety measures.
     I. Grace, Robert D.

   TN871.215.B56 2003
   622′.3382–dc21

                          2003044883

British Library Cataloguing-in-Publication Data

A catalogue record for this book is available from the British Library.

The publisher offers special discounts on bulk orders of this book.
For information, please contact:

Manager of Special Sales
Elsevier Science
200 Wheeler Road
Burlington, MA 01803
Tel: 781-313-4700
Fax: 781-313-4880

For information on all Gulf Professional Publishing titles available,
contact our World Wide Web home page at: http://www.gulfpp.com

10 9 8 7 6 5 4 3 2 1

CONTENTS

PREFACE

Well control problems are always interesting. The raw power that is released by nature in the form of an oil or gas well blowing out of control is awesome. Well control is one thing and WILD well control is something else.

There will be well control problems and wild wells as long as there are drilling operations anywhere in the world. There are some among us that think these problems are always the consequence of some error and can be eliminated. I don't think so. I've seen some that I don't think anyone could have avoided. These problems are part of the business and just go with the territory.

The consequences of failure are severe. Even the most simple blowout situation can result in the loss of millions of dollars in equipment and valuable natural resources. These situations can also result in the loss of something much more valuable—human life. Well control problems and blowouts are not particular. They occur in the operations of the very largest companies and the very smallest. They occur in the most complex operations such as deep, high-pressure gas wells, and they occur in the most simple, shallow operations. Men have lost their lives when things went wrong at surface pressures ranging from 12,000 psi to 15 psi. The potential for well control problems and blowouts is ever-present.

ACKNOWLEDGMENTS

I would like to acknowledge my contributors, Mr. Bob Cudd, Mr. Jerald L. Shursen, and Mr. Richard Carden. As a close personal friend and associate for more than 30 years, Bob Cudd contributed not only to this writing but also much more to the total experience in this work. I once reflected that Bob knew more than anyone about various aspects of well control. I later realized that he knew more than everyone else combined. Bob represents a wealth of experience, knowledge, and expertise.

I would also like to acknowledge Jerry Shursen for his contributions. As a close personal friend, business partner, and associate, Jerry and I worked very closely together to pioneer many of the concepts presented in this book. Jerry is the best drilling engineer in the industry today. There is no one I'd rather have with me in a tough situation than Jerry Shursen.

Richard Carden has been my friend and associate since his student days at Montana Tech. He is an outstanding engineer, and we have worked together on some very tough projects. Rich is technically solid and has worked diligently to contribute to this book and ensure the quality of the work.

For his inspiration, I would like to acknowledge my lifelong friend, Dr. Preston L. Moore. No one has contributed more to drilling than Preston. He was my inspiration in college more than 40 years ago. He continues to inspire me today. In the late 1960s, in Preston's world-renowned *Drilling Practices Seminars*, he and I were pioneering many of the well control concepts and techniques now considered classic in the industry.

Finally, I want to acknowledge the staff at GSM who has worked diligently and with professional pride to ensure the quality of this work. Particularly, I must mention and thank my friend, assistant, and secretary Mrs. Glennda Norman. Among many other things, she patiently read every word of this manuscript and found many mistakes after the rest of us were finished.

Blowout and Well Control Handbook

CHAPTER ONE

EQUIPMENT IN WELL CONTROL OPERATIONS

"... I could see that we were having a blowout!" Gas to the surface at 0940 hours.

0940 TO 1230 HOURS

Natural gas was at the surface on the casing side very shortly after routing the returning wellbore fluid through the degasser. The crew reported that most of the unions and the flex line were leaking. A $3\frac{1}{2}$-inch hammer union in the line between the manifold and the atmospheric-type "poor-boy" separator was leaking drilling mud and gas badly. The separator was mounted in the end of the first tank. Gas was being blown from around the bottom of the poor-boy separator. At about 1000 hours, the motors on the rig floor began to rev as a result of gas in the air intake. The crew shut down the motors.

At 1030 hours the annular preventer began leaking very badly. The upper pipe rams were closed.

1230 TO 1400 HOURS

Continuing to attempt to circulate the hole with mud and water.

1400 TO 1500 HOURS

The casing pressure continues to increase. The flow from the well is dry gas. The line between the manifold and the degasser is washing out and the leak is becoming more severe. The flow from the well is switched to the panic line. The panic line is leaking from numerous connections. Flow is to both the panic line and the separator.

The gas around the rig ignited at 1510 hours. The fire was higher than the rig. The derrick fell at 1520 hours.

This excerpt is from an actual drilling report. Well control problems are difficult without mechanical problems. With mechanical problems such as those described in this report, an otherwise routine well control problem escalates into a disastrous blowout. In areas where kicks are infrequent, it is common for contractors and operators to become complacent with poorly designed auxiliary systems. Consequently, when well control problems do occur, the support systems are inadequate, mechanical problems compound the situation, and a disaster follows.

Because this book is presented as an advanced blowout and well control operations manual, its purpose is not to present the routine discussion of blowout preventers and testing procedures. Rather, it is intended to discuss the role of equipment, which frequently contributes to the compounding of the problems. The components of the well control system and the more often encountered problems are discussed.

The saying "It will work great, if we don't need it!" applies to many well control systems. The fact is, if we don't ever need it, anything will suffice. And therein lies the root of many of the problems encountered. On a large number of rigs, the well control system has never been used and will never be needed.

Some rigs routinely encounter kicks and the crew is required to circulate out the kick using classical well control procedures. In these instances, the bare essentials will generally suffice. For most of these conditions, well site personnel need not be too concerned about how the equipment is rigged up or how tough it is.

In some parts of the industry, wells are routinely drilled underbalanced with the well flowing. In these cases, the well control system is much more critical and more attention must be paid to detail.

In a few instances, the kick gets out of control or the controlled blowout in the underbalanced operation becomes uncontrolled. Under these conditions, it is sometimes impossible to keep the best well control systems together. When it happens, every "i" must be dotted and every "t" crossed.

Unfortunately, it is not always possible to foretell when and where one of those rare instances will occur. It is easier and simpler to merely do it right the first time. Sometimes, the worst thing that can happen to us is that we get away with something we shouldn't. When we do, we are tempted to do things the same way over and over and even to see if we can get away with more. Sooner or later, it will catch up with the best of us. It is best to do it right the first time.

PRESSURE, EROSION, CORROSION, AND VIBRATION

When everything goes to hell in the proverbial hand basket, our first question should be, "How long is all this s— going to stay together?" The answer to that question is usually a function of the items listed above—pressure, erosion, corrosion, and vibration.

PRESSURE

If the well control system is rated to 10,000 psi and has been tested to 10,000 psi, I'm comfortable working up to that pressure provided none of the other three factors is contributing, though that is seldom the case. There is usually a large difference between the working pressure and test pressure for a given piece of equipment. For example, a 10,000 psi working pressure blowout preventer has a test pressure of 15,000 psi. That means the rams should operate with 10,000 psi, and under static conditions, everything should withstand 15,000 psi.

Wellheads, valves, and all other components are the same. It is easy to understand how a valve can have a "working" and a "test" pressure, but it is natural to wonder how a spool can have a "working" and a "test" pressure. Since a spool has no moving parts, it seems that the two should be the same.

VIBRATION

When things begin to vibrate, the working pressure goes down. There are no models available to predict the effect of vibration. All connections have a tendency to loosen when vibrated violently. As will be outlined in Chapter 9, at the E.N. Ross, a chicksan on a pump in line on the rig floor vibrated loose during the final kill attempt. Due to the presence of hydrogen sulfide, the gas was ignited and the rig was lost.

EROSION

Erosion of the well control system is the most serious problem normally encountered. When circulating out kicks and bubbles in routine drilling operations, erosion is not generally a factor. The exposure time is short and the velocities of the fluids are minimal. Therefore, almost any arrangement will suffice. It is usually under adverse conditions that things begin to fail. That is the reason most well control systems are inadequate for difficult conditions. Difficult conditions just do not happen that often.

In the time frame of most well control incidents, dry gas simply does not significantly erode; at least nothing harder than N-80 grade steel. At a production well blowout in North Africa, the flow rate was determined to be approximately 200 mmscf along with about 100,000 barrels of oil per day. The well flowed for almost six weeks with no significant erosion on the wellhead or flow lines. Thickness testers were used to monitor critical areas and showed no significant thickness reduction.

At a deep blowout in the southern United States, the flow rate was determined to be well over 100 mmscfpd through the $3\frac{1}{2}$-inch drillpipe by 7-inch casing annulus. The flowing surface pressure was less than 1000 psi. There was great concern about the condition of the drillpipe—would it be eroded or perhaps even severed by the flow? After about 10 days exposure, the drillpipe was recovered. At the flow cross, the drillpipe was shiny. Other than that, it was unaffected by the exposure to the flow.

Unfortunately, the industry has no guidelines for abrasion. Erosion in production equipment is well defined by API RP 14E. Although production equipment is designed for extended life and blowout systems are designed for extreme conditions over short periods of time, the API RP 14E offers insight into the problems and variables associated with the erosion of equipment under blowout conditions. This Recommended Practice relates a critical velocity to the density of the fluid being produced. The equations given by the API are as follows:

$$V_e = \frac{c}{\rho^{\frac{1}{2}}} \tag{1.1}$$

$$\rho = \frac{12{,}409\,S_l P + 2.7RS_g P}{198.7P + RTz} \tag{1.2}$$

$$A = \frac{9.35 + \frac{zRT}{21.25P}}{V_e} \tag{1.3}$$

Where:

V_e = Fluid erosional velocity, ft/sec
c = Empirical constant
   = 125 for non-continuous service
   = 100 for continuous service
ρ = Gas/liquid mixture density at operating temperature, lbs/ft^3
P = Operating pressure, psia
S_l = Average liquid specific gravity
R = Gas/liquid ratio, ft^3/bbl at standard conditions
T = Operating temperature, °Rankine
S_g = Gas specific gravity
z = Gas compressibility factor
A = Minimum cross-sectional flow area required,
   in^2/1000 bbls/day

Equations 1.1, 1.2, and 1.3 have been used to construct Figure 1.1, which has been reproduced from API RP 14E and offers insight into the factors affecting erosion. Because the velocity of a compressible fluid increases with decreasing pressure, it is assumed that the area required to avoid erosional velocities increases exponentially with decreasing pressure.

It is, however, interesting that pursuant to Figure 1.1 and Equations 1.1 through 1.3, a high gas/liquid ratio flow is more erosive than a low gas/liquid ratio flow.

The presence of solids causes the system to become unpredictable. Oil-field service companies specializing in fracture stimulation as well as those involved in slurry pipelines are very familiar with the erosional effects of solids in the presence of only liquids. Testing of surface facilities indicates that discharge lines, manifolds, and swivel joints containing elbows and short radius bends will remain intact for up to six months at a velocity of approximately 40 feet per second even at pressures up to 15,000 psi.

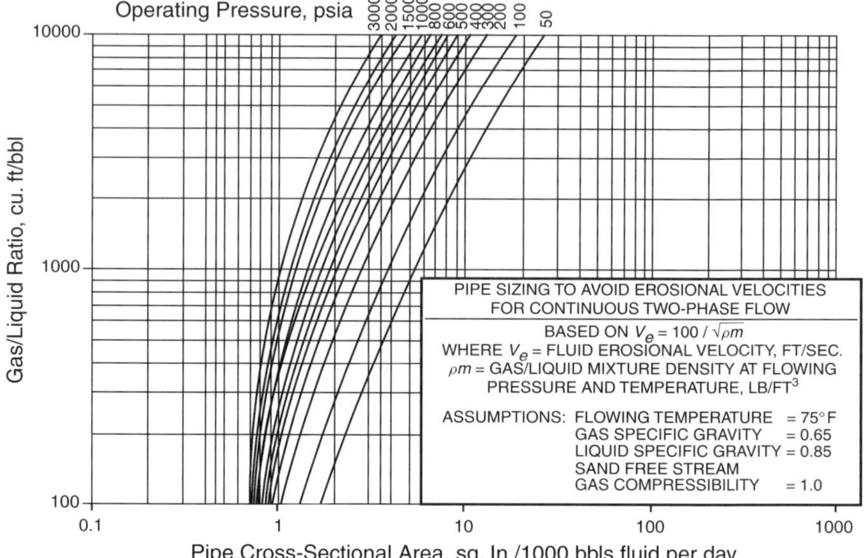

***Figure 1.1**   Erosional Velocity Chart.*

Further tests have shown that, in addition to velocity, abrasion is governed by the impingement angle or angle of impact of the slurry solids as well as the strength and ductility of the pipe and the hardness of the solids. At an impingement angle of 10 degrees or less, the erosion wear for a hard, brittle material is essentially zero. In these tests, the maximum wear rate occurred when the impingement angle was between 40 and 50 degrees. The wear rate increased when the solids in the slurry were harder than the tubular surface. Sand is slightly harder than steel. Barite is much less abrasive than hematite.

There is no authority for the erosion and wear rate when solids such as sand and barite are added to gas and drilling mud in a blowout or a well control situation. There can be little doubt that the steels are eroding under most circumstances. API RP 14E merely states that the empirical constant, c, should be reduced if sand production is anticipated.

Solids in the flow stream can be disastrous for the well control system. One case in point occurred at the Apache Key 1-11 in Wheeler County, Texas. Many called the blowout at the Key the biggest blowout in the history of the state of Texas. It was certainly one of the most baffling, spectacular, and unpredictable. With over 90 feet of Morrow sand and an open flow potential in excess of 90 mmscfpd, the Key was one of the best wells ever drilled in the Anadarko Basin. On October 4, 1981, while waiting on pipeline connection, the well inexplicably erupted, launching the well-head, 80 feet of $2\frac{7}{8}$-inch tubing, 80 feet of $7\frac{5}{8}$-inch casing, and 12 feet of $10\frac{3}{4}$-inch surface casing. The well flowed essentially unrestricted into the atmosphere.

The Key was routinely capped by the end of October. However, three days later the well cratered. All the vent lines were opened in an attempt to relieve the pressure. As illustrated in Figure 1.2, the 45-degree turns were cut out completely. In addition, a close look at Figure 1.2 shows that the $7\frac{1}{16}$-inch-by-10,000-psi valve body to the vent line had also cut out.

The capping stack was removed and the well continued to flow while subsequent well control operations were conducted. The particulate from the well had a distinctive color and was identified as coming from a zone that was originally separated from the flow stream by the tubing and two strings of casing.

At another point in the control operation at the Key, a 20-inch-by-10,000-psi stack was being rigged up to cap the well. Due to the size and weight, the stack had to be placed onto the well in sections. While bringing the second section into place, the crew noticed that the bolts in the first section were loose. The first section was removed and examined. As illustrated in Figure 1.3, the casing head was cut out beyond the rig groove. This severe erosion had taken place in the time required to make up the bolts on the spool and pick up the second section, which could have been no more than a couple of hours.

Time and again, the flow from the Key demonstrated its devastating nature. When the momentum kill procedure was implemented, several joints of $5\frac{1}{2}$-inch casing were run into the $7\frac{5}{8}$-inch casing. There was a 10,000-psi gate valve on top of the $5\frac{1}{2}$-inch casing. After the momentum kill was accomplished, the well remained dead for about one hour and then flow commenced again. The valve lasted only a few minutes. The connections of the $5\frac{1}{2}$-inch casing had been eroded to the extent that the threads on the pins were visible through the boxes. Imagine! All that from a cased hole!

The inside of all the equipment in the stack and the flow lines had to be protected with a special stellite material. Toward the end of

Figure 1.2

the well control operations, an average of 2000 cubic yards of particulate material was being separated from the flow stream and removed each month.

Wells do not have to be deep and high-pressured to demonstrate such fury. On March 12, 2000, a well located near Tabor, Alberta, Canada, got out of control, despite being only 3500 feet deep and normally pressured. The volume rate of flow was estimated to be between 20 and 40 mmscfpd.

Figure 1.3

Figure 1.4 *(Reproduced with permission of Conoco Canada Limited.)*

Within 30 minutes after the blowout, holes eroded in the choke line, filling the manifold house with gas. Later inspections revealed several holes in the choke manifold. One is shown in Figure 1.4. Within two hours, the drilling cross cut out (Figure 1.5).

Figure 1.5   *(Reproduced with permission of Conoco Canada Limited.)*

Figure 1.6   *(Reproduced with permission of Conoco Canada Limited.)*

When the wellhead was removed several days later, it was paper thin (Figure 1.6). After the well was killed by the relief well, the casing at the surface was found to be eroded and too thin to support a capping stack.

Formation solids are the usual culprits in problems with erosion. However, mud solids are also erosive. Barite will erode and hematite is even more erosive than barite.

At the aforementioned well in North Africa, a suction hose in the pump station forced a premature stoppage of a kill operation using 20-ppg mud weighted with barite and hematite. The well was dead when the hose failed. However, there was insufficient hydrostatic to control the bottomhole pressure and the well kicked off. Within minutes after the kill mud reached the surface, the side of the wellhead cut out.

Since erosion is a function of velocity, it should occur first at the perimeter of the well control system. Usually, turns in the flow path are the first to cut out. Anywhere the flow turns is a candidate for erosion. In blowout situations, it is prudent to begin monitoring wall thickness at all turns at the wellhead and downstream from the wellhead. Thickness testers are readily available to the industry and make the job very simple.

CORROSION

Most processes of corrosion require so much time they are not usually significant in well control operations. However, in well control operations, there are basically two corrosive elements that must be considered—hydrogen sulfide and carbon dioxide.

At the Lodge Pole blowout west of Edmonton, Alberta, Canada, in the mid 1980s, hydrogen sulfide played a disastrous role. The well blew out while on a trip. The decision was made to drop the drillpipe and close the blind rams. The hydrogen sulfide had caused the drillpipe to part not far beneath the rig floor. Consequently, when the pipe was cut, it didn't drop. Rather, it went out the derrick and sparks ignited the well.

The effect of hydrogen sulfide on steel is documented by NACE. The NACE guidelines define the working limits of oil field tubulars. These guidelines are very dependable. At a blowout in the southern part of the United States, hydrogen sulfide was a concern only if the surface temperature fell below 200 degrees Fahrenheit. Throughout the project, the surface temperature was maintained above 200 degrees Fahrenheit by adjusting the flow rate. No problems with the hydrogen sulfide were experienced.

Carbon dioxide corrosion takes place over a much longer period of time than hydrogen sulfide corrosion, which is usually very sudden. The exposure time required for CO_2 to cause problems is generally so long that it is not a factor in normal well control operations, which last only a few weeks at the most. However, the potential for CO_2 corrosion cannot be ignored. The environment required for carbon dioxide corrosion is not as predictable as that required for H_2S.

The general requirement for CO_2 corrosion is based on the partial pressure. The partial pressure is the mole fraction multiplied by the total pressure. For most gases, the mole fraction is approximately equal to the volume fraction. If the partial pressure to CO_2 is greater than 30 psi, corrosion is certain. If the partial pressure to CO_2 is between 7 and 30 psi, corrosion is likely. If the partial pressure to CO_2 is less than 7 psi, corrosion is unlikely.

At the TXO Marshall, the presence of CO_2 complicated the original work. The well control work was similarly plagued. When the blowout preventer was removed after the well was killed, the body was almost completely corroded and eroded away. One observer described its condition as resembling a spider web.

THREADED CONNECTIONS

I made a promise that I would address the issue of threaded connections in this book. So here it is. Threaded connections are okay—for sewer line connections, sprinkler systems, and household plumbing. But they have no place in the well control system. I have learned a lot of things from Bob Cudd over the last 35 years. One of the most important is this: In well control, Bob always advised to play only "aces, straights, and flushes." Threaded connections are none of these.

Sometime in the 1960s, I had the opportunity to investigate my first well control accident. A major company had a well control problem at a shallow well in the Oklahoma Panhandle. Two men were severely burned. One of them died. They were trying to kill the well by pumping into the annulus through a gate valve on a $2\frac{3}{8}$-inch double extra heavy nipple with 11V threads. My recollection is that the nipple was rated to 5000 psi. Due to casting problems, the threads only made up about halfway. The fact was that the nipple failed at about 1500 psi.

The 11V thread is not a good thread form. The "V" configuration allows stress concentrations at the base of the "V". In addition, the thread is tapered, which means that wall thickness is sacrificed to cut the thread. If not completely buried, the connection will be weakest at the last engaged thread. When I made all the calculations for strength reduction due to the reduced wall thickness, water hammering, etc., they indicated that the strength of the 5000-psi working pressure nipple was reduced by the conditions to approximately 1500 psi—almost exactly where it failed.

At another job, a 10V thread failed and several people were killed. More recently I investigated another accident where a threaded choke on the rig floor fill-up line failed and injured a man. It was the same scenario. The 11V thread was not fully made up and failed at far below the rated working pressure. Fortunately, no one was killed.

API Round threads are not much better. At a location on the edge of Perryton, Texas, the dual tree had 8 Rd connections. We were lubricating a plug into the wellhead one Saturday morning. The plug wedged in the tree above its intended setting location. The operator bled the lubricator. When he did, the plug came loose, and the impact severed the connection below the bottom master valve. That's as close as I ever came to being killed.

It is true that service companies routinely use threaded connections on high pressure equipment used in stimulation work. The difference is that trained personnel handle the connections. Around the rig, untrained roughnecks and roustabouts do the work.

It is also true that in the past I have personally built many choke manifolds using ball valves and threaded connections downstream from the chokes. There were two reasons. One, I never allowed pressure downstream from the choke, and two, I was there to supervise the installation. There should never be pressure downstream from the choke in a choke manifold. After all, the separator is usually a 300-psi-or-less vessel. But strange things happen when people are excited. I watched a roughneck try to close a ball valve downstream from a choke during a kill operation. Had he done so, it probably would have killed him. Fortunately for everyone, a ball valve is almost impossible to close if there is significant flow.

THE STACK

Interestingly, the industry doesn't experience many failures within the blowout preventer stack itself. There was one instance in Wyoming where a blowout preventer failed because of a casting problem. In another case, the 5000 psi annular failed at 7800 psi. In general, the stack components are very good and very reliable.

A problem that is continually observed is that the equipment doesn't function when needed. At a well at Canadian, Texas, the annular preventer had been closed on a blowout, but the accumulator would not maintain pressure. Shursen and another guy were standing on the rig floor when the accumulator lost pressure and the annular preventer opened unexpectedly. The annular opened so quickly that the floor was engulfed in a fireball. Fortunately, no one was seriously burned. The source of the fire was never determined. The rig had been completely shut down, but the accumulator system should have been in working order.

At another blowout in Arkansas, nothing worked. The accumulator wasn't rigged up properly, the ancient annular wouldn't work and, when the pipe rams were closed, the ram blocks fell off the transport arms. After that, the rams couldn't be opened. It's difficult to believe that this equipment was operated and tested as often as the reports indicated.

In another instance, the stack was to be tested prior to drilling the productive interval. The reports showed that the stack had been tested to the full working pressure of 5000 psi. After failing the test, it was found that the bolt holes had rusted out in the preventer!

These situations are not unique to one particular area of the world. Rather, they are common throughout the oil fields of the world. Operators should test and operate the components of the stack to be confident that they are functioning properly.

Having a remote accumulator away from any other part of the rig is a good idea. At one location, the accumulator was next to the mud pumps. The well pressured up and blew the vibrator hose between the mud pump and the stand pipe. The first thing that burned was the accumulator. A mud cross such as illustrated in Figure 1.7 would have saved that rig. It would have been possible to vent the well through the panic line and pump through the kill line and kill the well.

This chapter pertains to all operations, whether they are offshore, onshore, remote, or in the middle of a city. Some peculiarities persist that require special considerations. For example, all the equipment in an offshore operation is confined to a small space. However, it is important to

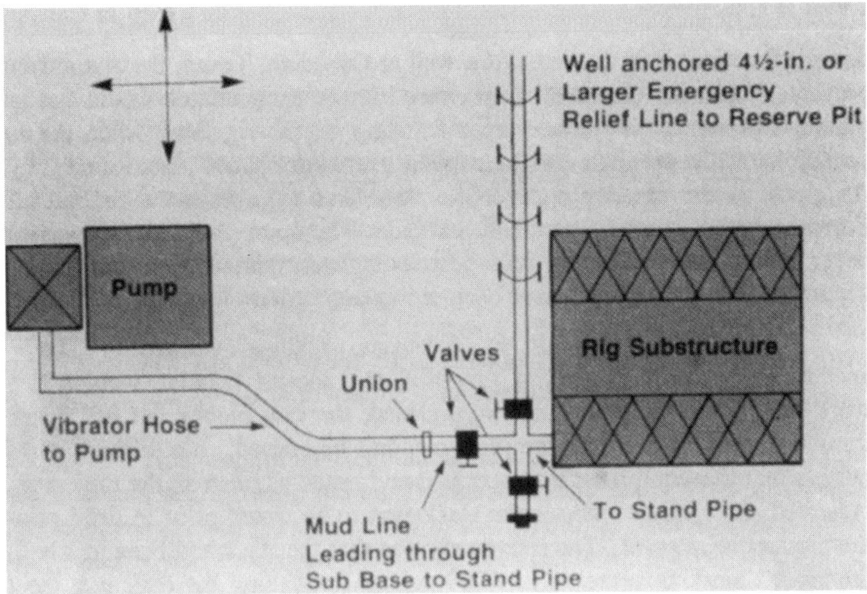

Figure 1.7

remember that a well is deep and dumb and doesn't know where it is or that there is some quantity of water below the rig floor. Therefore, when sacrifices are made and compromises are accepted due to self-imposed space limitations, serious consequences can result.

THE CHOKE LINE

Many well control problems begin in the choke line or downstream of the choke line. It is unusual to find a rig without the potential for a serious problem between the blowout preventer (BOP) stack and the end of the flare lines. In order to appreciate how a choke line must be constructed, it is necessary to remember that, in a well control situation, solids-laden fluids are extremely abrasive.

A typical choke line is shown in Figure 1.8. As illustrated, two valves are flanged to the drilling spool. There are outlets on the body of the blowout preventers. However, these outlets should not be used on a routine basis since severe body wear and erosion may result.

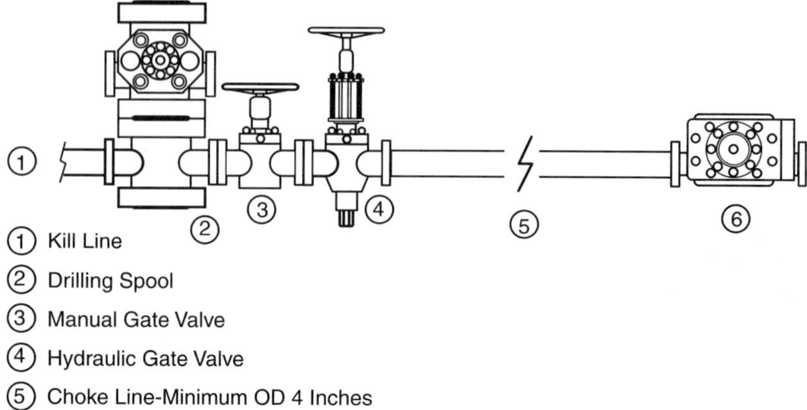

Figure 1.8   Choke Line.

One valve is hydraulically operated while the other is the backup or safety valve. The position of the hydraulic valve is important. Most often it is outboard with a safety valve next to the spool to be used only if the hydraulic valve fails to operate properly. Many operators put the hydraulic valve inboard of the safety valve. Experience has shown that the short interval between the wellbore and the valve can become plugged with drill solids or barite during the normal course of drilling. Therefore, when a problem does occur, the manifold is inoperable due to plugging. The problem can be minimized and often eliminated by placing the hydraulic valve next to the casing pool.

The outboard position for the hydraulic valve is the better choice under most circumstances since the inboard valve is always the safety valve. If the hydraulic valve is outboard, it is important that the system be checked and flushed regularly to insure that the choke line is not obstructed with drill solids.

In areas where underbalanced drilling is routine, such as West Texas, drilling with gas influx is normal and the wear on well control equipment can be a serious problem. In these areas, it is not uncommon to have more than one choke line to the manifold. The theory is sound. A backup choke line, in the event that the primary line washes out or is plugged, is an excellent approach.

A basic rule in well control is to have redundant systems where a failure in a single piece of equipment does not mean disaster for the operation. However, the second choke line must be as substantial and reliable as the primary choke line. In one instance, the secondary choke line was a 2-inch line from the braden head. The primary choke line failed and the secondary line failed even faster. Since the secondary line was on the braden head with no BOP below, the well blew out under the substructure, caught fire, and burned the rig.

Therefore, the secondary choke line should come from the kill-line side or from a secondary drilling spool below an additional pipe ram. In addition, it must meet the same specifications for dimension and pressure as the primary choke line.

The choke line from the valves to the choke manifold is a constant problem. This line must be flanged, have a minimum outside diameter of 4 inches, and should be STRAIGHT between the stack and the manifold. Any bends, curves, or angles are very likely to erode. When that happens, well control becomes very difficult, lost, extremely hazardous, or all of the aforementioned. Just remember, STRAIGHT and no threaded connections.

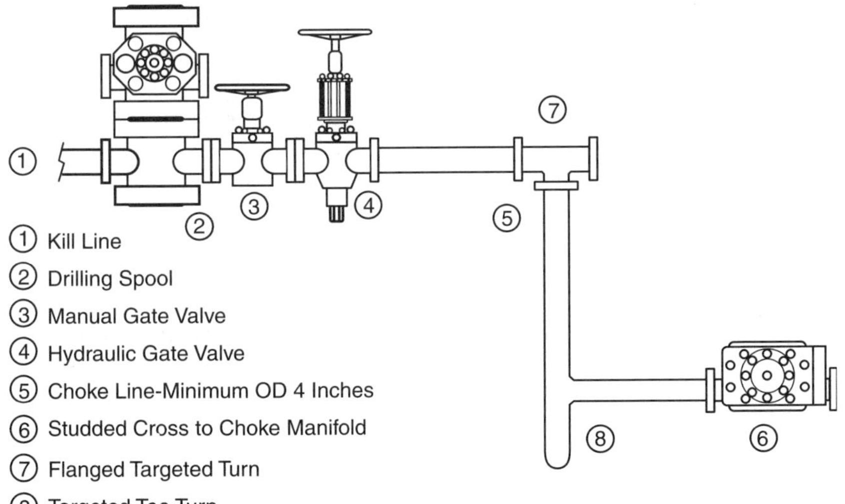

Figure 1.9   *Choke Line with Turns.*

Figure 1.10

If turns in the choke line are required, they should be made with T's and targets as illustrated in Figure 1.9. The targets must be filled with Babot and deep enough to withstand erosion. The direction of flow must be into the target.

Figure 1.10 illustrates an improperly constructed choke line. Note that the choke line is bent slightly. In addition, the targets are backwards or with the flow from the well. The direction of the targets is a common point of misunderstanding throughout the worldwide industry. These points should be checked on all operations.

Continuous, straight steel lines are the preferred choke lines. Swivel joints should only be used in fracturing and cementing operations and should not be used in a choke line or any well control operation. At a deep, high-pressure sour well in southern Mississippi, a hammer union failed and the rig burned.

Finally, the use of hoses has become more popular in recent years. Hoses are quick and convenient to install. However, hoses are recommended only in floating drilling operations which offer no alternative. Further, consider that, in the two most serious well control problems in the North Sea to date, hose failure was the root cause.

Hoses and swivel joints work well on many wells because serious well control problems do not occur on many wells. However, when serious well control problems do occur, solid equipment has better integrity. Swivel joints can be used on the pumping side in kill operations for short periods of time. As of this writing, hose use should be restricted to the suction side of the pumping equipment. Presently, hoses cannot be recommended to replace choke lines. Although the literature is compelling, it is illogical to conclude that rubber is harder than steel.

In summary, the choke line should be straight and no less than 4 inches in diameter. Hoses and threaded connections should not be used. Any turns should be targeted as illustrated. Finally, know this: If things get rough enough—that is, if there are lots of solids in the flow stream—there will be erosion everywhere the flow turns, including the drilling cross. The erosion will be on the outside of the turn immediately downstream.

THE CHOKE MANIFOLD

THE VALVES

The minimum requirement for a choke manifold is presented in Figure 1.11. I have seen some real pieces of garbage rigged up as choke manifolds. One such choke manifold is illustrated in Figure 1.12. It would be better, in most instances, to have only a panic line and no choke manifold than to have one like that shown in Figure 1.12.

With a manifold such as that shown in Figure 1.11, the outboard valves are routinely used and the inboard valves are reserved for an

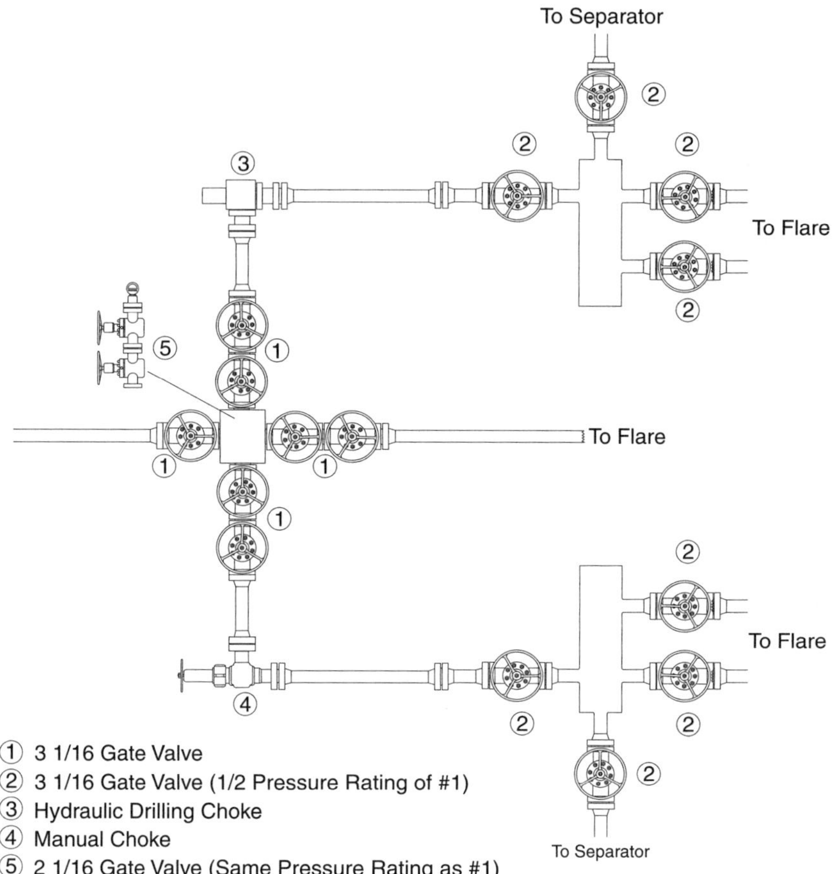

Figure 1.11   *The Choke Manifold.*

unexpected emergency or when the outboard valve simply fails to seal. In general, gate valves can endure considerable wear and tear and still function with sufficient seal to permit the installation of another valve, if necessary. Gate valves are designed to be open or closed. They are not intended to be used to merely restrict flow. That is the function of the choke. When used to restrict flow, the body of the valve is likely to cut out. Gate valves are often closed on flow. Absent unusually large volumes of sand or other abrasives in the flow, the valve will close and seal on almost any volume at any pressure.

Figure 1.12

I have rigged up many choke manifolds with gate valves upstream from the chokes and ball valves downstream from the chokes. I used these manifolds in rough conditions and they worked without problems or failures. However, a ball valve cannot be closed on any significant flow. For that reason alone, I no longer build manifolds using ball valves.

The general standard in the industry is for the valves downstream from the chokes to have a pressure rating of at least one half that of the valves upstream from the chokes. For example, by industry custom, a 10,000-psi choke manifold will have 10,000-psi working pressure valves upstream from the chokes and 5000-psi working pressure valves downstream from the chokes. That was not always the standard and I have never been able to determine any justification for that practice.

Downstream from the chokes, the pressure can never be more than a few hundred psi. The pressure is atmospheric in the mud pits, downstream from the separator, and at the end of the flare lines. The working pressure of the separator is never more than a couple hundred psi. Therefore, how could the pressure anywhere beyond the chokes ever reach 5000 psi? The answer is that it cannot. The choice of 5000 psi working pressure downstream from the chokes of a 10,000-psi manifold is arbitrary.

It is necessary for the plumbing downstream from the chokes to be substantial with respect to wall thickness. Remember, as the velocity increases, the tendency to erode also increases dramatically. Therefore, the most likely place for erosion is downstream from the choke. Everything downstream from the choke should be consistent with internal diameter. Any changes in internal diameter will increase the potential for erosion.

As the well becomes more complex and the probability of well control problems increases, redundancy in the manifold becomes a necessity. The manifold shown in Figure 1.13 was recently rigged up on a well control problem in the South Texas Gulf Coast. As illustrated in Figure 1.13, there were positions for four chokes on the manifold. Options were good. Either side of the manifold could be the primary side. Since each side was separately manifolded to individual separators, there was redundancy

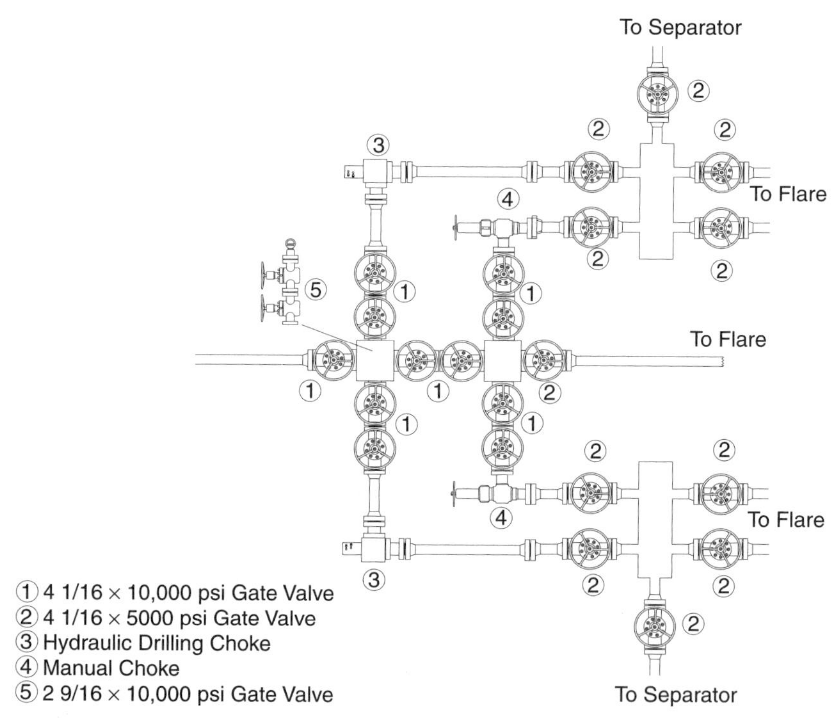

Figure 1.13 $4\text{-}\frac{1}{16}$-by-10,000-psi Choke Manifold.

for every system in the manifold. Failure of any single component of the manifold would not jeopardize the operation.

THE DRILLING CHOKE

Outboard of the gate valves are the drilling chokes. The drilling choke is the heart of the well control system. Well control was not routinely possible prior to the advent of the modern drilling choke. Positive chokes and production chokes were not designed for well control operations and are not tough enough. Production chokes and positive chokes should not be included in the choke manifold unless there is a specific production-well testing requirement. Under well control conditions, the best production choke can cut out in a few minutes. If the manifold is spaced for a production choke, a welder will be required to install an additional drilling choke if it is needed. Obviously, it is not good practice to weld on a manifold when well control operations are in progress. If a production choke is used in a manifold system, it is recommended that the system be spaced so that an additional drilling choke can readily fit. Even with this precaution, there remains the problem of testing newly installed equipment with the well control operation in progress.

In the late 1950s a company known as Drilling Well Control began controlling kicks with a series of skid-mounted separators. Prior to this technology, it was not possible to hold pressure on the annulus to control the well. It was awkward at best with the separators. The system was normally composed of two to three separators, depending on the anticipated annulus pressures. The annulus pressure was stepped down through the separators in an effort to maintain a specific annular pressure. It represented a significant step forward in technology but pointed to the need for a drilling choke that could withstand the erosion resulting from solids-laden multiphase flow.

The first effort was affectionately known as the "horse's ass" choke. The nickname will be apparent as its working mechanism is understood. The choke is illustrated in Figure 1.14. Basically, it was an annular blowout preventer turned on its side. The flow stream entered the rubber bladder. The pressure on the well was maintained by pressuring the back side of the bladder. The solids would cut out the bladder, requiring that the operator pay close attention to the pressures everywhere at the same time, making it an awkward choke to use.

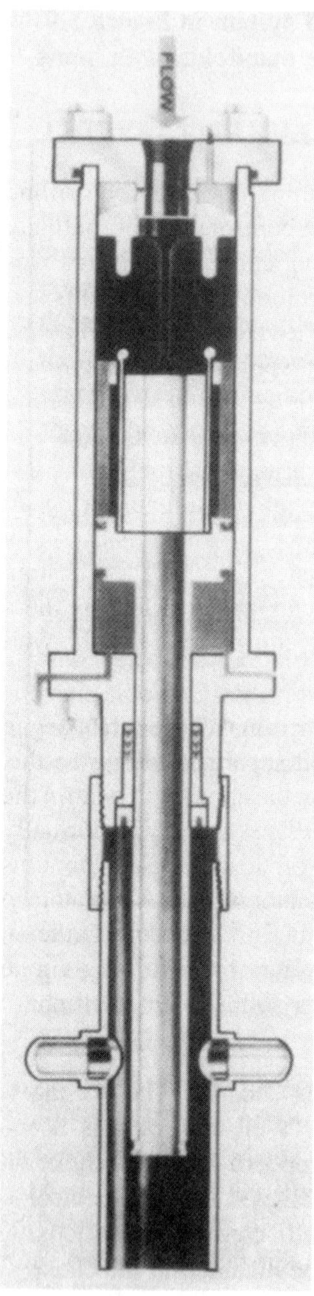

Figure 1.14

All modern drilling chokes fall into two categories according to the manner in which choking is achieved. In the first category, choking is achieved by rotating two surfaces with matching openings. Some use flat plates and others use cylinders or cages. The SWACO Super Choke shown in Figure 1.15 is typical of this category. Within the flow path of the SWACO Super Choke, there are two polished tungsten carbide plates. Both have half-moon-shaped openings. One plate is fixed and the other is hydraulically rotated. As one half-moon opening rotates over the opening in the fixed plate, flow is permitted. Choking is accomplished by reducing the size of the matched openings.

The other category is typified by the Cameron Drilling Choke shown in Figure 1.16. In this category, a tungsten carbide bean is hydraulically inserted into a tungsten carbide sleeve. The action is very similar to the typical production choke. The degree of choking is a function of the depth of the cylinder into the sleeve.

It is important to fully understand the operating system of the choke being used. Can it be operated without rig power? Can it be tested to full working pressure? Is there more than one operating station? If the hydraulic system in the operating station is connected to the choke by long hoses, is the choke compromised? Is it difficult and/or time-consuming to replace the choking mechanism?

Whichever choke one decides to use, most modern drilling chokes are very reliable.

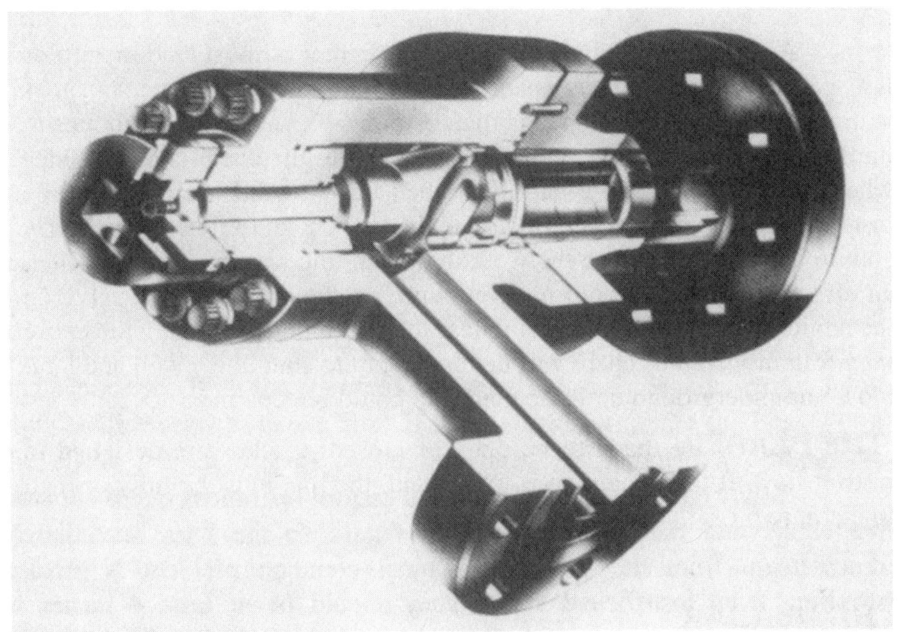

Figure 1.15

Figure 1.16

THE PANIC LINE

As illustrated in Figure 1.13, in the center of most land manifolds is a "panic line." This line is usually 4 inches or larger in diameter and goes straight to a flare pit. The idea is that, if the well condition deteriorates to intolerable conditions, the well can be vented to the pit. It is a good idea when properly used and a bad idea when misused. For example, in one instance the rig crew could not get the drilling choke to function properly and, in an effort to relieve the well, the panic line was opened. However, an effort was made to hold back pressure on the well with the valve on the panic line. The valve cut out in less than 30 minutes, making the entire manifold inoperable. There was no choice but to shut in the well and let it blow out underground until the manifold could be repaired.

Valves are made to be open or closed. Chokes are designed to restrict flow. If the panic line is to be used, the line must be fully opened to the flare pit.

THE HEADER

In the typical oil field choke manifold, the flow lines leaving the chokes are manifolded together at the "header." The idea is that from the "header" the flow can be directed from one or both chokes to the separator or to the flare pit.

Some headers are substantially constructed and are heavier than cannon barrels. However, most oil field headers are constructed from a discarded piece of casing. In general, most headers defeat the purpose and effectiveness of the well control system. In my opinion, the header is the heart of many problems and should be eliminated.

Consider Figures 1.17 and 1.18, which illustrate a header design common to many operations. This system failed for two reasons. The lines between the choke manifold and the header were 2 inches outside diameter, which resulted in excessive velocities, and the back side of the header was eroded away by the jetting action of the flow stream as it entered the header and expanded. Obviously, the well was out of control and had to be either vented or shut in to blow out underground, depending on the surface pressure and casing design. In most instances, the well is shut in and an underground blowout is the result.

Figure 1.17

Figure 1.18

If the "panic line" is also manifolded into the header, which is often the case, almost anything else that can happen is destined to be disastrous. Usually, the header is near the rig. If there is a failure, continuing to flow the well is obviously dangerous. If the flow is at the header, there is eminent danger of fire and total loss of the rig. If the decision is made to shut in and not flow the well, the casing could be ruptured or an underground blowout could ensue. There is no good reason to have the header when all the options can be maintained with alternative plumbing. Two manifolds without a common header are shown in Figures 1.11 and 1.13.

Many equipment failures in well control operations occur between the choke and the separator. With regard to the lines immediately downstream from the chokes, the most common problem is erosion resulting from insufficient size. They should be at least 4 inches in diameter in order to minimize the velocity. Any abrupt change in the diameter will cause an area of more severe erosion. These lines do not have to be high pressure; however, they should have a yield strength of 80,000 psi or greater in order that the steel will be harder than most of the particulate in the flow stream and resist erosion. Two-inch lines are too small for any operation and should not be used.

THE SEPARATOR

When well control operations are required, redundancy is a criterion for all components of the well control system. In most operations, there is no redundancy in the components downstream from the chokes, and if a failure occurs, the operation is in serious jeopardy. Therefore, the integrity of the components lacking redundancy is important and demands critical attention.

The lines connecting to the separator are often problematic. The fluid velocity downstream from the choke is higher. The drilling fluid, gas, barite, and drill solids at high velocities combine to produce a very effective cutter. Under severe conditions, sweeping turns can and will cut out in a matter of a few minutes. Even very slight bends, which are barely noticeable, have been known to erode completely in a few minutes.

These lines connecting the chokes to the separator must be straight if possible. If straight lines are possible, targeted turns are mandatory. A connection such as that shown in Figure 1.19, which was being used in a deep,

Figure 1.19

high pressure offshore operation, will erode in a few minutes under severe service. Hoses and swivel joints such as chicksans are not recommended.

The connecting lines should have an outside diameter equal to or greater than 4 inches. In the prolific Tuscaloosa trend, the outside diameter of the line from the header to the separator is routinely 8 inches. Even an 8-inch line must utilize targeted turns if turns are required. At one operation, a sweep turn in the 8-inch line from the manifold to the separator was completely washed out in only a few hours.

The separator routinely is a source of problems. The separator problems most frequently encountered are due to size, inability to adequately control the liquid level, and erosion of the body. The separator should be sized to adequately accommodate the anticipated gas volumes. The volume of gas the separator will accommodate is a function of the physical size of the vessel, maximum separator working pressure, and flare line size.

More often, the separator is too small and poorly designed. In offshore operations, space is always limited and the separator is easily

Figure 1.20

neglected. As illustrated in Figure 1.20, a typical offshore separator is small, improperly plumbed, and inaccessible. In land operations, where well control problems are rarely encountered, the vessels represented to be separators are commonly a bad joke, as is the manifold system as a whole, which is often incapable of performing as intended. A typical example is shown in Figure 1.21. These "separators" were simply constructed from a discarded piece of casing or drive pipe. In most of these cases, it would be better if there were no separator or manifold system and

Figure 1.21

the only alternative available to the crew was to shut in the well and wait for professionals.

All separators intended for use in well control operations should be BIG! No vessel should be smaller than 4 feet in diameter and 8 feet in height. Separators used in operations where production is prolific, such as the Tuscaloosa Trend, may be as large as 6 feet in diameter and 25 feet in length.

Separator size is also expressed as a function of the operating pressure of the vessel and the volume of gas and fluid the system can accommodate. Generally, the operating pressure of the vessel is approximately 125 psi. However, the operating pressure of the system may be limited by the liquid level control mechanism. A positive liquid level control is recommended and, if utilized, will make the system much more reliable.

However, it is often the case that the liquid level in the separator is simply controlled by a hydrostatic column of mud. Sometimes the separator vessel is simply immersed a few feet into the mud pit. In other cases, a hydrostatic riser controls the liquid level. Therefore, when the pressure in the vessel exceeds the hydrostatic pressure of the column controlling the liquid level, gas and reservoir fluids will pass out the bottom of the separator and into the mud pits. Such a scenario is common with separators designed as described and the results are unacceptably hazardous.

Consider the separator pictured in Figure 1.20. The liquid level in the separator is controlled by the hydrostatic loop in the left center of the figure. This loop is approximately 3 feet in length. Therefore, when the pressure in the vessel exceeds a maximum 3 psi, gas in the separator will pass directly to the mud room. In this case, the mud room had minimal ventilation. Any serious well control operation would not be possible with this arrangement.

The size and length of the flare line from the separator to the flare pit are important components of the system. The larger the flare line, the better. The flare line must be straight and not less than 6 inches in diameter. In the prolific Tuscaloosa Trend, 12-inch flare lines are common. The flare line should be no less than 100 yards in length, and the flare pit should not be adjacent to any roads or buildings. In one instance, the flare line terminated no more than 30 yards from the rig floor. A flare of any size would make access to the rig floor impossible. In another instance, the flare pit was adjacent to the only access road. When the well blew out, a new road had to be constructed before well control operations could be commenced.

Finally, fluid entry into the vessel is important. Some separators are designed to permit tangential entry. That is, the fluid enters tangential to the wall of the vessel. If the fluid is only gas and liquid, the design is satisfactory. However, if the fluid contains any solids, the vessel will erode rapidly. A perpendicular entry is best. An acceptable separator design is illustrated in Figure 1.22.

THE KILL LINE

The purpose of the kill line (see Figure 1.23) is to provide remote hydraulic access to the well. Usually, the kill line extends approximately 100 to 150 feet from the well head. Some operators have a valve at the end of the kill line.

The kill line access to the bore hole should never be used for any purpose other than that intended—emergency access. The line from the kill line check valve to the outer access should be straight. The valve at the end of the kill line is optional.

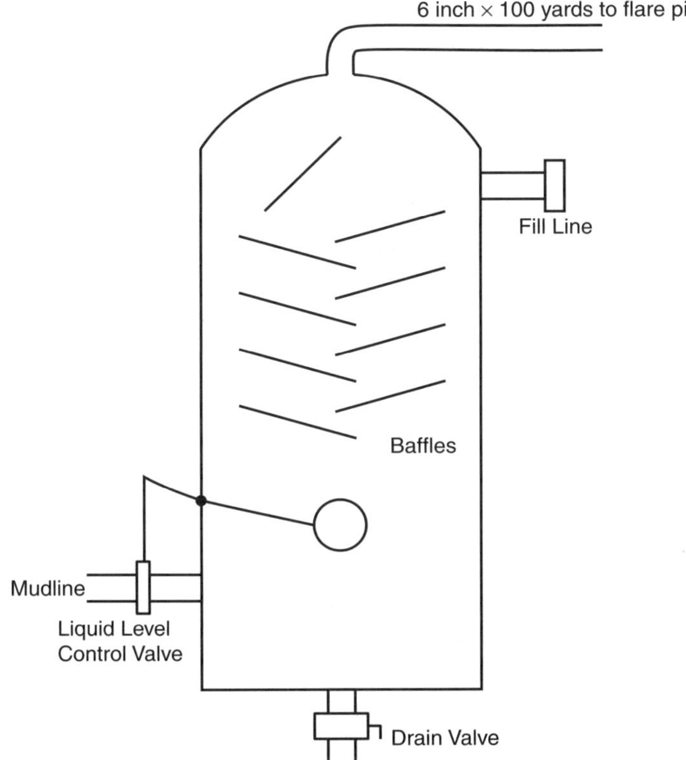

Figure 1.22   *Separator.*

The kill line access should never be used as a fill-up line. At one location, a fill-up line had been connected to the kill line access. When a kick was taken and the well was shut in, the subsequent pressure ruptured the fill-up line. The flow ignited and the rig was lost. At many locations, the kill line has provided the intended access to the wellbore and the well has been saved. The integrity of the kill line system can be assured by using the kill line only as intended.

THE STABBING VALVE

On all rigs and work-over operations there should be a valve readily accessible on the rig floor which is adapted to the tubulars. Such a valve

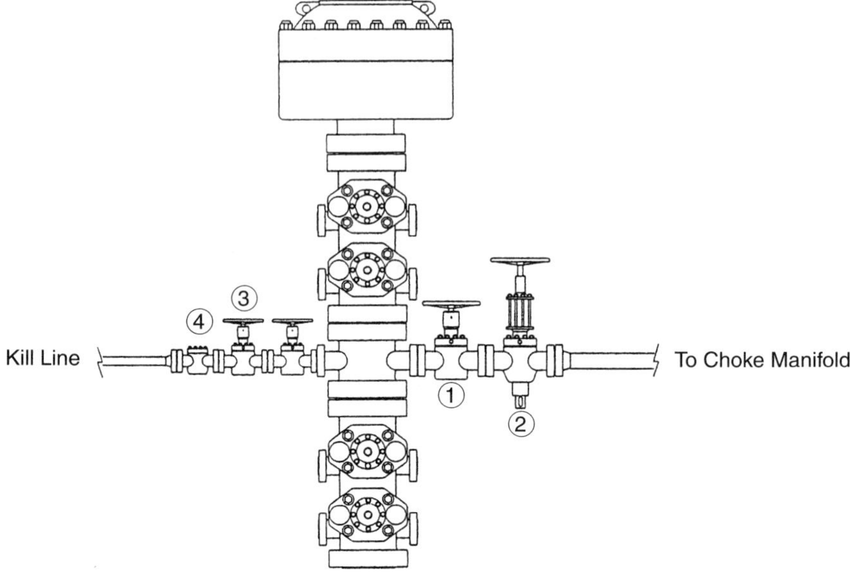

1. 4 1/16 × 10,000 psi Gate Valve
2. 4 1/16 × 10,000 psi Hydraulic Gate Valve
3. 2 9/16 × 10,000 psi Gate Valve
4. 10,000 psi Check Valve

Figure 1.23   10,000 psi Kill Line.

is routinely called a "stabbing valve." In the event of a kick during a trip, the stabbing valve is installed in the connection on the rig floor and closed to prevent flow through the drill string. Ordinarily, the stabbing valve is a ball valve, as illustrated in Figure 1.24.

The ball valve is probably the best alternative available under most circumstances. However, the valve is extremely difficult and problematic to operate under pressure. If the valve is closed and pressure is permitted to build under it, it becomes impossible to open the valve without equalizing the pressure. If the pressure below the valve is unknown, the pressure above the valve must be increased in small increments in order to equalize and open the valve.

The stabbing valve must have an internal diameter equal to or greater than the tubulars below the valve and the internal diameter of the

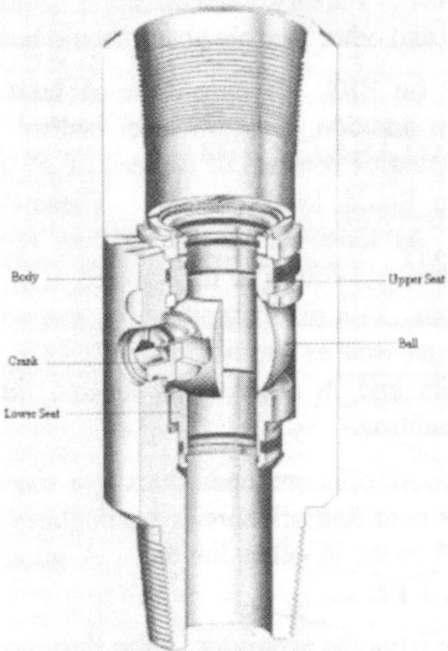

Figure 1.24   *TIW Valve.*

valve must be documented. Too often, the internal diameter is too small, unknown, and undocumented. Therefore, when a well control problem does occur, the first operation is to freeze the stabbing valve and replace it with a valve suitable for the job.

If the flow through the drill string is significant, it may not be possible to close the stabbing valve. In many instances, when a flow is observed, the blowout preventers are closed before the stabbing valve. With all of the flow through the drill string, it may not be possible to close the stabbing valve. In that instance, it is necessary to open the blowout preventers, attempt to close the stabbing valve, and re-close the blowout preventers. This discussion emphasizes the accessibility of the stabbing valve and the necessity of conducting drills to ensure that the stabbing valve can be readily installed and operated.

Finally, when the stabbing valve cannot be installed or closed and shear rams have been included in the stack, there is a tendency to activate the shear rams and sever the drill string. This option should be considered carefully. In most instances, shearing the drill string will result in the loss of the well. The energy released in a deep well, when the drill string is suddenly severed at the surface, is unimaginable and can be several times that required to cause the drill string to destroy itself along with the casing in the well. When that happens, chances for control from the surface are significantly reduced.

CHAPTER TWO

CLASSIC PRESSURE CONTROL PROCEDURES WHILE DRILLING

24 September

0600–0630	*Service rig.*
0630–2130	*Drilling 12,855 feet to 13,126 feet for 271 feet.*
2130–2200	*Pick up and check for flow. Well flowing. Shut well in.*
2200–2300	*Pump 35 barrels down drillpipe. Unable to fill drillpipe.*
2300–2330	*Observe well. 1000 psi on casing. 0 psi on drillpipe.*
2330–0030	*Pump 170 barrels down drillpipe using Barrel In–Barrel Out Method. Could not fill drillpipe. 0 psi on drillpipe. 1200 psi on casing at chokepanel. Shut pump off. Check gauge on choke manifold. Casing pressure 4000 psi at choke manifold. Choke panel gauge pegged at 1200 psi. Well out of control.*

This drilling report was taken from a recent blowout in the South Texas Gulf Coast. All of the men on the rig had been to well control school and were Minerals Management Service (MMS) certified. After the kick, it was decided to displace the influx using the Barrel In–Barrel Out Method. As a result, control was lost completely and an underground blowout followed.

Prior to 1960, the most common method of well control was known as the Constant Pit Level Method or the Barrel In–Barrel Out Method. However, it was realized that if the influx was anything other than water, this method would be catastrophic. Consequently, classical pressure control procedures were developed. It is incredible that even today there are those in the field who continue to use the older, antiquated methods.

Ironically, there are instances when these methods are appropriate and the classical methods are not. It is equally incredible that in some instances classical procedures are applied to situations which are completely inappropriate. If the actual situation is not approximated by the theoretical models used in the development of the classical procedures, the classical procedures are not appropriate. There is an obvious general lack of understanding. It is the purpose of this chapter to establish firmly the theoretical basis for the classical procedures as well as describe the classical procedures. The application of the theory must be strictly followed in the displacement procedure.

CAUSES OF WELL KICKS AND BLOWOUTS

A kick or blowout may result from one of the following:

1. Mud weight less than formation pore pressure
2. Failure to keep the hole full while tripping
3. Swabbing while tripping
4. Lost circulation
5. Mud cut by gas, water, or oil

MUD WEIGHT LESS THAN FORMATION PORE PRESSURE

There has been an emphasis on drilling with mud weights very near to and, in some instances, below formation pore pressures in order to maximize penetration rates. It has been a practice in some areas to take a kick to determine specific pore pressures and reservoir fluid composition. In areas where formation productivity is historically low (roughly less than 1 million standard cubic feet per day without stimulation), operators often drill with mud hydrostatics below the pore pressures.

Mud weight requirements are not always known for certain areas. The ability of the industry to predict formation pressures has improved in recent years and is sophisticated. However, a North Sea wildcat was recently 9 pounds per gallon overbalanced while several development wells in Central America were routinely 2 pounds per gallon underbalanced. Both used the very latest techniques to predict pore pressure while drilling. Many areas are plagued by abnormally pressured, shallow gas sands. Geologic correlation is always subject to interpretation and particularly difficult around salt domes.

FAILURE TO KEEP THE HOLE FULL AND SWABBING WHILE TRIPPING

Failure to keep the hole full and swabbing is one of the most frequent causes of well control problems in drilling. This problem is discussed in depth in Chapter 3.

LOST CIRCULATION

If returns are lost, the resulting loss of hydrostatic pressure will cause any permeable formation containing greater pressures to flow into the wellbore. If the top of the drilling fluid is not visible from the surface, as is the case in many instances, the kick may go unnoticed for some time. This can result in an extremely difficult well control situation.

One defense in these cases is to attempt to fill the hole with water in order that the well may be observed. Usually, if an underground flow is occurring, pressure and hydrocarbons will migrate to the surface within a few hours. In many areas it is forbidden to trip out of the hole without returns to the surface. In any instance, tripping out of the hole without mud at the surface should be done with extreme caution and care, giving consideration to pumping down the annulus while tripping.

MUD CUT

Gas-cut mud has always been considered a warning signal, but not necessarily a serious problem. Calculations demonstrate that severely gas-cut mud causes modest reductions in bottomhole pressures because of the compressibility of the gas. An incompressible fluid such as oil or water can cause more severe reductions in total hydrostatic and has caused serious well control problems when a productive oil or gas zone is present.

INDICATIONS OF A WELL KICK

Early warning signals are as follows:

1.   Sudden increase in drilling rate
2.   Increase in fluid volume at the surface, which is commonly termed a pit level increase or an increase in flow rate
3.   Change in pump pressure
4.   Reduction in drillpipe weight
5.   Gas, oil, or water-cut mud

SUDDEN INCREASE IN DRILLING RATE

Generally, the first indication of a well kick is a sudden increase in drilling rate or a "drilling break," which suggests that a porous formation may have been penetrated. Crews should be alerted that, in the potential pay interval, no more than some minimal interval (usually 2 to 5 feet) of any drilling break should be penetrated. This is one of the most important aspects of pressure control. Many multimillion-dollar blowouts could have been avoided by limiting the open interval.

INCREASE IN PIT LEVEL OR FLOW RATE

A variation of bit type may mask a drilling break. In that event, the first warning may be an increase in flow rate or pit level caused by the influx of formation fluids. Depending on the productivity of the formation, the influx may be rapid or virtually imperceptible. Therefore, the influx could be considerable before being noticed. No change in pit level or flow rate should be ignored.

CHANGE IN PUMP PRESSURE

A decrease in pump pressure during an influx is caused by the reduced hydrostatic in the annulus. Most of the time, one of the aforementioned indications will have manifested itself prior to a decrease in pump pressure.

REDUCTION IN DRILLPIPE WEIGHT

The reduction in string weight occurs with a substantial influx from a zone of high productivity. Again, the other indicators will probably have manifested themselves prior to or in conjunction with a reduction in drillpipe weight.

GAS, OIL, OR WATER-CUT MUD

Caution should be exercised when gas, oil, or water-cut mud is observed. Normally, this indicator is accompanied by one of the other indicators if the well is experiencing an influx.

SHUT-IN PROCEDURE

When any of these warning signals are observed, the crew must immediately proceed with the established shut-in procedure. The crew must be thoroughly trained in the procedure to be used and that procedure should be posted in the dog house. It is imperative that the crew be properly trained and react to the situation. Classic pressure control procedures cannot be used successfully to control large kicks. The success of the well control operation depends upon the response of the crew at this most critical phase.

A typical shut-in procedure is as follows:

1. Drill no more than 3 feet of any drilling break.
2. Pick up off bottom, space out, and shut off the pump.
3. Check for flow.
4. If flow is observed, shut in the well by opening the choke line, closing the pipe rams, and closing the choke, pressure permitting.
5. Record the pit volume increase, drillpipe, pressure, and annulus pressure. Monitor and record the drillpipe and annular pressures at 15-minute intervals.
6. Close annular preventer; open pipe rams.
7. Prepare to displace the kick.

The number of feet of a drilling break to be drilled prior to shutting in the well can vary from area to area. However, an initial drilling break of 2 to 5 feet is common. The drillpipe should be spaced out to insure that no tool joints are in the blowout preventers. This is especially important on offshore and floating operations. On land, the normal procedure would be to position a tool joint at the connection position above the rotary table to permit easy access for alternate pumps or wire-line operations. The pump should be left on while positioning the drillpipe. The fluid influx is distributed and not in a bubble. In addition, there is less chance of initial bit plugging.

When observing the well for flow, the question is "How long should the well be observed?" The obvious answer is that the well should be observed as long as necessary to satisfy the observer of the condition of the well. Generally, 15 minutes or less are required. If oil muds are being used, the observation period should be lengthened. If the well is deep, the observation period should be longer than for a shallow well.

If the drilling break is a potentially productive interval but no flow is observed, it may be prudent to circulate bottoms up before continuing drilling in order to monitor and record carefully such parameters as time, strokes, flow rate, and pump pressure for indications of potential well control problems. After it is determined that the well is under control, drill another increment of the drilling break and repeat the procedure. Again, there is flexibility in the increment to be drilled. The experience gained from the first increment must be considered. A second increment of 2 to 5 feet is common. Circulating out may not be necessary after each interval even in the productive zone; however, a short circulating period will disperse any influx. Repeat this procedure until the drilling rate returns to normal and the annulus is free of formation fluids.

Whether the annular preventer or the pipe rams are closed first is a matter of choice. The closing time for each blowout preventer must be considered along with the productivity of the formation being penetrated. The objective of the shut-in procedure is to limit the size of the kick. If the annular requires twice as much time to close as the pipe rams and the formation is prolific, the pipe rams may be the better choice. If both blowout preventers close in approximately the same time, the annular is the better choice since it will close on anything.

Shutting in the well by opening the choke, closing the blowout preventers, and closing the choke is known as a "soft shut-in." The alternative is known as a "hard shut-in," which is achieved by merely closing the blowout preventer on the closed choke line. The primary argument for the hard shut-in is that it minimizes influx volume, and influx volume is critical to success. The hard shut-in became popular in the early days of well control.

Before the advent of modern equipment with remote hydraulic controls, opening choke lines and chokes was time-consuming and could permit significant additional influx. With modern equipment, all hydraulic controls are centrally located and critical valves are hydraulically operated. Therefore, the shut-in is simplified and the time reduced. In addition, blowout preventers, like valves, are made to be open or closed while chokes are made to restrict flow. In some instances, during hard shut-in, the fluid velocity through closing blowout preventers has been sufficient to cut out the preventer before it could be closed effectively.

In the young rocks such as are commonly found in offshore operations, the consequences of exceeding the maximum pressure can be grave in that the blowout can fracture to the surface outside the casing. The blowout then becomes uncontrolled and uncontrollable. Craters can consume jack-up rigs and platforms. The plight of the floating rig can be even more grim due to the loss of buoyancy resulting from gas in the water.

The most infamous and expensive blowouts in industry history were associated with fracturing to the surface from under surface casing. It is often argued that fracturing to the surface can be avoided by observing the surface pressure after the well is closed in and opening the well if the pressure becomes too high. Unfortunately, in most instances there is insufficient time to avoid fracturing at the shoe. All things considered, the soft shut-in is the better procedure.

In the event the pressure at the surface reaches the maximum permissible surface pressure, a decision must be made either to let the well blow out underground or to vent the well to the surface. Either approach can result in serious problems. With only surface casing set to a depth of less than 3600 feet, the best alternative is to open the well and permit it to flow through the surface equipment. This procedure can result in the erosion of surface equipment. However, more time is made available for rescue operations and repairs to surface equipment. It also simplifies kill operations.

There is no history of a well fracturing to the surface with pipe set below 3600 feet. Therefore, with pipe set below 3600 feet, the underground blowout is an alternative. It is argued that an underground flow is not as hazardous as a surface flow in some offshore and land operations. When properly rigged up, flowing the well to the surface under controlled conditions is the preferred alternative. A shut-in well that is blowing out underground is difficult to analyze and often more difficult to control.

The maximum permissible shut-in surface pressure is the lesser of 80 to 90 percent of the casing burst pressure and the surface pressure required to produce fracturing at the casing shoe. The procedure for determining the maximum permissible shut-in surface pressure is illustrated in Example 2.1:

Example 2.1

Given:

Surface casing $= 2000$ feet $8\frac{5}{8}$-inch

Internal yield $= 2470$ psi

Fracture gradient, $F_g = 0.76$ psi/ft

Mud density, $\rho = 9.6$ ppg

Mud gradient, $\rho_m = 0.5$ psi/ft

Wellbore schematic $=$ Figure 2.1

Required:

Determine the maximum permissible surface pressure on the annulus, assuming that the casing burst is limited to 80 percent of design specification.

Solution:

$$80\% \text{ burst} = 0.8(2470 \text{ psi}) = 1976 \text{ psi}$$

$$P_f = P_a(Maximum) + \rho_m D_{sc} \tag{2.1}$$

Where:

$P_f =$ Fracture pressure, psi
$P_a =$ Annulus pressure, psi
$\rho_m =$ Mud gradient, psi/ft
$D_{sc} =$ Depth to the casing shoe

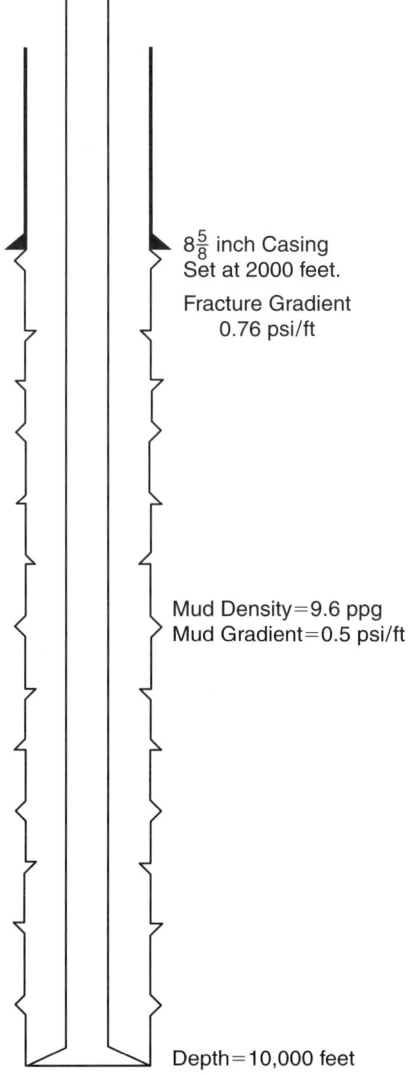

Figure 2.1   *Wellbore Schematic.*

Therefore:

$$P_a(Maximum) = P_f - \rho_m D_{sc}$$

$$= 0.76(2000) - 0.5(2000)$$

$$= \mathbf{520\ psi}$$

Therefore, the maximum permissible annular pressure at the surface is 520 psi, which is that pressure which would produce formation fracturing at the casing seat.

Recording the gain in pit volume, drillpipe pressure, and annulus pressure initially and over time is very important to controlling the kick. As will be seen in the discussion of special problems in Chapter 4, the surface pressures are critical for determining the condition of the well and the potential success of the well control procedure. Analysis of the gain in the surface volume in consideration of the casing pressure is critical in defining the potential for an underground blowout.

In some instances, due to a lack of familiarity with the surface equipment, crews have failed to shut in the well completely. When the pit volume continued to increase, the oversight was detected and the well shut in. Recording the surface pressures over time is extremely important. Gas migration, which is also discussed in Chapter 4, will cause the surface pressures to increase over time. Failure to recognize the resultant superpressuring can result in the failure of the well control procedure.

These procedures are fundamental to pressure control. They are the responsibility of the rig crew and should be practiced and studied until they become as automatic as breathing. The entire operation depends upon the ability of the driller and crew to react to a critical situation. Now, the well is under control and the kill operation can proceed to circulate out the influx.

CIRCULATING OUT THE INFLUX

THEORETICAL CONSIDERATIONS

Gas Expansion

Prior to the early 1960s, an influx was circulated to the surface by keeping the pit level constant. This was also known as the Barrel In–Barrel Out Method. Some insist on using this technique today although it is no more successful now than then. If the influx was mostly liquid, this technique was successful. If the influx was mostly gas, the results were disastrous. When a proponent of the Constant Pit Level Method was asked about the results, he replied, "Oh, we just keep pumping until something breaks!" Invariably, something did break, as illustrated in the drilling report at the beginning of this chapter.

In the late 1950s and early 1960s, some began to realize that this Barrel In–Barrel Out technique could not be successful. If the influx was gas, the gas had to be permitted to expand as it came to the surface. The basic relationship of gas behavior is given in Equation 2.2:

$$PV = znRT \tag{2.2}$$

Where:

$P =$ Pressure, psia
$V =$ Volume, ft^3
$z =$ Compressibility factor
$n =$ Number of moles
$R =$ Units conversion constant
$T =$ Temperature, °Rankine

For the purpose of studying gas under varying conditions, the general relationship can be extended to another form as given in Equation 2.3:

$$\frac{P_1 V_1}{z_1 T_1} = \frac{P_2 V_2}{z_2 T_2} \tag{2.3}$$

$1 =$ Denotes conditions at any point
$2 =$ Conditions at any point other than point 1

By neglecting changes in temperature, T, and compressibility factor, z, Equation 2.3 can be simplified into Equation 2.4 as follows:

$$P_1 V_1 = P_2 V_2 \tag{2.4}$$

In simple language, Equation 2.4 states that the pressure of a gas multiplied by the volume of the gas is constant. The significance of gas expansion in well control is illustrated by Example 2.2:

Example 2.2

Given:
Wellbore schematic $=$ Figure 2.2

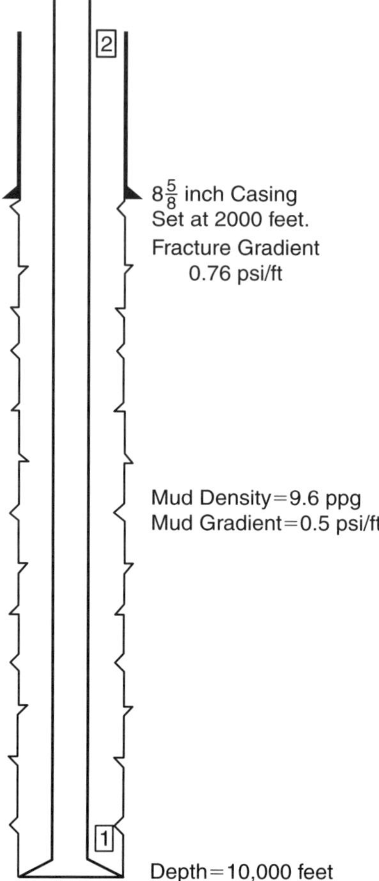

Figure 2.2   *Wellbore Schematic—Closed Container.*

Mud density, $\rho = 9.6$ ppg

Mud gradient, $\rho_m = 0.5$ psi/ft

Well depth, $D = 10,000$ feet

Conditions described in Example 2.1

The wellbore is a closed container.

1 cubic foot of gas enters the wellbore.

Gas enters at the bottom of the hole, which is point 1.

Required:

1. Determine the pressure in the gas bubble at point 1.

2. Assuming that the 1 cubic foot of gas migrates to the surface of the closed container (point 2) with a constant volume of 1 cubic foot, determine the pressure at the surface, the pressure at 2000 feet, and the pressure at 10,000 feet.

Solution:

1. The pressure of the gas, P_1, at point 1, which is the bottom of the hole, is determined by multiplying the gradient of the mud (psi/ft) by the depth of the well.

$$P_1 = \rho_m D \qquad (2.5)$$

$$P_1 = 0.5(10,000)$$

$$P_1 = \textbf{5000 psi}$$

2. The pressure in the 1 cubic foot of gas at the surface (point 2) is determined using Equation 2.4:

$$P_1 V_1 = P_2 V_2$$

$$(5000)(1) = P_2(1)$$

$$P_{surface} = \textbf{5000 psi}$$

Determine the pressure at 2000 feet:

$$P_{2000} = P_2 + \rho_m(2000)$$

$$P_{2000} = 5000 + 0.5(2000)$$

$$P_{2000} = \textbf{6000 psi}$$

Determine the pressure at the bottom of the hole.

$$P_{10,000} = \rho_m(10,000)$$

$$P_{10,000} = 5000 + 0.5(10,000)$$

$$P_{10,000} = \mathbf{10,000\ psi}$$

As illustrated in Example 2.2, the pressures in the well become excessive when the gas is not permitted to expand. The pressure at 2000 feet would build to 6000 psi if the wellbore was a closed container. However, the wellbore is not a closed container and the pressure required to fracture the wellbore at 2000 feet is 1520 psi. When the pressure at 2000 feet exceeds 1520 psi, the container will rupture, resulting in an underground blowout.

The goal in circulating out a gas influx is to bring the gas to the surface, allowing the gas to expand to avoid rupturing the wellbore. At the same time, there is the need to maintain the total hydrostatic pressure at the bottom of the hole at the reservoir pressure in order to prevent additional influx of formation fluids. As will be seen, classical pressure control procedures routinely honor the second condition of maintaining the total hydrostatic pressure at the bottom of the hole equal to the reservoir pressure and ignore any consideration of the fracture pressure at the shoe.

The U-Tube Model

All classical displacement procedures are based on the U-Tube Model illustrated in Figure 2.3. It is important to understand this model and premise. Too often, field personnel attempt to apply classical well control procedures to non-classical problems. If the U-Tube Model does not accurately describe the system, classical pressure control procedures cannot be relied upon.

As illustrated in Figure 2.3, the left side of the U-Tube represents the drillpipe while the right side of the U-Tube represents the annulus. Therefore, the U-Tube Model describes a system where the bit is on bottom and it is possible to circulate from bottom. If it is not possible to circulate from bottom, classical well control concepts are meaningless and not applicable. This concept is discussed in detail in Chapter 4.

As further illustrated in Figure 2.3, an influx of formation fluids has entered the annulus (right side of the U-Tube). The well has been shut in, which means that the system has been closed. Under these shut-in

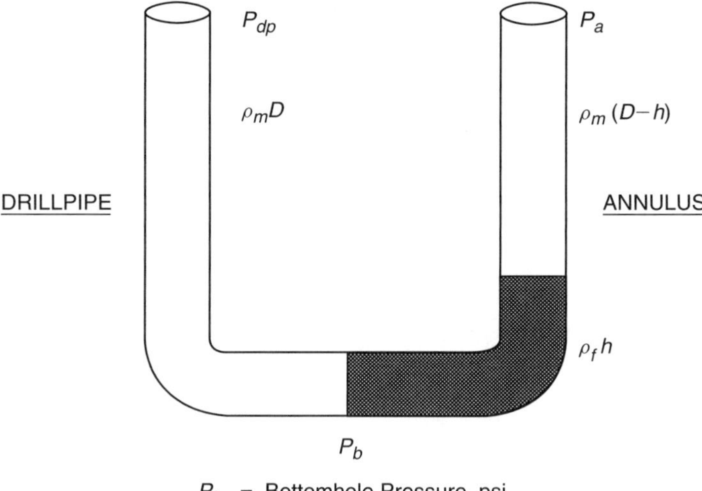

Figure 2.3 *The U-Tube Model.*

conditions, there is static pressure on the drillpipe, which is denoted by P_{dp}, and static pressure on the annulus, which is denoted by P_a. The formation fluid, ρ_f, has entered the annulus and occupies a volume defined by the area of the annulus and the height, h, of the influx.

An inspection of Figure 2.3 indicates that the drillpipe side of the U-Tube Model is more simple to analyze since the pressures are only influenced by mud of known density and pressure on the drillpipe that is easily measured. Under static conditions, the bottomhole pressure is easily determined utilizing Equation 2.6:

$$P_b = \rho_m D + P_{dp} \tag{2.6}$$

Where:

P_b = Bottomhole pressure, psi
ρ_m = Mud gradient, psi/ft
D = Well depth, feet
P_{dp} = Shut-in drillpipe pressure, psi

Equation 2.6 describes the shut-in bottomhole pressure in terms of the total hydrostatic on the drillpipe side of the U-Tube Model. The shut-in bottomhole pressure can also be described in terms of the total hydrostatic pressure on the annulus side of the U-Tube Model as illustrated by Equation 2.7:

$$P_b = \rho_f h + \rho_m (D - h) + P_a \tag{2.7}$$

Where:

P_b = Bottomhole pressure, psi
ρ_m = Mud gradient, psi/ft
D = Well depth, feet
P_a = Shut-in casing pressure, psi
ρ_f = Gradient of influx, psi/ft
h = Height of the influx, feet

Classic well control procedures, no matter what terminology is used, must keep the shut-in bottomhole pressure, P_b, constant to prevent additional influx of formation fluids while displacing the initial influx to the surface. Obviously, the equation for the drillpipe side (Equation 2.6) is the simpler and all of the variables are known; therefore, the drillpipe side is used to control the bottomhole pressure, P_b.

With the advent of pressure control technology, the necessity of spreading that technology presented an awesome task. Simplicity was in order and the classic Driller's Method for displacing the influx from the wellbore without permitting additional influx was developed.

THE DRILLER'S METHOD

The Driller's Method of displacement is simple and requires minimal calculations. The recommended procedure is as follows:

Step 1

On each tour, read and record the standpipe pressure at several rates in strokes per minute (spm), including the anticipated kill rate for each pump.

Step 2

After a kick is taken and prior to pumping, read and record the drillpipe and casing pressures. Determine the anticipated pump pressure at the kill rate using Equation 2.8:

$$P_c = P_{ks} + P_{dp} \tag{2.8}$$

Where:

P_c = Circulating pressure during displacement, psi
P_{ks} = Recorded pump pressure at the kill rate, psi
P_{dp} = Shut-in drillpipe pressure, psi

Important: If in doubt at any time during the entire procedure, shut in the well, read and record the shut-in drillpipe pressure and the shut-in casing pressure, and proceed accordingly.

Step 3

Bring the pump to a kill speed, keeping the casing pressure constant at the shut-in casing pressure. This step should require less than five minutes.

Step 4

Once the pump is at a satisfactory kill speed, read and record the drillpipe pressure. Displace the influx, keeping the recorded drillpipe pressure constant.

Step 5

Once the influx has been displaced, record the casing pressure and compare with the original shut-in drillpipe pressure recorded in Step 1. It is important to note that, if the influx has been completely displaced, the casing pressure should be equal to the original shut-in drillpipe pressure.

Step 6

If the casing pressure is equal to the original shut-in drillpipe pressure recorded in Step 1, shut in the well by keeping the casing pressure constant while slowing the pumps. If the casing pressure is greater than the original shut-in drillpipe pressure, continue circulating for an additional circulation, keeping the drillpipe pressure constant, and then shut in the well, keeping the casing pressure constant while slowing the pumps.

Step 7

Read, record, and compare the shut-in drillpipe and casing pressures. If the well has been properly displaced, the shut-in drillpipe pressure should be equal to the shut-in casing pressure.

Step 8

If the shut-in casing pressure is greater than the shut-in drillpipe pressure, repeat Steps 2 through 7.

Step 9

If the shut-in drillpipe pressure is equal to the shut-in casing pressure, determine the density of the kill-weight mud, ρ_1, using Equation 2.9 (Note that no "safety factor" is recommended or included):

$$\rho_1 = \frac{\rho_m D + P_{dp}}{0.052D} \tag{2.9}$$

Where:

ρ_1 = Density of the kill-weight mud, ppg
ρ_m = Gradient of the original mud, psi/ft
P_{dp} = Shut-in drillpipe pressure, psi
D = Well depth, feet

Step 10

Raise the mud weight in the suction pit to the density determined in Step 9.

Step 11

Determine the number of strokes to the bit by dividing the capacity of the drill string in barrels by the capacity of the pump in barrels per stroke according to Equation 2.10:

$$STB = \frac{C_{dp}l_{dp} + C_{hw}l_{hw} + C_{dc}l_{dc}}{C_p} \qquad (2.10)$$

Where:

$STB =$ Strokes to the bit, strokes
$C_{dp} =$ Capacity of the drillpipw, bbl/ft
$C_{hw} =$ Capacity of the heavy-weight drillpipe, bbl/ft
$C_{dc} =$ Capacity of the drill collars, bbl/ft
$l_{dp} =$ Length of the drillpipe, feet
$l_{hw} =$ Length of the heavy-weight drillpipe, feet
$l_{dc} =$ Length of the drill collars, feet
$C_p =$ Pump capacity, bbl/stroke

Step 12

Bring the pump to speed, keeping the casing pressure constant.

Step 13

Displace the kill-weight mud to the bit, keeping the casing pressure constant.

Warning: Once the pump rate has been established, no further adjustments to the choke should be required. The casing pressure should remain constant at the initial shut-in drillpipe pressure. If the casing pressure begins to rise, the procedure should be terminated and the well shut in.

Step 14

After pumping the number of strokes required for the kill mud to reach the bit, read and record the drillpipe pressure.

Step 15

Displace the kill-weight mud to the surface, keeping the drillpipe pressure constant.

Step 16

With kill-weight mud to the surface, shut in the well by keeping the casing pressure constant while slowing the pumps.

Step 17

Read and record the shut-in drillpipe pressure and the shut-in casing pressure. Both pressures should be 0.

Step 18

Open the well and check for flow.

Step 19

If the well is flowing, repeat the procedure.

Step 20

If no flow is observed, raise the mud weight to include the desired trip margin and circulate until the desired mud weight is attained throughout the system.

The discussion of each step in detail follows:

Step 1

On each tour, read and record the standpipe pressure at several flow rates in strokes per minute (spm), including the anticipated kill rate for each pump.

Experience has shown that one of the most difficult aspects of any kill procedure is bringing the pump to speed without permitting an additional influx or fracturing the casing shoe. This problem is compounded by attempts to achieve a precise kill rate. There is nothing magic about the kill rate used to circulate out a kick.

In the early days of pressure control, surface facilities were inadequate to bring an influx to the surface at a high pump speed. Therefore, one-half normal speed became the arbitrary rate of choice for circulating the influx to the surface. However, if only one rate such as the one-half speed is acceptable, problems can arise when the pump speed is slightly less or slightly more than the

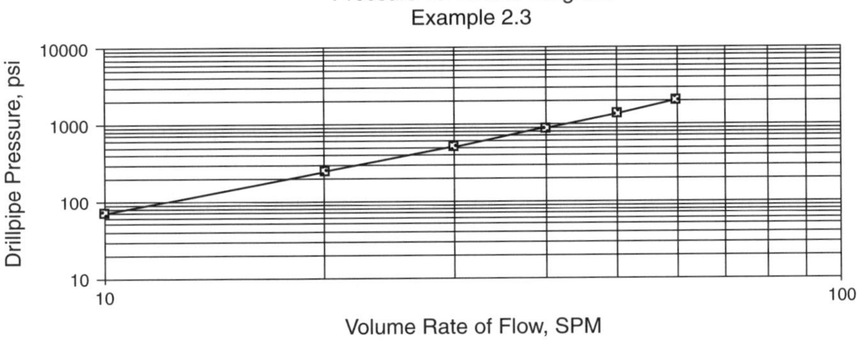

Figure 2.4

precise one-half speed. The reason for the potential problem is that the circulating pressure at rates other than the kill rate is unknown. Refer to further discussion after Step 4.

The best procedure is to record and graph several flow rates and corresponding pump pressures as illustrated in Figure 2.4. It is assumed in Examples 2.3 and 2.4 that the kill speed used is 30 strokes per minute. However, the actual pump speed used need not be exactly 30 strokes per minute. The drillpipe pressure corresponding to the actual pump speed being used could be verified using Figure 2.4.

Step 2

After a kick has been taken and prior to pumping, read and record the drillpipe and casing pressures. Determine the pump pressure at the kill speed.

Important: If in doubt at any time during the entire procedure, shut in the well, read and record the shut-in drillpipe pressure and the shut-in casing pressure, and proceed accordingly.

It is not uncommon for the surface pressures to fluctuate slightly due to temperature, gas migration, or gauge problems. Therefore, it is important to record the surface pressures immediately prior to commencing pumping operations.

The second statement is extremely important to keep in mind. When in doubt, shut in the well! It seems that the prevailing impulse is to continue circulating regardless of the consequences. If the condition of the well has deteriorated since it was shut in, it deteriorated during the pumping phase. When in doubt, shut in the well, read the surface pressures, compare with the original pressures, and evaluate the situation prior to further operations. If something is wrong with the displacement procedure being used, the situation is less likely to deteriorate while shut in and more likely to continue to deteriorate if pumping is continued.

Step 3

Bring the pump to a kill speed, keeping the casing pressure constant. This step should require less than five minutes.

As previously stated, bringing the pump to speed is one of the most difficult problems in any well control procedure. Experience has shown that the most practical approach is to keep the casing pressure constant at the shut-in casing pressure while bringing the pump to speed. The initial gas expansion is negligible over the allotted time of five minutes required to bring the pump to speed.

It is not important that the initial volume rate of flow be exact. Any rate within 10 percent of the kill rate is satisfactory. This procedure will establish the correct drillpipe pressure to be used to displace the kick. Figure 2.4 can be used to verify the drillpipe pressure being used.

Practically, the rate can be lowered or raised at any time during the displacement procedure. Simply read and record the circulating casing pressure and hold that casing pressure constant while adjusting the pumping rate and establishing a new drillpipe pressure. No more than one to two minutes can be allowed for changing the rate when the gas influx is near the surface because the expansion near the surface is quite rapid.

Step 4

Once the pump is at a satisfactory kill speed, read and record the drillpipe pressure. Displace the influx, keeping the recorded drillpipe pressure constant.

Actually, all steps must be considered together and are integral to each other. The correct drillpipe pressure used to circulate out the influx will be that drillpipe pressure established by Step 4. The pump rates and pressures established in Step 1 are to be used as a confirming reference only once the operation has commenced. Consideration of the U-Tube Model in Figure 2.3 clearly illustrates that, by holding the casing pressure constant at the shut-in casing pressure while bringing the pump to speed, the appropriate drillpipe pressure will be established for the selected rate.

All adjustments to the circulating operation must be performed considering the casing annulus pressure. In adjusting the pressure on the circulating system, the drillpipe pressure response must be considered secondarily because there is a significant lag time between any choke operation and the response on the drillpipe pressure gauge. This lag time is caused by the time required for the pressure transient to travel from the choke to the drillpipe pressure gauge.

Pressure responses travel at the speed of sound in the medium. The speed of sound is 1088 feet per second in air and about 4800 feet per second in most water-based drilling muds. Therefore, in a 10,000-foot well, a pressure transient caused by opening or closing the choke would not be reflected on the standpipe pressure gauge until four seconds later. Utilizing only the drillpipe pressure and the choke usually results in large cyclical variations which cause additional influxes or unacceptable pressures at the casing shoe.

Step 5

Once the influx has been displaced, record the casing pressure and compare with the original shut-in drillpipe pressure recorded in Step 1. It is important to note that, if the influx has been completely displaced, the casing pressure should be equal to the original shut-in drillpipe pressure.

Consider the U-Tube Model presented in Figure 2.5 and compare with the U-Tube Model illustrated in Figure 2.3. If the influx has been properly and completely displaced, the conditions in the annulus side of Figure 2.5 are exactly the same as the conditions in

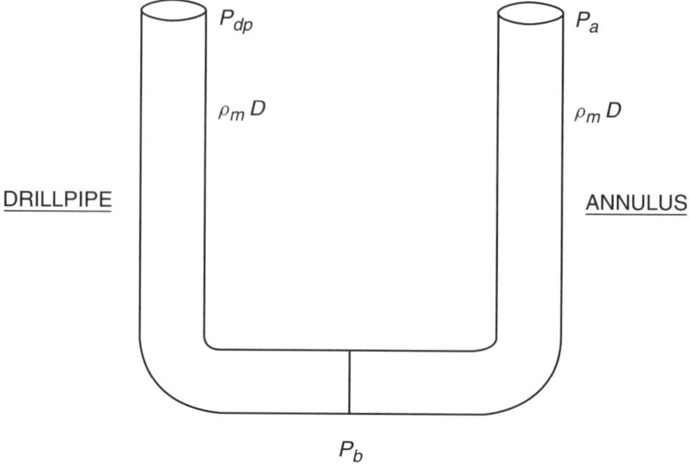

Figure 2.5   *Influx Displaced—The U-Tube Model.*

the drillpipe side of Figure 2.3. If the frictional pressure losses in the annulus are negligible, the conditions in the annulus side of Figure 2.5 will be approximately the same as the drillpipe side of Figure 2.3. Therefore, once the influx is displaced, the circulating annulus pressure should be equal to the initial shut-in drillpipe pressure.

Step 6

If the casing pressure is equal to the original shut-in drillpipe pressure recorded in Step 1, shut in the well by keeping the casing pressure constant while slowing the pumps. If the casing pressure is greater than the original shut-in drillpipe pressure, continue circulating for an additional circulation, keeping the drillpipe pressure constant, and then shut in the well, keeping the casing pressure constant while slowing the pumps.

Step 7

Read, record, and compare the shut-in drillpipe and casing pressures. If the well has been properly displaced, the shut-in drillpipe pressure should be equal to the shut-in casing pressure.

Again consider Figure 2.5. Assuming that the influx has been completely displaced, conditions in both sides of the U-Tube Model are exactly the same. Therefore, the pressures at the surface on both the drillpipe and casing should be exactly the same.

Often, pressure is trapped in the system during the displacement procedure. If the drillpipe pressure and casing pressure are equal after displacing the influx but greater than the original shut-in drillpipe pressure or that drillpipe pressure recorded in Step 2, the difference between the two values is probably due to trapped pressure.

If the surface pressures recorded after displacement are equal but greater than the initial shut-in drillpipe pressure and formation influx is still present in the annulus, this discussion is not valid. These conditions are discussed in the special problems in Chapter 4.

Step 8

If the shut-in casing pressure is greater than the shut-in drillpipe pressure, repeat Steps 2 through 7.

If, after displacing the initial influx, the shut-in casing pressure is greater than the shut-in drillpipe pressure, it is probable that an additional influx was permitted at some point during the displacement procedure. Therefore, it will be necessary to displace that second influx.

Step 9

If the shut-in drillpipe pressure is equal to the shut-in casing pressure, determine the density of the kill-weight mud, ρ_1, using Equation 2.9 (Note that no "safety factor" is recommended or included):

$$\rho_1 = \frac{\rho_m D + P_{dp}}{0.052 D}$$

Safety factors are discussed in detail in Chapter 4.

Step 10

Raise the mud weight in the suction pit to the density determined in Step 9.

Step 11

Determine the number of strokes to the bit by dividing the capacity of the drill string in barrels by the capacity of the pump in barrels per stroke according to Equation 2.10:

$$STB = \frac{C_{dp}l_{dp} + C_{hw}l_{hw} + C_{dc}l_{dc}}{C_p}$$

Sections or different weights of drillpipe, drill collars, or heavy-weight drillpipe may be added or deleted from Equation 2.10 simply by adding to or subtracting from the numerator of Equation 2.10 the product of the capacity and the length of the section.

Step 12

Bring the pump to speed, keeping the casing pressure constant.

Step 13

Displace the kill-weight mud to the bit, keeping the casing pressure constant.

Warning: Once the pump rate has been established, no further adjustments to the choke should be required. The casing pressure should remain constant at the initial shut-in drillpipe pressure. If the casing pressure begins to rise, the procedure should be terminated and the well shut in.

It is vital to understand Step 13. Again, consider the U-Tube Model in Figure 2.5. While the kill-weight mud is being displaced to the bit on the drillpipe side, under dynamic conditions no changes are occurring in any of the conditions on the annulus side. Therefore, once the pump rate has been established, the casing pressure should not change and it should not be necessary to adjust the choke to maintain the constant drillpipe pressure.

If the casing pressure does begin to increase, with everything else being constant, in all probability there is some gas in the annulus. If there is gas in the annulus, this procedure must be terminated. Since the density of the mud at the surface has been increased to the kill, the proper procedure under these conditions would be the Wait and Weight Method, which is further described on page 70. The Wait and Weight Method would be used to circulate the gas in the annulus to the surface and control the well.

Step 14

After pumping the number of strokes required for the kill mud to reach the bit, read and record the drillpipe pressure.

Step 15

Displace the kill-weight mud to the surface, keeping the drillpipe pressure constant.

Referring to Figure 2.5, once kill-weight mud has reached the bit and the displacement of the annulus begins, conditions on the drillpipe side of the U-Tube Model are constant and do not change. Therefore, the kill-weight mud can be displaced to the surface by keeping the drillpipe pressure constant. Some change in casing pressure and adjustment in the choke size can be expected during this phase. If the procedure has been executed properly, the choke size will be increased to maintain the constant drillpipe pressure and the casing pressure will decline to 0 when the kill-weight mud reaches the surface.

Step 16

With kill-weight mud to the surface, shut in the well by keeping the casing pressure constant while slowing the pumps.

Step 17

Read and record the shut-in drillpipe pressure and the shut-in casing pressure. Both pressures should be 0.

Step 18

Open the well and check for flow.

Step 19

If the well is flowing, repeat the procedure.

Step 20

If no flow is observed, raise the mud weight to include the desired trip margin and circulate until the desired mud weight is attained throughout the system.

The Driller's Method is illustrated in Example 2.3:

Example 2.3

Given:

Wellbore schematic = Figure 2.6

Well depth, $D = 10{,}000$ feet

Hole size, $D_h = 7\frac{7}{8}$ inches

Drillpipe size, $D_p = 4\frac{1}{2}$ inches

$8\frac{5}{8}$-inch surface casing = 2000 feet

Casing internal diameter, $D_{ci} = 8.017$ inches

Fracture gradient, $F_g = 0.76\,\text{psi/ft}$

Mud weight, $\rho = 9.6\,\text{ppg}$

Mud gradient, $\rho_m = 0.50\,\text{psi/ft}$

A kick is taken with the drill string on bottom and:

Shut-in drillpipe pressure, $P_{dp} = 200\,\text{psi}$

Shut-in annulus pressure, $P_a = 300\,\text{psi}$

Pit level increase = 10 barrels

Normal circulation rate = 6 bpm at 60 spm

Kill rate = 3 bpm at 30 spm

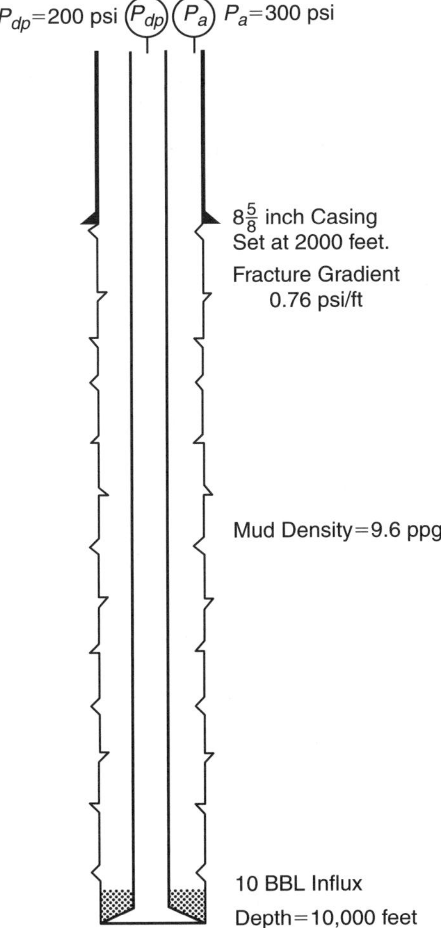

Figure 2.6   *Wellbore Schematic.*

Circulating pressure at kill rate, $P_{ks} = 500\,\text{psi}$

Pump capacity, $C_p = 0.1\,\text{bbl/stk}$

Capacity of the:

Drillpipe, $C_{dpi} = 0.0142\,\text{bbl/ft}$

Drillpipe casing annulus, $C_{dpca} = 0.0428\,\text{bbl/ft}$

Drillpipe hole annulus, $C_{dpha} = 0.0406\,\text{bbl/ft}$

Note: For simplicity in calculation and illustration, no drill collars are assumed. The inclusion of drill collars adds only an intermediate calculation.

Required:

Describe the kill procedure using the Driller's Method.

Solution:

1.   Establish pressure versus volume diagram (Figure 2.4).

2.   Record the shut-in drillpipe pressure and shut-in casing pressure.

$$P_{dp} = 200 \text{ psi}$$

$$P_a = 300 \text{ psi}$$

3.   Establish the pumping pressure at the kill rate of 30 spm using Equation 2.8:

$$P_c = P_{ks} + P_{dp}$$

$$Pc = 500 + 200$$

$$P_c = \textbf{700 psi}$$

4.   Bring the pump to 30 spm, maintaining 300 psi on the casing annulus.

5.   Read and record the drillpipe pressure equal to 700 psi. Confirm the drillpipe pressure using Figure 2.4.

6.   Displace the annulus and all gas which has entered the wellbore, keeping the drillpipe pressure constant at 700 psi.

7.   Read and record the drillpipe pressure equal to 700 psi and casing pressure equal to 200 psi.

8.   Shut in the well, keeping the casing pressure constant at 200 psi. Allow the well to stabilize.

9.   Determine that all gas is out of the mud.

$$P_a = P_{dp} = \textbf{200 psi.}$$

10.   Determine the kill weight, ρ_1, using Equation 2.9:

$$\rho_1 = \frac{\rho_m D + P_{dp}}{0.052 D}$$

$$\rho_1 = \frac{0.5(10{,}000) + 200}{0.052(10{,}000)}$$

$$\rho_1 = \textbf{10 ppg}$$

11.   Determine the strokes to the bit using Equation 2.10:

$$STB = \frac{C_{dp} l_{dp} + C_{hw} l_{hw} + C_{dc} l_{dc}}{C_p}$$

$$STB = \frac{(0.0142)(10{,}000)}{0.1}$$

$$STB = \textbf{1420 strokes}$$

12.   Raise the mud weight at the surface to 10 ppg.

13.   Bring the pump to 30 spm, keeping the casing pressure constant at 200 psi.

14.   Displace the 10-ppg mud to the bit with 1420 strokes, keeping the annulus pressure, P_a, constant at 200 psi. The choke size must not change.

15.   At 1420 strokes, observe and record the circulating pressure on the drillpipe. Assume that the observed pressure on the drillpipe is 513 psi.

16.   Circulate the 10-ppg mud to the surface, keeping the drillpipe pressure constant at 513 psi.

17.   Shut in and check $P_a = P_{dp} = 0$ psi. The well is dead.

18.   Circulate and raise the mud weight to some acceptable trip margin, generally between 150 to 500 psi above the formation pressure, P_b, or 10.3 to 11.0 ppg.

19.   Continue drilling ahead.

THE WAIT AND WEIGHT METHOD

The alternative classical method is commonly known as the Wait and Weight Method. As the name implies, the well is shut in while the mud density is increased to the kill weight as determined by Equation 2.9. Therefore, the primary difference is operational in that the kill-weight mud, ρ_1, is pumped while the gas is being displaced. The result is that the well is killed in one circulation with the Wait and Weight Method whereas, with the Driller's Method, two circulations are required.

In the early days of pressure control, the time required to increase the density of the mud in the surface system to the kill-weight was significant. During that time, it was not uncommon for the gas to migrate or for the drillpipe to become stuck. However, modern mud mixing systems have eliminated the time factor from most operations in that most systems can raise the density of the surface system as fast as the mud is pumped. There are other important comparisons, which will be presented after the Wait and Weight Method is presented, illustrated, and discussed.

While each step will be subsequently discussed in detail, the displacement procedure for the Wait and Weight Method is as follows:

Step 1

On each tour, read and record the standpipe pressure at several rates in strokes per minute (spm), including the anticipated kill rate for each pump.

Step 2

Prior to pumping, read and record the drillpipe and casing pressures. Determine the anticipated pump pressure at the kill rate using Equation 2.8:

$$P_c = P_{ks} + P_{dp}$$

Step 3

Determine the density of the kill-weight mud, ρ_1, using Equation 2.9 (note that no "safety factor" is recommended or included):

$$\rho_1 = \frac{\rho_m D + P_{dp}}{0.052D}$$

Step 4

Determine the number of strokes to the bit by dividing the capacity of the drill string in barrels by the capacity of the pump in barrels per stroke according to Equation 2.10:

$$STB = \frac{C_{dp}l_{dp} + C_{hw}l_{hw} + C_{dc}l_{dc}}{C_p}$$

Step 5

Determine the new circulating pressure, P_{cn}, at the kill rate with the kill-weight mud at the bit utilizing Equation 2.11:

$$P_{cn} = P_{dp} - 0.052(\rho_1 - \rho)D + \left(\frac{\rho_1}{\rho}\right)P_{ks} \qquad (2.11)$$

Where:

$\rho_1 =$ Density of the kill-weight mud, ppg
$\rho =$ Density of the original mud, ppg
$P_{ks} =$ Original circulating pressure at kill rate, psi
$P_{dp} =$ Shut-in drillpipe pressure, psi
$D =$ Well depth, feet

Step 6

For a complex drill string configuration, determine and graph the pumping schedule for reducing the initial circulating pressure, P_c, determined in Step 2 to the final circulating pressure, P_{cn}, determined in Step 5. Using Equations 2.12 and 2.13, calculate Table 1 and create the corresponding graph.

Note: A "section" of drill string is the length where all the diameters remain the same. A new section would start any time the hole size or pipe diameter changes. As long as the diameters remain the same, it is one section. Therefore, each section has one annular capacity. The calculations begin from the surface.

For example, if the hole size does not change and the string consists of two weights of drillpipe, heavy-weight drillpipe, and drill collars, four calculations would be required.

The table would look like the following:

Strokes	Pressure
0	700
STKS 1	P_1
STKS 2	P_2
STKS 3	P_3
$\ldots$	$\ldots$
STB	P_{cn}

$$STKS\ 1 = \frac{C_{ds1}l_{ds1}}{C_p} \tag{2.12a}$$

$$STKS\ 2 = \frac{C_{ds1}l_{ds1} + C_{ds2}l_{ds2}}{C_p} \tag{2.12b}$$

$$STKS\ 3 = \frac{C_{ds1}l_{ds1} + C_{ds2}l_{ds2} + C_{ds3}l_{ds3}}{C_p} \tag{2.12c}$$

$$STB = \frac{C_{ds1}l_{ds1} + C_{ds2}l_{ds2} + C_{ds3}l_{ds3} + \ldots + C_{dc}l_{dc}}{C_p} \tag{2.12d}$$

$$P_1 = P_c - 0.052(\rho_1 - \rho)(l_{ds1})$$
$$+ \left(\frac{\rho_1 P_{ks}}{\rho} - P_{ks} \right) \left(\frac{STKS\ 1}{STB} \right) \tag{2.13a}$$

$$P_2 = P_c - 0.052(\rho_1 - \rho)(l_{ds1} + l_{ds2})$$
$$+ \left(\frac{\rho_1 P_{ks}}{\rho} - P_{ks} \right) \left(\frac{STKS\ 2}{STB} \right) \tag{2.13b}$$

$$P_3 = P_c - 0.052(\rho_1 - \rho)(l_{ds1} + l_{ds2} + l_{ds3})$$

$$+ \left(\frac{\rho_1 P_{ks}}{\rho} - P_{ks}\right)\left(\frac{STKS\ 3}{STB}\right) \tag{2.13c}$$

$$P_{cn} = P_{dp} - 0.052(\rho_1 - \rho) \times (l_{ds1} + l_{ds2} + l_{ds3} + \ldots + l_{dc})$$

$$+ \left(\frac{\rho_1 P_{ks}}{\rho} - P_{ks}\right) \tag{2.13d}$$

Where:

STKS 1 = Strokes to end of section 1 of drill string
STKS 2 = Strokes to end of section 2 of drill string
STKS 3 = Strokes to end of section 3 of drill string
STB = Strokes to the bit as determined in Step 4

Where:

ρ_1 = Density of kill-weight mud, ppg
ρ = Density of original mud, ppg
$l_{ds1,2,3}$ = Length of section of drill string, feet
$C_{ds1,2,3}$ = Capacity of section of drill string, bbl/ft
$P_{1,2,3}$ = Circulating pressure with kill-weight mud to the
        end of section 1,2,3 psi
P_{dp} = Shut-in drillpipe pressure, psi
P_{ks} = Circulating pressure at kill speed determined
        in Step 1, psi
C_p = Pump capacity, bbl/stroke
P_{cn} = New circulating pressure, psi
P_c = Initial displacement pressure determined in Step 2
        using Equation 2.8, psi

For a drill string composed of only one weight of drillpipe and one string of heavy-weight drillpipe or drill collars, the pumping schedule can be determined using Equation 2.14:

$$\frac{STKS}{25\ psi} = \frac{25(STB)}{P_e - P_{cn}} \tag{2.14}$$

Step 7

Raise the density of the mud in the suction pit to the kill weight determined in Step 3.

Step 8

Bring the pump to a kill speed, keeping the casing pressure constant at the shut-in casing pressure. This step should require less than five minutes.

Step 9

Once the pump is at a satisfactory kill speed, read and record the drillpipe pressure. Adjust the pumping schedule accordingly. Verify the drillpipe pressure using the diagram established in Step 1. Displace the kill-weight mud to the bit pursuant to the pumping schedule established in Step 6 as revised in this step.

Step 10

Displace the kill-weight mud to the surface, keeping the drillpipe pressure constant.

Step 11

Shut in the well, keeping the casing pressure constant, and observe that the drillpipe pressure and the casing pressure are 0 and the well is dead.

Step 12

If the surface pressures are not 0 and the well is not dead, continue to circulate, keeping the drillpipe pressure constant.

Step 13

Once the well is dead, raise the mud weight in the suction pit to provide the desired trip margin.

Step 14

Drill ahead.

Discussion of each step in detail follows:

Step 1

On each tour, read and record the standpipe pressure at several rates in strokes per minute (spm), including the anticipated kill rate for each pump.

This is the same discussion as presented after Step 1 of the Driller's Method. Experience has shown that one of the most difficult aspects of any kill procedure is bringing the pump to speed without permitting an additional influx or fracturing the casing shoe. This problem is compounded by attempts to achieve a precise kill rate. There is nothing magic about the kill rate used to circulate out a kick.

In the early days of pressure control, surface facilities were inadequate to bring an influx to the surface at a high pump speed. Therefore, one-half normal speed became the arbitrary rate of choice for circulating the influx to the surface. However, if only one rate such as the one-half speed is acceptable, problems can arise when the pump speed is slightly less or slightly more than the precise one-half speed. The reason for the potential problem is that the circulating pressure at rates other than the kill rate is unknown. Refer to further discussion after Step 4.

The best procedure is to record and graph several flow rates and corresponding pump pressures as illustrated in Figure 2.4. It is assumed in Examples 2.3 and 2.4 that the kill speed used is 30 strokes per minute. However, the actual pump speed used need not be exactly 30 strokes per minute. The drillpipe pressure corresponding to the actual pump speed being used could be verified using Figure 2.4.

Step 2

Prior to pumping, read and record the drillpipe and casing pressures. Determine the anticipated pump pressure at the kill rate using Equation 2.8:

$$P_c = P_{ks} + P_{dp}$$

This is the same discussion as presented after Step 2 of the Driller's Method. It is not uncommon for the surface pressures to fluctuate slightly due to temperature, migration, or gauge problems. Therefore, it is important to record the surface pressures immediately prior to commencing pumping operations.

When in doubt, shut in the well! It seems that the prevailing impulse is to continue circulating regardless of the consequences. If the condition of the well has deteriorated since it was shut in, it deteriorated during the pumping phase. When in doubt, shut in the well, read the surface pressures, compare with the original pressures and evaluate the situation prior to further operations. If something is wrong with the displacement procedure being used, the situation is less likely to deteriorate while shut in and more likely to continue to deteriorate if pumping is continued.

Step 3

Determine the density of the kill-weight mud, ρ_1, using Equation 2.9 (note that no "safety factor" is recommended or included):

$$\rho_1 = \frac{\rho_m D + P_{dp}}{0.052 D}$$

Safety factors are discussed in Chapter 4.

Step 4

Determine the number of strokes to the bit by dividing the capacity of the drill string in barrels by the capacity of the pump in barrels per stroke according to Equation 2.10:

$$STB = \frac{C_{dp} l_{dp} + C_{hw} l_{hw} + C_{dc} l_{dc}}{C_p}$$

Sections of different weights of drillpipe, drill collars, or heavy-weight drillpipe may be added or deleted from Equation 2.10 simply by adding to or subtracting from the numerator of Equation 2.10 the product of the capacity and the length of the section.

Step 5

Determine the new circulating pressure, P_{cn}, at the kill rate with the kill-weight mud utilizing Equation 2.11:

$$P_{cn} = P_{dp} - 0.052(\rho_1 - \rho)D + \left(\frac{\rho_1}{\rho}\right)P_{ks}$$

The new circulating pressure with the kill-weight mud will be slightly greater than the recorded circulating pressure at the kill speed since the frictional pressure losses are a function of the density of the mud. In Equation 2.11 the frictional pressure loss is considered a direct function of the density. In reality, the frictional pressure loss is a function of the density to the 0.8 power. However, the difference is insignificant.

Step 6

For a complex drill string configuration, determine and graph the pumping schedule for reducing the initial circulating pressure, P_c, determined in Step 2 to the final circulating pressure, P_{cn}, determined in Step 5. Using Equations 2.12 and 2.13, calculate Table 1 and create the corresponding graph.

Note: A "section" of drill string is the length where all the diameters remain the same. A new section would start any time the hole size or pipe diameter changed. As long as the diameters remain the same, it is one section. Therefore, each section has one annular capacity. The calculations begin from the surface. For example, if the hole size does not change and the string consists of two weights of drillpipe, heavy-weight drillpipe, and drill collars, four calculations would be required.

Determining this pump schedule is a most critical phase. Use of these equations is illustrated in Example 2.4. Basically, the circulating drillpipe pressure is reduced systematically to offset the increase in hydrostatic introduced by the kill-weight mud and ultimately to keep the bottomhole pressure constant.

The systematic reduction in drillpipe pressure must be attained by reducing the casing pressure by the scheduled amount and waiting

4 to 5 seconds for the pressure transient to reach the drillpipe pressure gauge. Efforts to control the drillpipe pressure directly by manipulating the choke are usually unsuccessful due to the time lag.

The key to success is to observe several gauges at the same time. The sequence is usually to observe the choke position, the casing pressure, and drillpipe pressure. Then concentrate on the choke position indicator while slightly opening the choke. Next, check the choke pressure gauge for the reduction in choke pressure. Continue that sequence until the designated amount of pressure has been bled from the annulus pressure gauge.

Finally, wait 10 seconds and read the result on the drillpipe pressure gauge. Repeat the process until the drillpipe pressure has been adjusted appropriately.

Step 7

Raise the density of the mud in the suction pit to the kill weight determined in Step 3.

Step 8

Bring the pump to a kill speed, keeping the casing pressure constant at the shut-in casing pressure. This step should require less than five minutes.

Step 9

Once the pump is at a satisfactory kill-speed, read and record the drillpipe pressure. Adjust the pumping schedule accordingly. Verify the drillpipe pressure using the diagram established in Step 1. Displace the kill-weight mud to the bit pursuant to the pumping schedule established in Step 6 as revised in this step.

As discussed in the Driller's Method, the actual kill speed used is not critical. Once the actual kill speed is established at a constant casing pressure equal to the shut-in casing pressure, the drillpipe pressure read is correct. The pumping schedule must be adjusted to reflect a pump speed different from the pump speed used to construct the table and graph.

The adjustment of the table is accomplished by reducing arithmetically the initial drillpipe pressure by the shut-in drillpipe pressure and remaking the appropriate calculations. The graph is more easily adjusted. The circulating drillpipe pressure marks the beginning point. Using that point, a line is drawn which is parallel to the line drawn in Step 6. The new line becomes the correct pumping schedule. The graph of pump pressure versus volume constructed in Step 1 is used to confirm the calculations.

If doubt arises during the pumping procedure, the well should be shut in by keeping the casing pressure constant while slowing the pump. The shut-in drillpipe pressure, shut-in casing pressure, and volume pumped should be used to evaluate the situation. The pumping procedure can be continued by bringing the pump to speed keeping the casing pressure constant, reading the drillpipe pressure, plotting the point on the pumping schedule graph, and establishing a new line parallel to the original. These points are clarified in Examples 2.3 and 2.4.

Keeping the casing pressure constant in order to establish the pump speed and correct circulating drillpipe pressure is an acceptable procedure provided that the time period is short and the influx is not near the surface. The time period should never be more than five minutes. If the influx is near the surface, the casing pressures will be changing very rapidly. In that case, the time period should be one to two minutes.

Step 10

Displace the kill-weight mud to the surface, keeping the drillpipe pressure constant.

Step 11

Shut in the well, keeping the casing pressure constant, and observe that the drillpipe pressure and the casing pressure are 0 and the well is dead.

Step 12

If the surface pressures are not 0 and the well is not dead, continue to circulate, keeping the drillpipe pressure constant.

Step 13

Once the well is dead, raise the mud weight in the suction pit to provide the desired trip margin.

Step 14

Drill ahead.

The Wait and Weight Method is illustrated in Example 2.4:

Example 2.4

Given:

Wellbore schematic = Figure 2.6

Well depth, $D = 10,000$ feet

Hole size, $D_h = 7\frac{7}{8}$ inches

Drillpipe size, $D_p = 4\frac{1}{2}$ inches

$8\frac{5}{8}$-inch surface casing = 2000 feet

Casing internal diameter, $D_{ci} = 8.017$ inches

Fracture gradient, $F_g = 0.76$ psi/ft

Mud weight, $\rho = 9.6$ ppg

Mud gradient, $\rho_m = 0.50$ psi/ft

A kick is taken with the drill string on bottom and:

Shut-in drillpipe pressure, $P_{dp} = 200$ psi

Shut-in annulus pressure, $P_a = 300$ psi

Pit level increase = 10 barrels

Normal circulation rate = 6 bpm at 60 spm

Kill rate $= 3$ bpm at 30 spm

Circulating pressure at kill rate, $P_{ks} = 500$ psi

Pump capacity, $C_p = 0.1$ bbl/stk

Capacity of the:

Drillpipe, $C_{dpi} = 0.0142$ bbl/ft

Drillpipe casing annulus, $C_{dpca} = 0.0428$ bbl/ft

Drillpipe hole annulus, $C_{dpha} = 0.0406$ bbl/ft

Note: For simplicity in calculation and illustration, no drill collars are assumed. The inclusion of drill collars adds only an intermediate calculation.

Required:

Describe the kill procedure using the Wait and Weight Method.

Solution:

1. Establish pressure versus volume diagram using Figure 2.4.

2. Record the shut-in drillpipe pressure and shut-in casing pressure.

$$P_{dp} = 200 \text{ psi}$$

$$P_a = 300 \text{ psi}$$

3. Establish the pumping pressure at the kill rate of 30 spm using Equation 2.8:

$$P_c = P_{ks} + P_{dp}$$

$$P_c = 500 + 200$$

$$P_c = \textbf{700 psi}$$

4.   Determine the kill weight, ρ_1, using Equation 2.9:

$$\rho_1 = \frac{\rho_m D + P_{dp}}{0.052 D}$$

$$\rho_1 = \frac{0.5(10{,}000) + 200}{0.052(10{,}000)}$$

$$\rho_1 = \textbf{10 ppg}$$

5.   Determine the strokes to the bit using Equation 2.10:

$$STB = \frac{C_{dp}l_{dp} + C_{hw}l_{hw} + C_{dc}l_{dc}}{C_p}$$

$$STB = \frac{(0.0142)(10{,}000)}{0.1}$$

$$STB = \textbf{1420 strokes}$$

6.   Determine the new circulating pressure, P_{cn}, at the kill rate with the kill-weight mud utilizing Equation 2.11:

$$P_{cn} = P_{dp} - 0.052(\rho_1 - \rho)D + \left(\frac{\rho_1}{\rho}\right) P_{ks}$$

$$P_{cn} = 200 - 0.052(10 - 9.6)(10{,}000)$$

$$+ \left(\frac{10}{9.6}\right) 500$$

$$P_{cn} = \textbf{513 psi}$$

7.   Determine the pumping schedule for a simple drill string pursuant to Equation 2.14:

$$\frac{STKS}{25\,psi} = \frac{25(STB)}{P_c - P_{cn}}$$

$$\frac{STKS}{25\,psi} = \frac{25(1420)}{700 - 513}$$

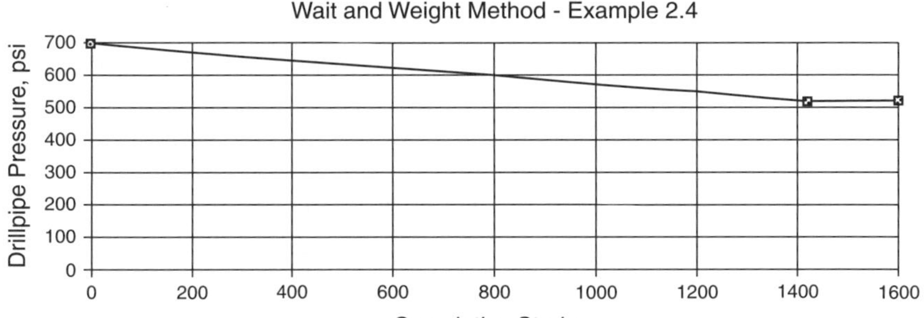

Figure 2.7

$$\frac{STKS}{25\,psi} = 190 \text{ strokes}$$

Strokes	Pressure
0	700
190	675
380	650
570	625
760	600
950	575
1140	550
1330	525
1420	513
1600	513

8.   Construct Figure 2.7—graph of pump schedule.

9.   Bring pump to 30 spm, keeping the casing pressure constant at 300 psi.

10.   Displace the kill-weight mud to the bit (1420 strokes) according to the pump schedule developed in Steps 7 and 8.

After 190 strokes, reduce the casing pressure observed at that moment by 25 psi. After 10 seconds, observe that the drillpipe pressure has dropped to 675 psi. After

380 strokes, reduce the casing pressure observed at that moment by 25 psi. After 10 seconds, observe that the drillpipe pressure has dropped to 650 psi. Continue in this manner until the kill-weight mud is at the bit and the drillpipe pressure is 513 psi.

11.   With the kill-weight mud to the bit after 1420 strokes, read and record the drillpipe pressure equal to 513 psi.

12.   Displace the kill-weight mud to the surface, keeping the drillpipe pressure constant at 513 psi.

13.   Shut in the well, keeping the casing pressure constant, and observe that the drillpipe and pressure are 0.

14.   Check for flow.

15.   Once the well is confirmed dead, raise the mud weight to provide the desired trip margin and drill ahead.

Obviously, the most potentially confusing aspect of the Wait and Weight Method is the development and application of the pumping schedule used to circulate the kill-weight mud properly to the bit while maintaining a constant bottomhole pressure. The development and application of the pump schedule is further illustrated in Example 2.5 to provide additional clarity:

Example 2.5

Given:

Example 2.4

Required:

1.   Assume that the kill speed established in Step 9 of Example 2.4 was actually 20 spm instead of the anticipated 30 spm. Determine the effect on the pump schedule and demonstrate the application of Figure 2.7.

2.   Assume that the drill string is complex and composed of 4000 feet of 5-inch 19.5 #/ft, 4000 feet of $4\frac{1}{2}$-inch 16.6 #/ft,

1000 feet of $4\frac{1}{2}$-inch heavy-weight drillpipe, and 1000 feet of 6-inch-by-2-inch drill collars. Illustrate the effect of the complex string configuration on Figure 2.7. Compare the pump schedule developed for the complex string with that obtained with the straight line simplification.

Solution:

1. Pursuant to Figure 2.4, the surface pressure at a kill speed of 20 spm would be 240 psi. The initial displacement surface pressure would be given by Equation 2.8 as follows:

$$P_c = P_{ks} + P_{dp}$$

$$P_c = 240 + 200$$

$$P_c = \textbf{440 psi}$$

Therefore, simply locate 440 psi on the Y axis of Figure 2.7 and draw a line parallel to that originally drawn. The new line is the revised pumping schedule. This concept is illustrated as Figure 2.8.

As an alternative, merely subtract 260 psi (700−440) from the values listed in the table in Step 7.

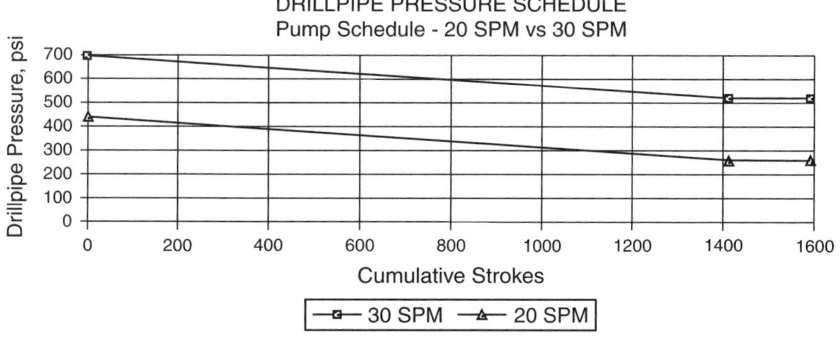

Figure 2.8

Strokes	Pressure
0	440
190	415
380	390
570	365
760	340
950	315
1140	290
1330	265
1420	253
1600	253

2.   Equations 2.10, 2,12, and 2.13 are used to graph the new pump schedule presented as Figure 2.9:

$$STB = \frac{C_{dp}l_{dp} + C_{hw}l_{hw} + C_{dc}l_{dc}}{C_p}$$

$$STB = \{(0.01776)(4000) + (0.01422)(4000)$$

$$+ (0.00743)(1000)$$

$$+ (0.00389)(1000)\} \div 0.10$$

$$STB = \textbf{1392 strokes}$$

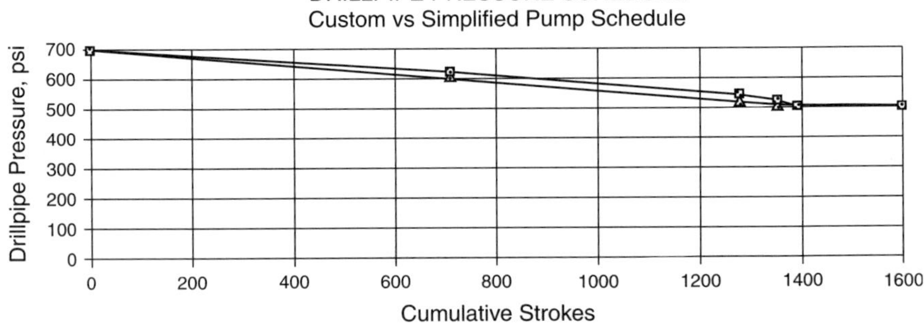

Figure 2.9

The graph is determined from Equations 2.12 and 2.13:

$$STKS\ 1 = \frac{C_{ds1}l_{ds1}}{C_p}$$

$$STKS\ 1 = \frac{(0.01776)(4000)}{0.1}$$

$$STKS\ 1 = \textbf{710 strokes}$$

$$STKS\ 2 = \frac{C_{ds1}l_{ds1} + C_{ds2}l_{ds2}}{C_p}$$

$$STKS\ 2 = \frac{(0.01776)(4000) + (0.01422)(4000)}{0.1}$$

$$STKS\ 2 = \textbf{1279 strokes}$$

$$STKS\ 3 = \frac{C_{ds1}l_{ds1} + C_{ds2}l_{ds2} + C_{ds3}l_{ds3}}{C_p}$$

$$STKS\ 3 = \{(0.01776)(4000)$$

$$+ (0.01422)(4000)$$

$$+ (0.00743)(1000)\} \div 0.10$$

$$STKS\ 3 = \textbf{1353 strokes}$$

Similarly, $STKS\ 4 = \textbf{1392 strokes}$

The circulating pressure at the surface at the end of each section of drill string is given by Equation 2.13:

$$P_1 = P_c - 0.052(\rho_1 - \rho)(l_{ds1})$$

$$+ \left(\frac{\rho_1 P_{ks}}{\rho} - P_{ks}\right)\left(\frac{STKS\ 1}{STB}\right)$$

$$P_1 = 700 - 0.052(10 - 9.6)(4000)$$

$$+ \left(\frac{(10)(500)}{(9.6)} - 500\right)\left(\frac{710}{1392}\right)$$

$$P_1 = \textbf{627 psi}$$

$$P_2 = P_c - 0.052(\rho_1 - \rho)(l_{ds1} + l_{ds2})$$

$$+ \left(\frac{\rho_1 P_{ks}}{\rho} - P_{ks}\right)\left(\frac{STKS\,2}{STB}\right)$$

$$P_2 = 700 - 0.052(10 - 9.6)(4000 + 4000)$$

$$+ \left(\frac{(10(500))}{(9.6)} - 500\right)\left(\frac{1279}{1392}\right)$$

$$P_2 = \mathbf{552\ psi}$$

$$P_3 = P_c - 0.052(\rho_1 - \rho)(l_{ds1} + l_{ds2} + l_{ds3})$$

$$+ \left(\frac{\rho_1 P_{ks}}{\rho} - P_{ks}\right)\left(\frac{STKS\,3}{STB}\right)$$

$$P_3 = 700 - 0.052(10 - 9.6)$$

$$\times (4000 + 4000 + 1000)$$

$$+ \left(\frac{(10)(500)}{(9.6)} - 500\right)\left(\frac{1353}{1392}\right)$$

$$P_3 = \mathbf{534\ psi}$$

Adding the final section of drill collars to the above equation,

$$P_4 = \mathbf{513\ psi}$$

As illustrated in Figure 2.8, changing the pump speed at which the kick is displaced merely moves the pump schedule to a parallel position on the graph. As illustrated in Figure 2.9, the complex pump schedule is slightly more difficult to construct. The simplified straight line pump schedule will underbalance the well during the period that the kill-weight mud is being displaced to the bit. In this example, the underbalance is only as much as 25 psi.

In reality, in most cases the annular frictional pressure losses, which are considered negligible in classical pressure control analysis, would more than compensate and an additional influx would not occur. However, that may not be the case in any specific instance, and an additional influx could occur. In most instances, the simplified pump schedule would suffice. In significantly complex drill strings, this comparison should be made.

SUMMARY

The Driller's Method was the first and most popular displacement procedure. The crew proceeded immediately to displace the influx. The required calculations were not difficult. The calculations were made, the kill-weight mud was easily displaced, and the drilling operation was resumed. One disadvantage of the Driller's Method is that at least two circulations are required to control the well.

The Wait and Weight Method is slightly more complicated but offers some distinct advantages. First, the well is killed in half the time. Modern mud-mixing facilities permit barite to be mixed at rates up to 600 sacks per hour with dual mixing systems; therefore, time required to weight up the suction pit is minimized and kill rate is not penalized. The Wait and Weight Method results in kill mud reaching the well sooner, and that is always an advantage.

In addition, as discussed and illustrated in Chapter 4, the annulus pressures are lower when the Wait and Weight Method is used. The primary disadvantage is the potential for errors and problems while displacing the kill-weight mud to the bit. With the Driller's Method, the procedure can be stopped and started easily. Stopping and starting when using the Wait and Weight Method is not as easy, especially during the period when the kill-weight mud is being displaced to the bit. It is not uncommon for good drilling men to get confused during displacement using the Wait and Weight Method.

In view of all considerations, the Wait and Weight Method is the preferred technique.

CHAPTER THREE

PRESSURE CONTROL PROCEDURES WHILE TRIPPING

22 June

Trip out of the hole. Well began flowing. Trip in the hole with the bottomhole assembly. Gained 60 bbls. Circulated the hole. Pressure continued to increase. Shut in well with 3000 psi. Drillpipe started coming through the rotary table. Opened choke, closed pipe rams above tool joint, and closed choke. Total gain was 140 barrels. Attempt to close safety valve without success. Open choke, close safety valve, close choke. Had 200 barrel total gain. Rigged up snubbing unit.

14 July

Began snubbing in the hole.

15 July

Continued snubbing to 4122 feet. Circulated 18.5-ppg mud. Surface pressure is 3700 psi.

16 July

Snubbed to 8870 feet. Began to circulate 18.5-ppg mud. Surface pressure increased to 5300 psi on casing. Mud in is 540 bbls versus mud out of 780 bbls. Increased pump rate from 3 bpm to 6 bpm. Drillpipe pressure increased to 6700 psi. Hammer union on rig floor washed out. Unable to close safety valves on rig floor due to excessive pressure and flow. Hydrogen sulfide monitors sounded. Ignited rig with flare gun.

The drilling report above is a good illustration of a disaster resulting from complications during tripping operations at a well that was under control only a few hours prior to the trip. It must be reasoned that any well that is tripped is under control when the bit leaves bottom. Therefore, operations subsequent to the beginning of the trip precipitate the well control problem. As is often the case, due to operations subsequent to the time that a well control problem was detected, the condition of the well deteriorated and the well control problem became progressively more complicated.

The reason for this familiar unfortunate chain of events is that the industry has been and still is inconsistent under these circumstances. A recent survey of the major well control schools revealed substantial inconsistencies under the same given circumstances.

Actually, classical pressure control procedures apply only to problems which occur during drilling operations. Unfortunately, there is no widely accepted procedure to be followed when a kick occurs during a trip. Further, procedures and instructions that apply to problems during drilling are routinely posted on the rig floor. However, just as routinely, there are often no posted instructions that apply to procedures to be followed if the kick occurs during a trip. A federal court in Pecos, Texas, found a major oil company grossly negligent because procedures for problems occurring while drilling were posted and procedures for problems occurring while tripping were not. One purpose of this chapter is to suggest that classical trip procedures be adopted, taught, and posted.

CAUSES OF KICKS WHILE TRIPPING

Any well control problem that occurs during a trip is generally the result of a failure on the part of the rig crew to keep the hole full or the failure of the crew to recognize that the hole is not filling properly. The problem of keeping the hole full of fluid has been emphasized for many years. Pressure control problems and blowouts associated with trips continue to be a major occurrence. A lack of training and understanding contributes to these circumstances.

Classical pressure control procedures apply to drilling operations, not to tripping operations. All of the modeling and technology used in pressure control was developed based on a drilling model as opposed to a tripping model. Therefore, the technology that applies to pressure control problems that occur during drilling operations does not apply to pressure control problems that occur during tripping operations. As a result, when pressure control problems occur while tripping, drilling procedures are often applied, confusion reigns, and disaster results.

TRIP SHEETS AND FILLING PROCEDURES

Prior to a trip, it is assumed that the well is under control and that a trip can be made safely if the full hydrostatic is maintained. Pressure control problems that occur during a trip are generally the result of swabbing or a simple failure to keep the hole full. In either case, recognition and prevention of the problem is much easier than the cure.

Accurate "trip sheets" must be kept whenever productive horizons have been penetrated or on the last trip before entering the transition zone or pay interval. The trip sheet is simply a record of the actual amount of mud used to keep the hole full while the drill string is being pulled, compared to the theoretical quantity required to replace the pipe that is being removed. Properly monitoring the tripping operation and utilizing the trip sheet will forewarn the crew of potential well control problems.

In order to fill and monitor the hole properly, the drillpipe must be slugged dry with a barite pill. Difficulty in keeping the pipe dry is not an acceptable excuse for failure to fill the hole properly. If the first pill fails to dry the drillpipe, pump a second pill heavier than the first. If the pipe is dry for some time and then pulls wet, pump another pill. A common question is "How frequently should the hole be filled?" The basic factors determining frequency are regulations, critical nature of the well, and the wellbore geometry.

Often, special field rules or regulatory commissions will specify the method to be used to maintain the hydrostatic in particular fields or areas. Certainly, these rules should be observed. It is acceptable to deviate from the established procedure with appropriate cause or when the procedure to be used is widely accepted as being more definitive than that established by regulation. Certainly, the condition of some wells is more critical than that of others. The critical nature may be related to location, depth, pressure, hydrocarbon composition, or toxic nature of the formation fluids, to name a few.

To illustrate the significance of the wellbore geometry, consider Example 3.1:

Example 3.1

Given:

Well depth, $D = 10,000$ feet

Mud density, $\rho = 15$ ppg

Mud gradient, $\rho_m = 0.78$ psi/ft

Length of stand, $L_{std} = 93$ feet

Displacement of:

$4\frac{1}{2}$-inch drillpipe, $DSP_{4\frac{1}{2}} = 0.525$ bbl/std

$2\frac{7}{8}$-inch drillpipe, $DSP_{2\frac{7}{8}} = 0.337$ bbl/std

Capacity of:

$12\frac{1}{4}$-inch hole less pipe displacement, $C_1 = 0.14012$ bbl/ft

$4\frac{3}{4}$-inch hole less pipe displacement, $C_2 = 0.01829$ bbl/ft

Stands of pipe to be pulled $= 10$ stands

Wellbore configuration 1:

$4\frac{1}{2}$-inch drillpipe in $12\frac{1}{4}$-inch hole

Wellbore configuration 2:

$2\frac{7}{8}$-inch drillpipe in $4\frac{3}{4}$-inch hole

Required:

Compare the loss in hydrostatic resulting from pulling 10 stands without filling from wellbore configuration 1 and wellbore configuration 2.

Solution:

Determine the displacement for 10 stands for wellbore configuration 1.

$$Displacement = (DSP_{ds})(\text{number of stands}) \qquad (3.1)$$

Where:

$$DSP_{ds} = \text{Displacement of the drill string, bbls/std}$$
$$Displacement = (0.525)(10)$$
$$Displacement = \textbf{5.25 bbls}$$

Determine hydrostatic loss for wellbore configuration 1.

$$Loss = \frac{\rho_m(Displacement)}{C_1} \qquad (3.2)$$

Where:

$$\rho_m = \text{Mud gradient, psi/ft}$$
$$C_1 = \text{Hole capacity less pipe displacement, bbl/ft}$$

$$Loss = \frac{(0.78)(5.25)}{0.14012}$$

$$Loss = \textbf{29 psi}$$

Determine the displacement for 10 stands for wellbore configuration 2.

$$Displacement = (DSP_{ds})(\text{number of stands})$$

$$Displacement = (0.337)(10)$$

$$Displacement = \textbf{3.37 bbls}$$

Determine hydrostatic loss for wellbore configuration 2.

$$Loss = \frac{\rho_m(Displacement)}{C_2}$$

$$Loss = \frac{(0.78)(3.37)}{0.01829}$$

$$Loss = \textbf{144 psi}$$

The loss in hydrostatic for the first wellbore configuration is obviously insignificant for most drilling operations. The loss of almost 150 psi in hydrostatic for the second case is much more significant. The trip margin, which is the difference between the mud hydrostatic and the formation pore pressure, is often no more than 150 psi. Therefore, all things being equal, filling after 10 stands would be acceptable in the first instance while continuous filling would be in order in the second case.

PERIODIC FILLING PROCEDURE

Periodic filling, which is filling the hole after pulling a specified number of stands, is the minimum requirement and is usually accomplished utilizing a pump stroke counter according to a schedule. The periodic filling procedure is as follows:

Periodic Filling Procedure:

Step 1

Determine the pump capacity, C_p, bbls/stk.

Step 2

Determine the drill string displacement, DSP_{ds}, bbls/std, for each section of drill string.

Step 3

Determine the number of stands of each section of drill string to be pulled prior to filling the hole.

Step 4

Determine the theoretical number of pump strokes required to fill the hole after pulling the number of stands determined in Step 3.

Step 5

Prior to reaching the critical interval, begin maintaining a record of the number of pump strokes required to fill the hole after pulling the number of stands determined in Step 3.

Maintain the data in tabular form, comparing the number of stands pulled with the actual strokes required to fill the hole, the number of strokes theoretically required and the number of strokes required on the previous trip. This data, in tabular form, is the trip sheet. The trip sheet should be posted and maintained on the rig floor.

Step 6

Mix and pump a barite slug in order to pull the drill string dry. Wait for the hydrostatic inside the drill string to equalize.

Step 7

Pull the specified number of stands.

Step 8

Zero the stroke counter. Start the pump. Observe the return of mud to the surface and record in the appropriate column the actual number of strokes required to bring mud to the surface.

Step 9

Compare the actual number of strokes required to bring mud to the surface to the number of strokes required on the previous trip and the number of strokes theoretically determined.

The trip sheet generated by the periodic filling procedure is illustrated in Example 3.2:

Example 3.2

Given:

Pump capacity, $C_p = 0.1$ bbl/stk

Drillpipe displacement, $DSP_{ds} = 0.525$ bbl/std

Drillpipe $= 4\frac{1}{2}$-inch 16.60 #/ft

Drillpipe pulled between filling $= 10$ stands

Actual strokes required as illustrated in Table 3.1

Strokes required on previous trip as illustrated in Table 3.1

Required:

Illustrate the proper trip sheet for a periodic filling procedure.

Solution:

Strokes per 10 stands

$$Strokes = \frac{DSP_{ds}(\text{number of stands})}{C_p} \qquad (3.3)$$

Where:

$DSP_{ds} = $ Displacement of the drill string, bbl/std

$C_p = $ Pump capacity, bbl/stk

$$Strokes = \frac{(0.525)(10)}{0.1}$$

$Strokes = $ **52.5 per 10 stands**

The proper trip sheet for periodic filling is illustrated as Table 3.1.

Table 3.1

Trip Sheet Periodic Filling Procedure

Cumulative Stands Pulled	Actual Strokes Required	Theoretical Strokes	Strokes Required on Previous Trip
10	55	52.5	56
20	58	52.5	57
30	56	52.5	56
...	...	...	...

The periodic filling procedure represents the minimum acceptable filling procedure. The "flo-sho," or drilling fluid return indicator, should not be used to indicate when circulation is established. It is preferable to zero the stroke counter, start the pump, and observe the flow line returns. As a matter of practice, the displacement should be determined for each different section of the drill string. The number of stands of each section to be pulled between fillings may vary. For example, the hole should be filled after each stand of drill collars since the drill collar displacement is usually approximately five times drillpipe displacement.

CONTINUOUS FILLING PROCEDURE

In critical well situations, continuous filling is recommended using a trip tank. A trip tank is a small-volume tank (usually less than 60 barrels) which permits the discerning of fractions of a barrel. The better arrangement is with the trip tank in full view of the driller or floor crew and rigged with a small centrifugal pump for filling the tank and continuously circulating the mud inside the tank through the bell nipple or drillpipe annulus and back into the trip tank. When that mechanical arrangement is used, the hydrostatic will never drop.

The procedure would be for the hole to be filled continuously. After each 10 (or some specified number of) stands, the driller would observe, record, and compare the volume pumped from the trip tank into the hole with the theoretical volume required. The procedure for continuously filling the hole using the trip tank would be as follows:

Step 1

Determine the drill string displacement, DSP_{ds}, bbl/std, for each section of drill string.

Step 2

Determine the number of stands of each section of drill string to be pulled prior to checking the trip tank.

Step 3

Determine the theoretical number of barrels required to replace the drill string pulled from the hole. Fill the hole after pulling the number of stands determined in Step 2.

Step 4

Prior to reaching the critical interval, begin maintaining a record of the number of barrels required to maintain the hydrostatic during the pulling of the number of stands determined in Step 2.

Maintain the data in tabular form, comparing the number of stands pulled with the cumulative volume required to maintain the hydrostatic, the volume required as theoretically determined, and the volume required on the previous trip. This data, in tabular form, is the trip sheet. It should be posted and maintained on the rig floor.

Step 5

Mix and pump a barite slug in order to pull the drill string dry. Wait for the hydrostatic inside the drill string to equalize.

Step 6

Fill the trip tank and isolate the hole from the mud pits.

Step 7

Start the centrifugal pump and observe the return of mud to the trip tank.

Step 8

With the centrifugal pump circulating the hole, pull the number of stands specified in Step 2.

Step 9

After pulling the number of stands specified in Step 2, observe and record the number of barrels of mud transferred from the trip tank to the hole.

Step 10

Compare the number of barrels of mud transferred from the trip tank to the hole during this trip with the same volume transferred during the previous trip and the volume theoretically determined.

The trip sheet generated by the continuous filling procedure is illustrated in Example 3.3:

Example 3.3

Given:

Drillpipe $= 4\frac{1}{2}$-inch 16.60 #/ft

Drillpipe displacement, $DSP_{ds} = 0.525$ bbl/std

Stands pulled between observations $= 10$ stands

Trip tank capacity is 60 barrels in $\frac{1}{4}$-barrel increments

Actual volume required as illustrated in Table 3.2

Volume required on previous trip as illustrated in Table 3.2

Required:

Illustrate the proper trip sheet for a continuous filling procedure.

Solution:

Determine volume per 10 stands.

$$Displacement = (DSP_{ds})(\text{number of stands})$$

$$Displacement = (0.525)(10)$$

$$Displacement = \textbf{5.25 barrels per 10 stands}$$

The proper trip sheet for periodic filling is illustrated in Table 3.2.

Generally, the actual volume of mud required to keep the hole full exceeds the theoretical calculations. The excess can be as much as 50 percent. On rare occasions, however, the actual volume requirements to keep the hole full are consistently less than those theoretically determined. Therefore, it is vitally important that the trip sheets from previous trips be kept for future reference. Whatever the fill pattern, it must be recorded faithfully for future comparison.

Table 3.2

Trip Sheet Continuous Filling Procedure

Cumulative Number of Stands	Cumulative Volume Required	Cumulative Theoretical Volume	Previous Trip
10	5.50	5.25	5.40
20	11.50	10.05	11.00
30	17.00	15.75	16.50
40	23.25	21.00	22.75
. . .	. . .	. . .	. . .

TRIPPING INTO THE HOLE

In specific instances, though not always, it is prudent to monitor displacement while tripping in the hole to insure that fluid displacement is not excessive. The best means of measuring the displacement going in the hole is to displace directly into the isolated trip tank. A trip sheet exactly like Table 3.2 should be maintained. All too often crews are relaxed and not as diligent as necessary on the trip in the hole. As a result, industry history has recorded several instances where excessive displacement went unnoticed and severe pressure control problems resulted.

Calculations and experience prove that swabbing can occur while tripping out or in. Swab pressures should be calculated, as additional trip time can be more costly and hazardous than insignificant swab pressures. Swab pressures are real and should be considered. If a well is swabbed in on a trip in the hole, the influx will most probably be inside the drillpipe rather than the annulus because the frictional pressure is greater inside the drill string.

Further, the potential for problems does not disappear once the bit is on bottom. The pit level should be monitored carefully during the first circulation after reaching bottom. The evolution of the trip gas from the mud as it is circulated to the surface may reduce the total hydrostatic sufficiently to permit a kick.

Special attention is due when using inverted oil-emulsion systems. Historically, influxes into oil muds are difficult to detect. Because gas is infinitely soluble in oil, significant quantities of gas may pass undetected by the usual means until the pressure is reduced to the bubble point for that particular hydrocarbon mixture. At that point, the gas can flash out of solution, unload the annulus, and result in a kick.

SHUT-IN PROCEDURE

WELL KICKS WHILE TRIPPING

When a trip sheet is maintained and the well fails to fill properly, the correct procedure is as follows:

Step 1

The hole is observed not to be filling properly. Discontinue the trip and check for flow.

Step 2

If the well is observed to be flowing, space out as may be necessary.

Step 3

Stab a full opening valve in the drillpipe and shut in the drillpipe.

Step 4

Open the choke line, close the blowout preventers, and close the choke, pressure permitting.

Step 5

Observe and record the shut-in drillpipe pressure, the shut-in annulus pressure, and the volume of formation fluid that has invaded the wellbore.

Step 6

Repeat Step 5 at 15-minute intervals.

Step 7

Prepare to strip or snub back to bottom.

The following is a discussion of each step:

Step 1

The hole is observed not to be filling properly. Discontinue the trip and check for flow.

If the hole is not filling properly, it should be checked for flow. The observation period is a function of experience in the area, the productivity of the productive formation, the depth of the well and the mud type. Under most conditions, 15 minutes is sufficient. In a deep well below 15,000 feet, or if oil-based mud is being used, the observation period should be extended to a minimum of 30 minutes.

If the well is not observed to be flowing, the trip can be continued with the greatest caution. If a periodic filling procedure is being used, the hole should be filled after each stand and checked for flow until the operation returns to normal. If, after pulling another designated number of stands, the well continues to fill improperly, the trip should be discontinued.

If the well is not flowing when the trip is discontinued, the bit may be returned cautiously to bottom. In the event that the bit is returned to bottom, the displacement of each stand should be monitored closely and the well should be checked for flow after each stand.

If at any time the well is observed to be flowing, the trip should be discontinued. It is well known that in many areas of the world trips are made with the well flowing; however, these should be considered isolated instances and special cases.

Step 2

If the well is observed to be flowing, space out as may be necessary.

The drill string should be spaced out to insure that there is not a tool joint in the rams. If that is not a consideration, a tool joint is normally spotted at the connection position.

Step 3

Stab a full opening valve in the drillpipe and shut in the drillpipe.

The drillpipe should be shut in first. It is well known that the ball valves normally used to shut in the drillpipe are difficult to close under flow or pressure.

Step 4

Open the choke line, close the blowout preventers, and close the choke, pressure permitting.

This step represents a soft shut-in. For a discussion of the soft shut-in as opposed to the hard shut-in, refer to the shut-in procedure for the Driller's Method in Chapter 2.

Step 5

Observe and record the shut-in drillpipe pressure, the shut-in annulus pressure, and the volume of formation fluid that has invaded the wellbore.

Step 6

Repeat Step 5 at 15-minute intervals.

If the well has been swabbed in, the bubble should be below the bit, as illustrated in Figure 3.1. In that event, the shut-in drillpipe pressure will be equal to the shut-in casing pressure. The pressures and influx must be monitored at 15-minute intervals in order to insure that the well is effectively shut in, to establish the true reservoir pressure, and to monitor bubble rise. Bubble rise is discussed in Chapter 4.

Step 7

Prepare to strip or snub back to bottom.

Stripping is a simple operation. However, stripping does require a means of bleeding and accurately measuring small volumes of mud as the drill string is stripped in the hole. A trip tank is adequate for measuring the mud volumes bled. Alternatively, a service company pump truck can be rigged up to the annulus and the displacement can be measured into its displacement tanks. If the gain is large and the pressures are high, stripping through the rig equipment may not be desirable. This point is more thoroughly discussed in Snubbing Operations in Chapter 6.

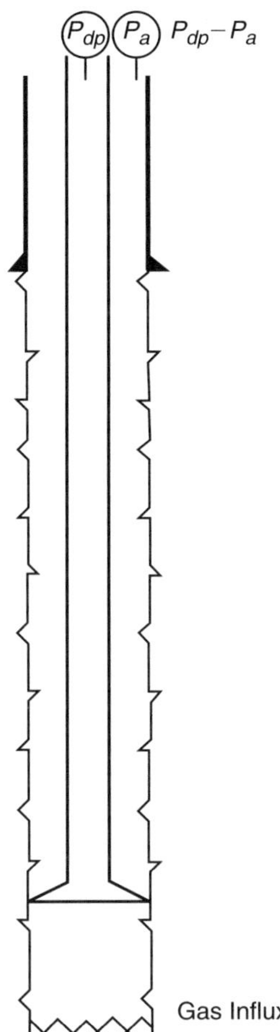

Figure 3.1   *Influx Swabbed In.*

Once shut in, the well is under control. Properly done, the surface pressure on the well should be less than 500 psi, which is almost insignificant. With the well shut in, there is adequate time to consider alternatives without danger of disaster. The well that has been swabbed in is a well that was under control when the bit started off bottom. Properly done, it should be a simple procedure to return the well to full control.

Once the well is shut in and under control, the options are follows:

STRIPPING IN THE HOLE

Stripping in is not complicated if a few simple rules and concepts are observed. The procedure is as follows:

Step 1

Install a back-pressure valve on top of the safety valve in the drill string. Open the safety valve.

Step 2

Determine the displacement in barrels per stand of the drill string to be stripped into the hole. Consider that the inside of the drill string will be void.

Step 3

Determine the anticipated increase in surface pressure when the bit enters the influx according to Equation 3.4:

$$\Delta P_{incap} = \frac{DSP_{ds}}{C_{dsa}}(\rho_m - \rho_f) \tag{3.4}$$

Where:

DSP_{ds} = Displacement of drill string, bbl/std

$$C_{dsa} = \text{Capacity of drill string annulus, bbl/ft}$$
$$\rho_m = \text{Mud gradient, psi/ft}$$
$$\rho_f = \text{Influx gradient, psi/ft}$$

$$\rho_f = \frac{S_g P_b}{53.3 z_b T_b} \tag{3.5}$$

Where:

$$S_g = \text{Specific gravity of gas}$$
$$P_b = \text{Bottomhole pressure, psi}$$
$$T_b = \text{Bottomhole temperature, }^\circ\text{Rankine}$$
$$z_b = \text{Compressibility factor}$$

Step 4

Determine the anticipated top of the influx, TOI, pursuant to Equation 3.6:

$$TOI = D - \frac{Influx\ Volume}{C_h} \tag{3.6}$$

Where:

$$D = \text{Well depth, feet}$$
$$C_h = \text{Hole capacity, bbl/ft}$$

Step 5

Prepare to lubricate the drill string with water as it passes through the surface equipment.

Step 6

Lower one stand into the hole.

Step 7

At the same time, bleed and precisely measure the displacement determined in Step 2.

Step 8

Shut in the well.

Step 9

Read and record in tabular form the shut-in casing pressure. Compare the shut-in casing pressures before and after the stand was lowered into the hole.

Note: The shut-in casing pressure should remain constant until the bit reaches the influx.

Step 10

Repeat Step 9 until the bit reaches the top of the influx as determined in Step 4.

Step 11

When the bit reaches the top of the influx as determined in Step 4, the shut-in surface pressure will increase even after the proper volume of mud is released.

Read and record the new shut-in casing pressure and compare with the original shut-in casing pressure and the anticipated increase in shut-in casing pressure as determined in Step 2.

The new shut-in casing pressure should not be greater than the original shut-in casing pressure plus the anticipated increase.

Step 12

Repeat Step 11 until the bit is on bottom.

Step 13

Once the bit is on bottom, circulate out the influx using the Driller's Method as outlined in Chapter 2.

Step 14

If necessary, circulate and raise the mud weight and trip out of the hole.

Each step will be discussed as appropriate:

Step 1

Install a back-pressure valve on top of the safety valve in the drill string. Open the safety valve.

Step 2

Determine the displacement in barrels per stand of the drill string to be stripped into the hole. Consider that the inside of the drill string will be void.

Step 3

Determine the anticipated increase in surface pressure when the bit enters the influx according to Equation 3.4:

$$\Delta P_{incap} = \frac{DSP_{ds}}{C_{dsa}}(\rho_m - \rho_f)$$

When the bit enters the influx, the influx will become longer because it will then occupy the annular area between the drill string and the hole as opposed to the open hole. Therefore, the volume of mud which is bled from the well will be replaced by the increased length of the influx. Provided that the volume of mud bled from the well is exactly as determined, the result is that the surface pressure will increase automatically by the difference between the hydrostatic of the mud and the hydrostatic of the influx, and the bottomhole pressure will remain constant. No additional influx will be permitted.

Step 4

Determine the anticipated top of the influx, TOI, pursuant to Equation 3.6:

$$TOI = D - \frac{Influx\ Volume}{C_h}$$

It is necessary to anticipate the depth at which the bit will enter the influx. As discussed, the annular pressure will suddenly increase when the bit enters the influx.

Step 5

Prepare to lubricate the drill string with water as it passes through the surface equipment.

Lubricating the string as it is lowered into the hole will reduce the weight required to cause the drill string to move. In addition, the lubricant will reduce the wear on the equipment used for stripping.

Step 6

Lower one stand into the hole.

It is important that the drill string be stripped into the hole, stand by stand.

Step 7

At the same time, bleed and precisely measure the displacement determined in Step 2.

It is vital that the volume of mud removed is exactly replaced by the drill string that is stripped into the well.

Step 8

Shut in the well.

Step 9

Read and record in tabular form the shut-in casing pressure. Compare the shut-in casing pressures before and after the stand was lowered into the hole.

Note: The shut-in casing pressure should remain constant until the bit reaches the influx.

It is important to monitor the surface pressure after each stand. Prior to the bit entering the influx, the surface pressure should remain constant. However, if the influx is gas and begins to migrate to the surface, the surface pressure will slowly begin to increase. The rate of increase in surface pressure indicates whether the increase is caused by influx migration or penetration of the influx.

If the increase is due to influx penetration, the pressure will increase rapidly during the stripping of one stand. If the increase is due to influx migration, the increase is almost imperceptible for a single stand. Bubble migration and the procedure for stripping in the hole with influx migration is discussed in Chapter 4.

Step 10

Repeat Step 9 until the bit reaches the top of the influx as determined in Step 4.

Step 11

When the bit reaches the top of the influx as determined in Step 4, the shut-in surface pressure will increase even after the proper volume of mud is released.

Read and record the new shut-in casing pressure and compare with the original shut-in casing pressure and the anticipated increase in shut-in casing pressure as determined in Step 2.

The new shut-in casing pressure should not be greater than the original shut-in casing pressure plus the anticipated increase.

Step 12.

Repeat Step 11 until the bit is on bottom.

Step 13

Once the bit is on bottom, circulate out the influx using the Driller's Method as outlined in Chapter 2.

Step 14

If necessary, circulate and raise the mud weight and trip out of the hole.

The procedure for stripping is illustrated in Example 3.4:

Example 3.4

Given:

Wellbore $=$ Figure 3.2

Number of stands pulled $=$ 10 stands

Length per stand, $L_{std} = 93$ ft/std

Stands stripped into the hole $=$ 10 stands

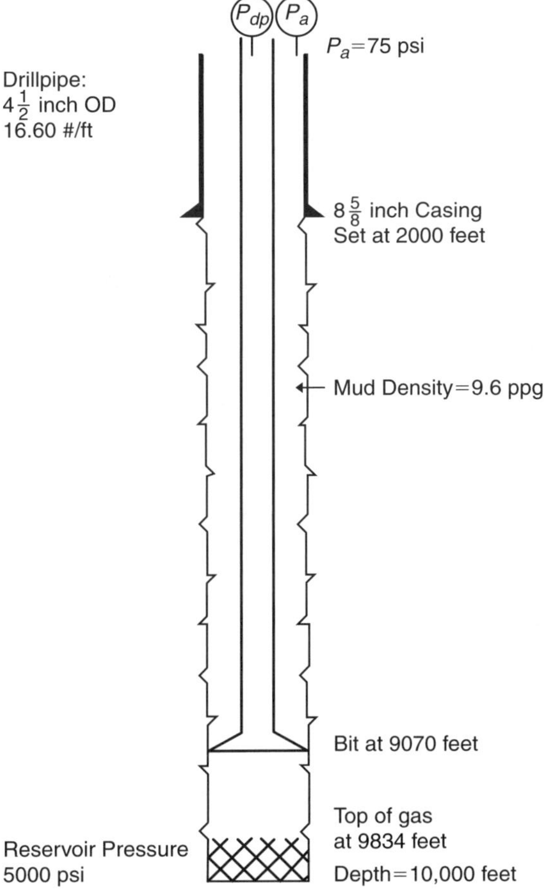

Figure 3.2   *Wellbore Schematic.*

Drill string to be stripped $= 4\frac{1}{2}$-inch 16.60 #/ft

Drill string displacement, $DSP_{ds} = 2$ bbl/std

Mud density, $\rho = 9.6$ ppg

Influx $= 10$ bbls of gas

Annular capacity, $C_{dsa} = 0.0406$ bbl/ft

Depth, $D = 10,000$ feet

Hole diameter, $D_h = 7\frac{7}{8}$ inches

Hole capacity, $C_h = 0.0603$ bbl/ft

Bottomhole pressure, $P_b = 5000$ psi

Bottomhole temperature, $T_b = 620\ °$Rankine

Gas specific gravity, $S_g = 0.6$

Shut-in casing pressure, $P_a = 75$ psi

Required:

Describe the procedure for stripping the 10 stands back to bottom.

Solution:

Determine influx height, h_b:

$$h_b = \frac{Influx\ Volume}{C_h} \tag{3.7}$$

Where:

$$C_h = \text{capacity, bbl/ft}$$

$$h_b = \frac{10}{0.0603}$$

$$h_b = \mathbf{166\ feet}$$

Determine the top of the influx using Equation 3.6:

$$TOI = D - \frac{Influx\ Volume}{C_h}$$

$$TOI = 10,000 - \frac{10}{0.0603}$$

$$TOI = \mathbf{9834\ feet}$$

The bit will enter the influx on the 9th stand.

Determine the depth to the bit:

$$Depth\ to\ Bit = D - (\text{number of stands})(L_{std}) \qquad (3.8)$$

Where:

$D = $ Well depth, feet

$L_{std} = $ Length of a stand, feet

$Depth\ to\ Bit = 10{,}000 - (10)(93)$

$Depth\ to\ Bit = \textbf{9070 feet}$

Determine ΔP_{incap} using Equation 3.4:

$$\rho_f = \frac{S_g(P_b)}{53.3 z_b T_b}$$

$$\rho_f = \frac{0.6(5000)}{53.3(1.1)(620)}$$

$$\rho_f = \textbf{0.0825 psi/ft}$$

$$\Delta P_{incap} = \frac{DSP_{ds}}{C_{dsa}}(\rho_m - \rho_f)$$

$$\Delta P_{incap} = \frac{2}{0.0406}(0.4992 - 0.0825)$$

$$\Delta P_{incap} = \textbf{20.53 psi/std}$$

$$\Delta P_{incap} = \textbf{0.22 psi/ft}$$

Therefore, lower 8 stands, bleeding and measuring 2 barrels with each stand. The shut-in casing pressure will remain constant at 75 psi.

Table 3.3

Stripping Procedure Example 3.4

Stand Number	Beginning Shut-in Time	Initial Annulus Pressure	Barrels Bled	Final Shut-in Annulus Pressure
1	0800	75	2	75
2	0810	75	2	75
3	0820	75	2	75
4	0830	75	2	75
5	0840	75	2	75
6	0850	75	2	75
7	0860	75	2	75
8	0900	75	2	75
9	0910	75	2	91
10	0920	91	2	112

Lower the 9th stand, bleeding 2 barrels. Observe that the shut-in casing pressure increases to 91 psi:

$$P_{an} = P_a + \Delta P_{incap}(\textit{feet of influx penetrated}) \qquad (3.9)$$

Where:

P_a = Shut-in casing pressure, psi

ΔP_{incap} = Increase in pressure with bit penetration

$$P_{an} = 75 + 0.22(166 - 93)$$

$$P_{an} = \textbf{91 psi}$$

Lower the 10th stand, bleeding 2 barrels. Observe that the shut-in casing pressure increases to 112 psi.

Maintain the results as in Table 3.3. Table 3.3 summarizes the stripping procedure. Note that if the procedure is properly done, the shut-in casing pressure remains constant until the bit penetrates the influx. This is true only if the influx is not migrating. The stripping procedure must be modified to accommodate the case of migrating influx. The proper stripping procedure, including migrating influx, is presented in Chapter 4.

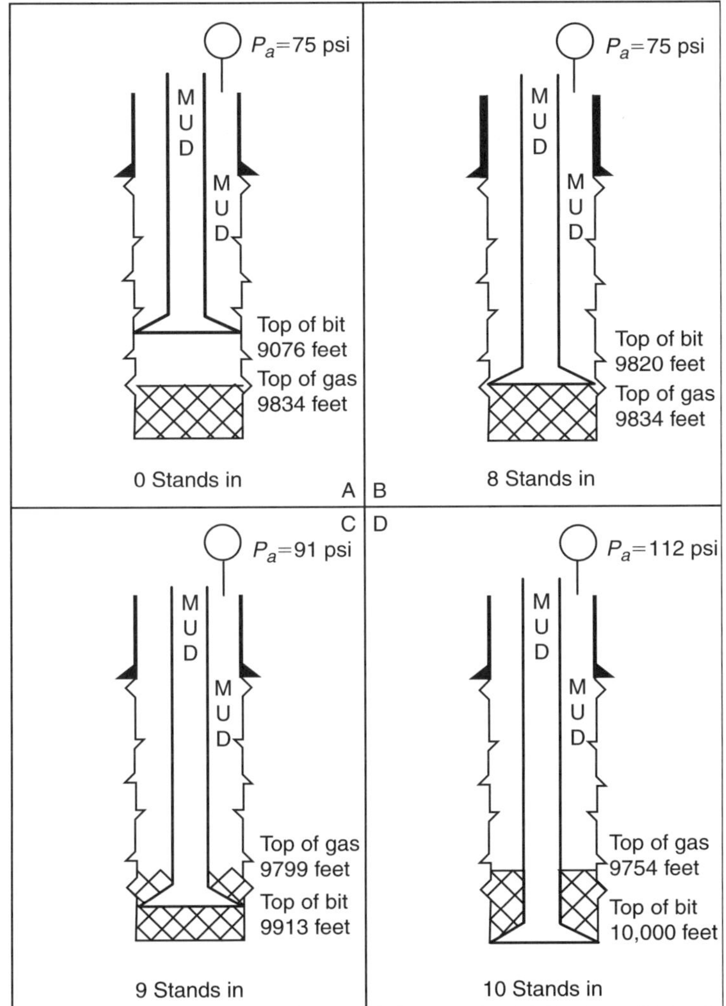

Figure 3.3

Example 3.4 is further summarized in Figure 3.3. Figure 3.3 illustrates the relative positions of the bit with respect to the influx during stages of the stripping operation. With 8 stands stripped into the hole, the bit is at 9820 feet, or 14 feet above the top of the influx. Until then, the shut-in surface pressure has remained constant at 75 psi. When the 9th stand is run, the bit enters the influx and the shut-in surface pressure increases to 91 psi.

On the last stand, the bit is in the influx and the shut-in surface pressure increases to 112 psi. Throughout the stripping procedure, the bottomhole pressure has remained constant at 5000 psi.

One should not conclude that stripping is performed while keeping the shut-in surface pressure constant. As illustrated in Example 3.4, had keeping the shut-in surface pressure constant been the established procedure, the well would have been underbalanced during the time that the last 2 stands were run, and additional influx would have resulted. Certainly, the shut-in surface pressure must be considered. However, it is only one of many important factors.

Once the bit is back on bottom, the influx can easily be circulated to the surface using the Driller's Method as outlined in Chapter 2. With the influx circulated out, the well is under control and in the same condition as before the trip began.

CHAPTER FOUR

SPECIAL CONDITIONS, PROBLEMS, AND PROCEDURES IN WELL CONTROL

9 November

Took kick of approximately 100 barrels at 1530 hours. Shut in well with 2400 psi on drillpipe and casing. At 1730 pressure increased to 2700 psi. At 1930 pressure increased to 3050 psi.

10 November

0200 pressure is 2800 psi. Determined total gain was 210 barrels. At 0330 pressure is 3800 psi. Bled 10 barrels of mud and pressure dropped to 3525 psi. At 0445 pressure is 3760 psi and building 60 psi every 15 minutes. At 0545 pressure is 3980 psi and building 40 psi every 15 minutes. Pressure stabilized at 4400 psi at 1000 hours.

As illustrated in this drilling report, the circumstances surrounding a kick do not always fit the classic models. In this drilling report from southeast New Mexico, the bit was at 1500 feet and total depth was below 14,000 feet. In addition, the surface pressures were changing rapidly. These conditions are not common to classical pressure control procedures and must be given special consideration.

This chapter is intended to discuss non-classical situations. It is important to understand classical pressure control. However, for every well control situation that fits within the classical model, there is a well control situation which bears no resemblance to the classical. According to statistics reported by the industry to the UK Health and Safety Executive, classic kicks are uncommon. For the three-year period from 1990 to 1992, of the 179 kicks reported, only 39 (22%) were classic.

The student of well control must be aware of the situations in which classical procedures are appropriate and be capable of distinguishing those non-classic situations where classical procedures have no application. In addition, when the non-classic situation occurs, it is necessary to know and understand the alternatives and which one has the greatest potential for success. In the non-classical situation, the use of classic procedures may result in the deterioration of the well condition to the point that the well is lost or the rig is burned.

SIGNIFICANCE OF SURFACE PRESSURES

In any well control situation, the pressures at the surface reflect the heart of the problem. A well out of control must obey the laws of physics and chemistry. Therefore, it is for the well control specialists to analyze and understand the problem. The well will always accurately communicate its condition. It is for us to interpret the communication properly.

A KICK IS TAKEN WHILE DRILLING

As discussed in Chapter 2 on classical pressure control, when a kick is taken while drilling and the well is shut in, the shut-in drillpipe pressure and the shut-in casing pressure are routinely recorded. The relationship between these two pressures is very important. The applicability of the classic Driller's or Wait and Weight Method must be considered in light of the relationship between the shut-in drillpipe pressure and the shut-in casing pressure.

Consider the classical U-Tube Model presented as Figure 4.1. In this figure, the left side of the U-Tube represents the drillpipe while the right side represents the annulus. When the well is first shut in, the possible relationships between the shut-in drillpipe pressure, P_{dp}, and the shut-in annulus pressure, P_a, are described in Inequalities 4.1 and 4.2 and Equation 4.3 as follows:

$$P_a > P_{dp} \tag{4.1}$$

$$P_a < P_{dp} \tag{4.2}$$

$$P_a \cong P_{dp} \tag{4.3}$$

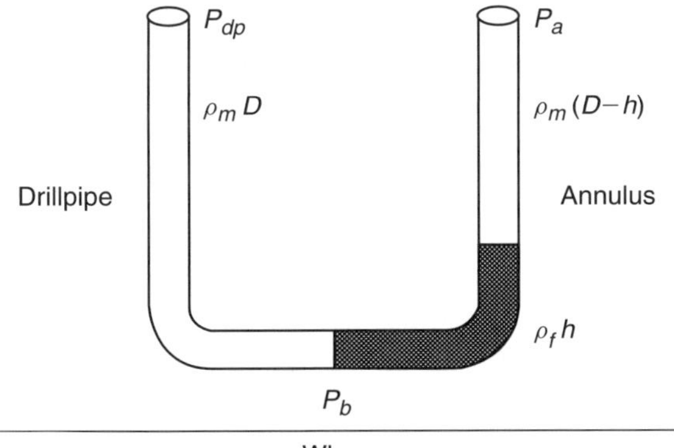

Where:

P_{dp} = Shut-in drillpipe pressure, psi
P_a = Shut-in annulus pressure, psi
D = Depth of well, feet
ρ_m = Mud density, psi/ft
ρ_f = Influx density, psi/ft
h = Height of influx, psi/ft
P_b = Bottomhole pressure, psi

Figure 4.1   *Classic U-Tube Model.*

With respect to the U-Tube Model in Figure 4.1, the bottomhole pressure, P_b, may be defined by the conditions on the drillpipe side and the conditions on the annulus side pursuant to Equations 2.6 and 2.7.

On the drillpipe side, P_b is given by Equation 2.6 as follows:

$$P_b = \rho_m D + P_{dp}$$

On the annulus side, P_b is given by Equation 2.7 as follows:

$$P_b = \rho_f h + \rho_m (D - h) + P_a$$

Equations 2.6 and 2.7 may be rearranged to give the following expressions for P_{dp} and P_a, respectively:

$$P_{dp} = P_b - \rho_m D$$

and

$$P_a = P_b - \rho_m (D - h) - \rho_f h$$

To illustrate the significance of the relationship between the shut-in drillpipe pressure and the shut-in annulus pressure, it is first assumed that a kick is taken, the well is shut in, and the shut-in annulus pressure is greater than the shut-in drillpipe pressure as expressed in Inequality 4.1:

$$P_a > P_{dp}$$

Substituting Equation 2.6 for the right side of the inequality and Equation 2.7 for the left side of the inequality results in the following expression:

$$P_b - \rho_m(D - h) - \rho_f h > P_b - \rho_m D$$

Expanding the terms gives

$$P_b - \rho_m D + \rho_m h - \rho_f h > P_b - \rho_m D$$

Adding $\rho_f h$ and $\rho_m D$ to both sides and subtracting P_b from both sides gives

$$P_b - P_b + \rho_m D - \rho_m D + \rho_m h + \rho_f h - \rho_f h >$$
$$P_b - P_b + \rho_m D - \rho_m D + \rho_f h$$

Simplifying the above equation results in the following:

$$\rho_m h > \rho_f h$$

Finally, dividing both sides of the inequality by h gives the consideration that

$$\rho_m > \rho_f \tag{4.4}$$

The significance of the analysis is this: When a well is on bottom drilling, a kick is taken and the well is shut in pursuant to the condition illustrated by the U-Tube Model in Figure 4.1. One of the conditions presented as expressions 4.1, 4.2, or 4.3 must describe the relationship between the shut-in drillpipe pressure and the shut-in casing pressure.

For the purpose of illustration in this analysis, it was assumed that the shut-in casing pressure was greater than the shut-in drillpipe pressure as described in Inequality 4.1, which results in Inequality 4.4. Therefore, it is a certainty that, when a well kicks and is shut in, if the shut-in annulus pressure is greater than the shut-in drillpipe pressure, the density of the fluid that has entered the wellbore must be less than the density of the mud that is in the wellbore.

If the mud density is 15 ppg, the pressure on the annulus, P_a, must be greater than the pressure on the drillpipe, P_{dp}, since the heaviest naturally occurring salt water is about 9 ppg. If the mud density is 9 ppg and P_a is greater than P_{dp}, the fluid entering the wellbore is gas or some combination of gas and oil or water. (Determining the density of the fluid that entered the wellbore is illustrated later in this chapter.)

Consider that by similar analysis it can be shown that if the Inequality 4.2 or Equation 4.3 describes the relationship between the shut-in drillpipe pressure and the shut-in annulus pressure, the following must be true:

If

$$P_a < P_{dp}$$

it must follow that

$$\rho_m < \rho_f \tag{4.5}$$

or if

$$P_a \cong P_{dp}$$

it must follow that

$$\rho_m \cong \rho_f \tag{4.6}$$

That is, if the shut-in annulus pressure is less than or equal to the shut-in drillpipe pressure, the density of the fluid that has entered the wellbore must be greater than or equal to the density of the mud in the wellbore!

Further, when the density of the drilling mud in the wellbore is greater than 10 ppg or when the influx is known to be significantly hydrocarbon, it is theoretically not possible for the shut-in casing pressure to be equal to or less than the shut-in drillpipe pressure.

Now, here is the point and it is vitally important that it be understood. If the reality is that the well is shut in, the density of the mud exceeds 10 ppg, or the fluid which has entered the wellbore is known to be significantly hydrocarbon, and the shut-in annulus pressure is equal to or less than the shut-in drillpipe pressure, the mathematics and reality are incompatible.

When the mathematics and reality are incompatible, the mathematics have failed to describe reality. In other words, something is wrong downhole. Something is different downhole than assumed in the mathematical U-Tube Model, and that something is usually that lost circulation has occurred and the well is blowing out underground. The shut-in annulus pressure is influenced by factors other than the shut-in drillpipe pressure.

As a test, pump a small volume of mud down the drillpipe with the annulus shut in and observe the shut-in annulus pressure. No response indicates lost circulation and an underground blowout.

Whatever the cause of the incompatibility between the shut-in casing pressure and the shut-in drillpipe pressure, the significance is that under these conditions the U-Tube Model is not applicable and **CLASSIC PRESSURE CONTROL PROCEDURES ARE NOT APPLICABLE.** The Driller's Method will not control the well! The Wait and Weight Method will not control the well! "Keep the drillpipe pressure constant" has no more meaning under these conditions than any other five words in any language. "Pump standing on left foot" has as much significance and as much chance of success as "Keep the drillpipe pressure constant"!

Under these conditions, non-classical pressure control procedures must be used. There are no established procedures for non-classical pressure control operations. Each instance must be analyzed considering the unique and individual conditions, and the procedure must be detailed accordingly.

INFLUX MIGRATION

To suggest that a fluid of lesser density will migrate through a fluid of greater density should be no revelation. However, in drilling operations there are many factors that affect the rate of influx migration. In some instances, the influx has been known not to migrate.

In recent years there has been considerable research related to influx migration. In the final analysis the variables required to predict the rate of influx migration are simply not known in field operations. The old field rule of migration of approximately 1000 feet per hour has proven to be as reliable as many much more theoretical calculations.

Some interesting and revealing observations and concepts have resulted from the research which has been conducted. Whether or not the influx will migrate depends upon the degree of mixing which occurs when the influx enters the wellbore. If the influx that enters over a relatively long period of time is significantly distributed as small bubbles in the mud and the mud is viscous, the influx may not migrate. If the influx enters the wellbore in a continuous bubble, as is the case when the influx is swabbed into the wellbore, it is more likely to migrate. As the viscocity approaches that of water, the probability of migration increases.

Researchers have observed many factors which will influence the rate of migration of an influx. For example, a migrating influx in a vertical annulus will travel up one side of the annulus with liquid backflow occupying an area opposite the influx. In addition, the migrating velocity of an influx is affected by annular clearances. The smaller the annular clearances, the slower the influx will migrate. The greater the density difference between the influx and the drilling mud, the faster the influx will migrate.

Therefore, the composition of the influx will affect the rate of migration, as will the composition of the drilling fluid. Further, the rate of migration of an influx is reduced as the viscosity of the drilling mud is increased. Finally, an increase in the velocity of the drilling fluid will increase the migration velocity of the influx. Obviously, without specific laboratory tests on the drilling fluid, the influx fluid, and the resulting mixture of the fluids in question, predictions concerning the behavior of an influx would be virtually meaningless.

As previously stated, the surface pressures are a reflection of the conditions in the wellbore. Influx migration can be observed and analyzed from the changes in the shut-in surface pressures. Basically, as the influx migrates toward the surface, the shut-in surface pressure increases provided that the geometry of the wellbore does not change. An increase in the surface pressure is the result of the reduction in the drilling mud hydrostatic above the influx as it migrates through the drilling mud toward the surface.

As the influx migrates and the surface pressure increases, the pressure on the entire wellbore also increases. Thereby, the system is super-pressured until the fracture gradient is exceeded or until mud is released at the surface, permitting the influx to expand properly. The procedures for controlling the pressures in the wellbore as the influx migrates are discussed later in this section.

At this point it is important to understand that, even under ideal conditions, the surface annular pressure will increase as the influx migrates, provided that the geometry of the wellbore does not change. If the casing is larger in the upper portion of the wellbore and the influx is permitted to expand properly, the surface pressure will decrease as the length of the influx shortens in the larger diameter casing. After decreasing as the influx enters the larger casing, the surface pressure will increase as the influx continues to migrate toward the surface.

A few field examples illustrate the points discussed. At the well in southeastern New Mexico from which the drilling report at the beginning of this chapter was excerpted, a 210-barrel influx was taken while on a trip at 14,000 feet. The top of the influx in the $6\frac{1}{2}$-inch hole was calculated to be at 8326 feet. The kick was taken at 1530 hours and migrated to the surface through the 11.7-ppg water-base mud at 1000 hours the following morning. The average rate of migration was 450 feet per hour.

At the Pioneer Corporation Burton #1 in Wheeler County, Texas, all gas was circulated out of the wellbore at the end of each day for more than one month. In this case, the gas migrated from 13,000 feet to the surface in the 7-inch casing by $2\frac{7}{8}$-inch tubing annulus in approximately 8 hours. The average rate of migration under those conditions was calculated to be 1600 feet per hour. However, the pressure was recorded and the rate of rise was exponential.

It is generally considered that an influx will migrate faster in a directional well with all other factors being equal (Figure 4.2). At a directional well in New Zealand, 7-inch casing had been set to 2610 meters and a 6-inch core was being recovered from 2777 meters measured depth, 2564 meters true vertical depth. The wellbore was deviated 38 degrees from vertical at an azimuth of 91 degrees. The gel polymer mud density was

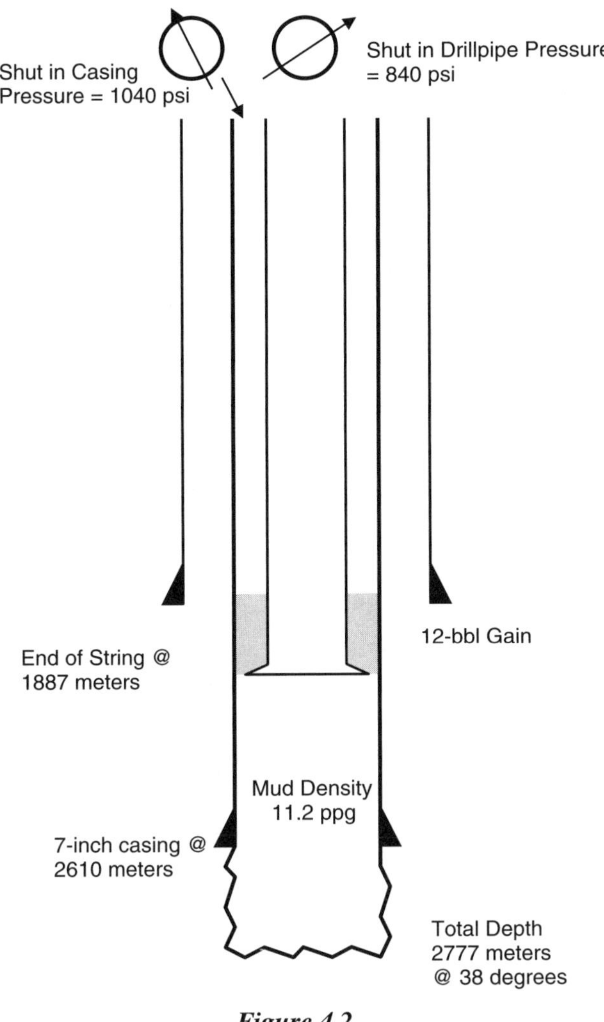

Figure 4.2

11.2 ppg, the plastic viscosity was 29 centipoise, and the yield point was 29 pounds force per 100 square feet.

At 0700 hours with the core bit at 1884 meters on the trip out of the hole to recover the core, an influx was taken and the well was shut in. A 12-barrel influx was recorded. The shut-in drillpipe and casing pressures were 830 psi and 1040 psi, respectively.

Analysis of the pressure data indicated that the top of the influx was at 1428 meters or 4685 feet. Preparations were being made to strip the core bit back to bottom. During the preparations, the influx migrated, reaching the surface before stripping operations could be commenced. Gas was detected at the surface at 1030 hours for a migration rate of 1330 feet per hour.

More often than one would think, the influx will not migrate. Consider another well in New Zealand where 7-inch casing had been set at 2266 meters (Figure 4.3). After coring to a total depth of 2337 meters, a trip was commenced to retrieve the core. The gel polymer mud had a density of 10.1 ppg, a plastic viscosity of 14 centipoise, a yield point of 16 pounds force per 100 square feet, and a funnel viscosity of 40 seconds per liter.

With the core bit at 757 meters on the trip out, a gain of three barrels was observed. The well was shut in with a total gain of six barrels. The shut-in drillpipe and annulus pressures were equal at 350 psi. The kelly was picked up. The choke was opened and the well was circulated. When the well was shut in the second time, the total gain was 115 barrels and the shut-in surface pressures were equal at 1350 psi.

Calculations indicated that the top of the influx was only 1335 meters from the surface. With these conditions, intuition and experience would strongly suggest that the influx would rise rapidly to the surface. The operator decided to permit the influx to migrate to the surface. Accordingly, the well was observed for three days. Ironically, the surface pressures remained constant at 1320 psi, indicating no influx movement.

During the next 24 hours, the drill string was stripped back to the bottom of the hole. As the drill string was being stripped, seven barrels of gas were randomly bled from the annulus along with the appropriate quantity of mud, which was replaced by the drill string. The remainder of the original 115 barrels of gas was in a bubble on the bottom of the hole and had to be circulated to the surface.

At the E. N. Ross No. 2 near Jackson, Mississippi, a 260-barrel kick was taken inside a $7\frac{5}{8}$-inch liner while out of the hole on a 19,419-foot sour gas well. The top of the influx was calculated to be at 13,274 feet. A 17.4-ppg oil-base mud was being used. The initial shut-in surface pressure was 3700 psi. The pressure remained constant for the next 17 days while snubbing equipment was being rigged up, indicating that the influx did not move during that 17 days. After 17 days, the influx began to migrate into the

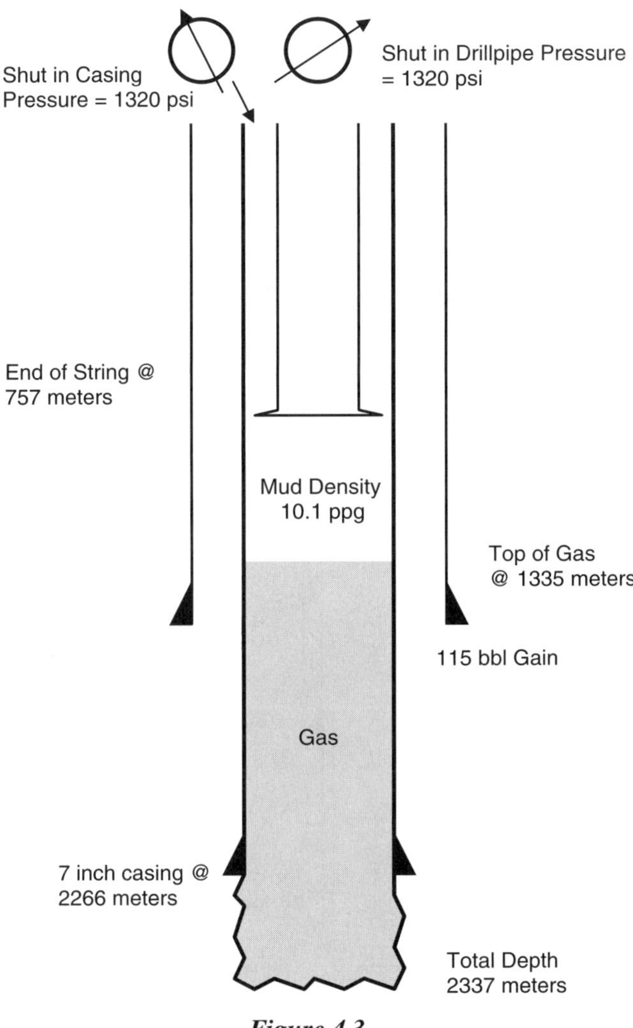

Figure 4.3

$9\frac{5}{8}$-inch intermediate casing and the surface pressure declined accordingly. Six days later the influx was encountered during snubbing operations at 10,000 feet. The influx had migrated only 3274 feet in six days. Consider Example 4.1:

Example 4.1

Given:

Wellbore schematic $=$ Figures 4.4 and 4.5

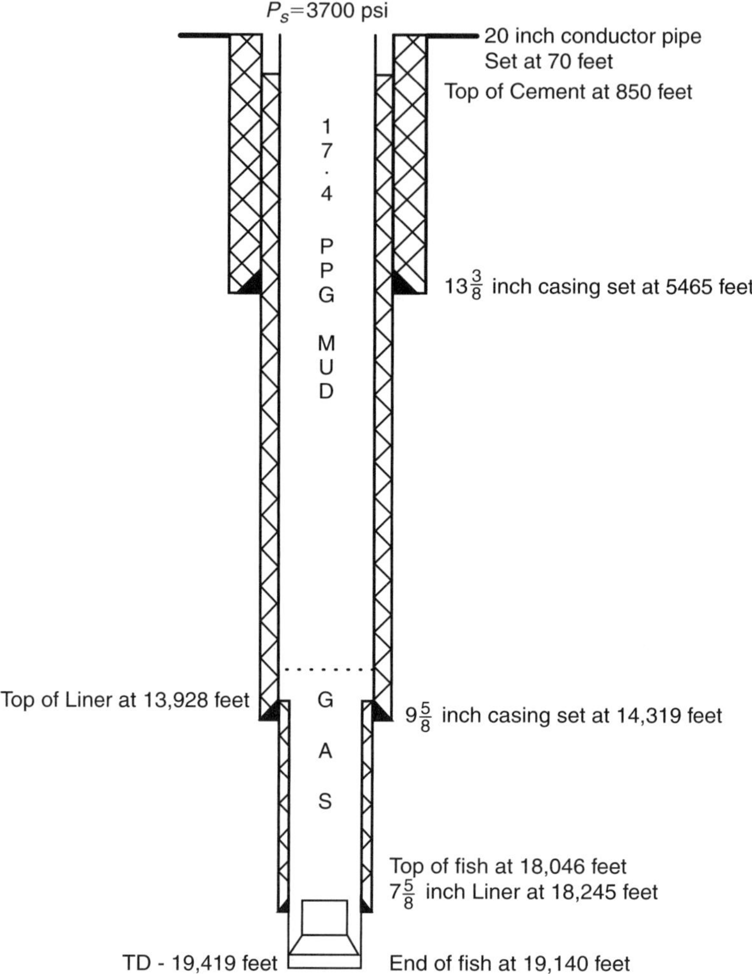

Figure 4.4   *E.N. Ross No. 2 Conditions after Initial Kick.*

Top of $7\frac{5}{8}$-inch liner, $D_t = 13{,}928$ ft

Well depth, $D = 19{,}419$ ft

Influx volume $= 260$ bbl

Capacity of casing, $C_{dpca} = 0.0707$ bbl/ft

Mud weight, $\rho = 17.4$ ppg OBM

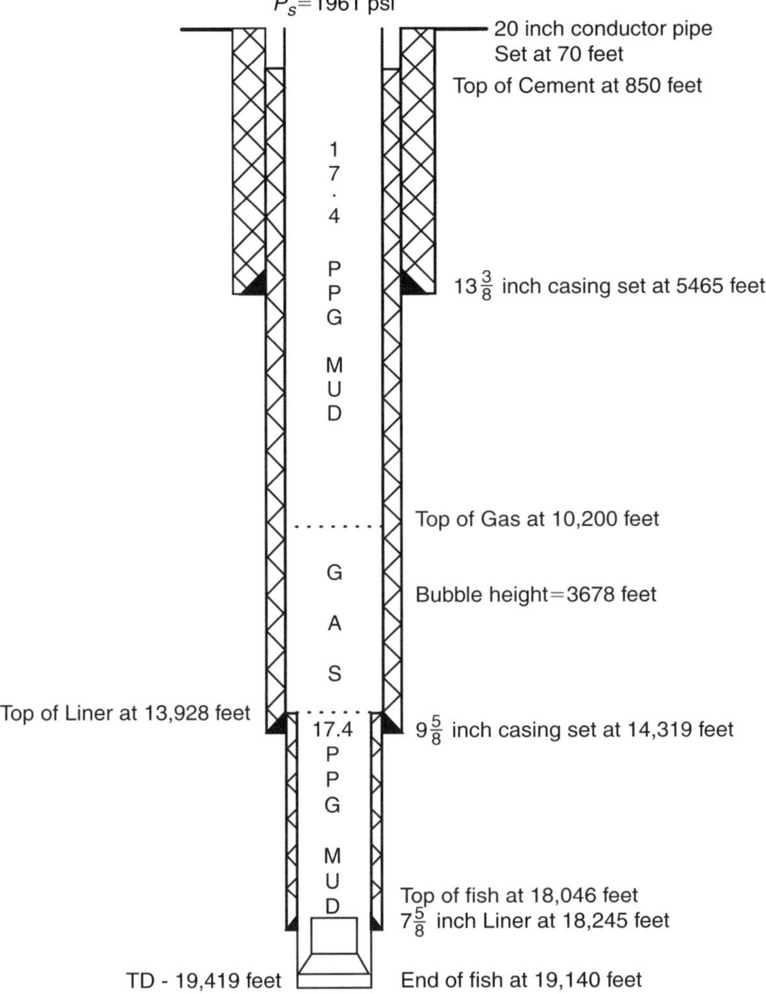

Figure 4.5 *E.N. Ross No. 2 Bubble Migration.*

Mud gradient, $\rho_m = 0.9048$ psi/ft

Bottomhole pressure, $P_b = 16,712$ psi

Bottomhole temperature, $T_b = 772\,°$Rankine

Temperature at 10,200 feet, $T_x = 650\,°$Rankine

Influx gradient, $\rho_f = 0.163$ psi/ft on bottom

$$\rho_f = 0.158 \text{ psi/ft inside } 9\tfrac{5}{8}$$

Compressibility factor at:

10,200 feet, $z_x = 1.581$

19,419 feet, $z_b = 1.988$

Required:
Determine the surface pressure when the influx has migrated to 10,200 feet and is completely inside the $9\tfrac{5}{8}$-inch intermediate casing.

Solution:
From the Ideal Gas Law:

$$\frac{P_b V_b}{z_b T_b} = \frac{P_x V_x}{z_x T_x}$$

Since the influx has migrated without expansion,

$$V_b = V_x = 260 \text{ bbls}$$

Therefore,

$$P_x = \frac{Z_x T_x P_b}{Z_b T_b}$$

or

$$P_x = \frac{16,712(1.581)(650)}{(1.988)(772)}$$

$$P_x = \mathbf{11,190 \ psi}$$

and

$$P_s = 11,190 - 0.9048(10,200)$$

$$P_s = \mathbf{1961 \ psi}$$

Analysis of the surface pressure data was in good agreement with the actual condition encountered. The surface pressure at the time that the influx was encountered at 10,200 feet was 2000 psi. The calculated surface pressure under the given conditions was 1961 psi. It is important to note that this influx was not expanded as it migrated and the surface pressure decreased significantly. Instinctively it is anticipated that the surface pressure will increase significantly as an unexpanded influx migrates. However, under these unusual conditions, the opposite was true.

Recent research has suggested that migration analysis based upon pressure interpretations is limited due to the fact that the compressibilities of the mud, hole filter cake, and formations are not routinely considered in field analyses. However, field application of the techniques described in this chapter has proven generally successful. An increase in surface pressure is the result of the reduction in drilling mud hydrostatic above the influx as the influx migrates through the drilling mud. Field observations have generally proven consistent with the theoretical considerations described herein.

Consider the wellbore schematic presented as Figure 4.6 which depicts a blowout near Albany, Texas. As illustrated, there was a hole in the drillpipe at 980 feet and an open hole from the $8\frac{5}{8}$-inch casing shoe at 720 feet to total depth at 4583 feet. After the well was killed, gas from supercharged shallow zones was migrating to the surface through the hole in the drillpipe. As a result, an opportunity to analyze the reliability of traditional calculations was presented. Further, since the substantial open hole interval contained a variety of fluids including mud, gas, water, and cement, it was possible to observe the error-producing effects described in literature related to compressibility.

On one of many pumping operations, the gas migrated to the surface through water in the $4\frac{1}{2}$-inch drillpipe in one hour and fifteen minutes for a migration rate of 784 feet per hour. On another occasion, as the gas migrated, water was pumped in four 2-barrel increments and the resulting change in surface pressure was recorded. The volume of water was carefully measured from the calibrated tank of a service company cement pump truck. In the $4\frac{1}{2}$-inch drillpipe, two barrels of fresh water represents 142 feet of hydrostatic or 122 psi. On each occasion, the surface pressure declined by a measured 120 psi after two barrels were pumped. The increase in surface pressure was the result of the reduction in drilling mud hydrostatic above the influx as the influx migrated through the water. Therefore, it must be concluded in this case that the compressibility of the formation

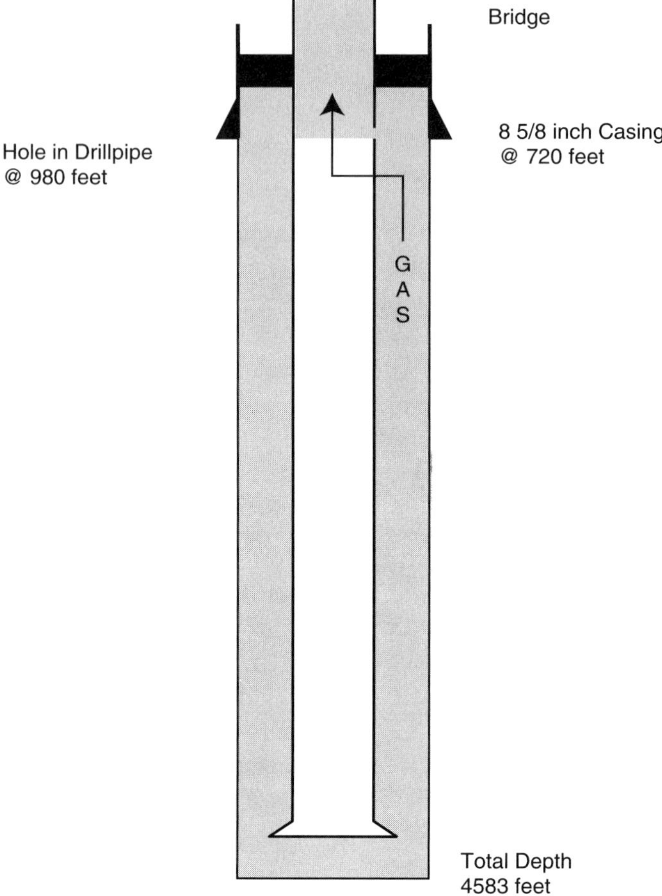

Figure 4.6

and fluids had no apparent effect on the ability to predict and analyze the behavior of the influx as reflected by the surface pressures.

The concepts of influx migration and rate of migration are further illustrated in Example 4.2:

Example 4.2

Given:

Wellbore schematic $=$ Figure 4.7

Well depth, $D = 10,000$ feet

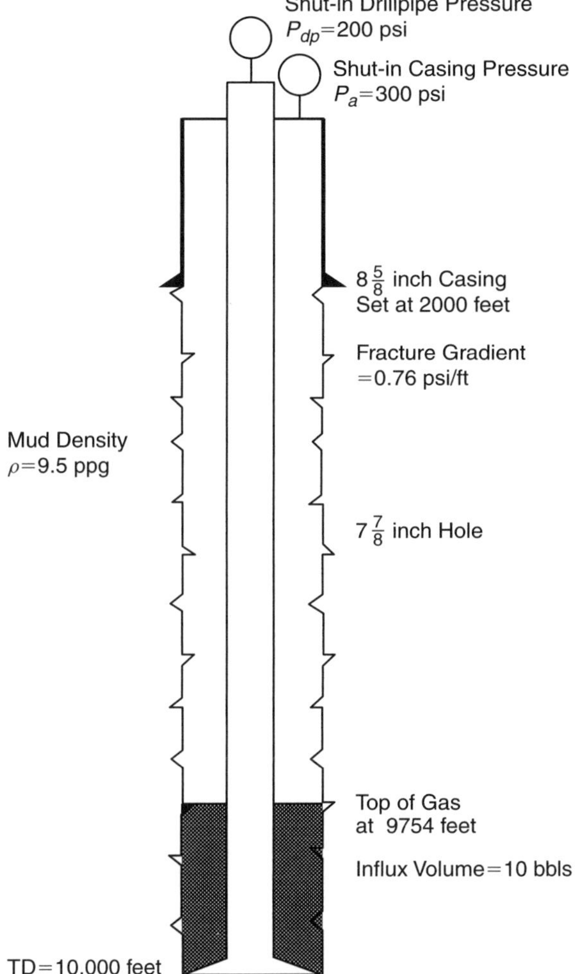

Figure 4.7 *Influx Migration.*

Hole size, $D_h = 7\frac{7}{8}$ inches

Drillpipe size, $D_p = 4\frac{1}{2}$ inches

$8\frac{5}{8}$-inch surface casing = 2000 feet

Casing internal diameter, $D_{ci} = 8.017$ inches

Fracture gradient, $F_g = 0.76$ psi/ft

Mud weight, $\rho = 9.6$ ppg

Mud gradient, $\rho_m = 0.50$ psi/ft

A kick is taken with the drill string on bottom and

Shut-in drillpipe pressure, $P_{dp} = 200$ psi

Shut-in annulus pressure, $P_a = 300$ psi

Pit level increase$= 10$ barrels

Capacity of the:

Drillpipe casing annulus, $C_{dpca} = 0.0428$ bbl/ft

Drillpipe hole annulus, $C_{dpha} = 0.0406$ bbl/ft

Depth to top of influx $= 9754$ feet

Further: The rig has lost all power and is unable to displace the influx. After one hour, the shut-in drillpipe pressure has increased to 300 psi and the shut-in annulus pressure has increased to 400 psi.

Required:
The depth to the top of the influx after one hour and the rate of influx migration.

Solution:
The distance of migration, D_{mgr}, is given by Equation 4.7:

$$D_{mgr} = \frac{\Delta P_{inc}}{0.052\rho} \tag{4.7}$$

Where:
$$\Delta P_{inc} = \text{Pressure increase, psi}$$
$$\rho = \text{Mud weight, ppg}$$

$$D_{mgr} = \frac{100}{(0.052)(9.6)}$$

$$D_{mgr} = \textbf{200 feet}$$

The depth to the top of the influx, D_{toi}, after one hour is

$$D_{toi} = TOI - D_{mgr} \qquad (4.8)$$

Where:

$TOI =$ Initial top of influx, feet

$D_{mgr} =$ Distance of migration, feet

$D_{toi} = 9754 - 200$

$D_{toi} = \textbf{9554 feet}$

Velocity of migration, V_{mgr}, is given by

$$V_{mgr} = \frac{D_{mgr}}{\text{Time}} \qquad (4.9)$$

$$V_{mgr} = \frac{200}{1}$$

$V_{mgr} = \textbf{200 feet per hour}$

The condition of the well after one hour is schematically illustrated as Figure 4.8. After one hour, the shut-in surface pressures have increased by 100 psi. The shut-in drillpipe pressure has increased from 200 psi to 300 psi and the shut-in casing pressure has increased from 300 psi to 400 psi. Therefore, the drilling mud hydrostatic equivalent to 100 psi has passed from above the influx to below the influx or the influx has migrated through the equivalent of 100 psi mud hydrostatic, which is equivalent to 200 feet of mud hydrostatic. The loss in mud hydrostatic of 100 psi has been replaced by additional shut-in surface pressure of 100 psi. Therefore, the rate of migration for the first hour is 200 feet per hour.

It should not be anticipated that the rate of migration will remain constant. As the influx migrates toward the surface, the velocity normally will increase. The influx will expand, the diffused bubbles will accumulate into one large bubble, and the migration velocity will increase.

The influx could be permitted to migrate to the surface. The methodology would be exactly the same as the Driller's Method at 0-barrels-per-minute circulation rate. The drillpipe pressure would be kept constant

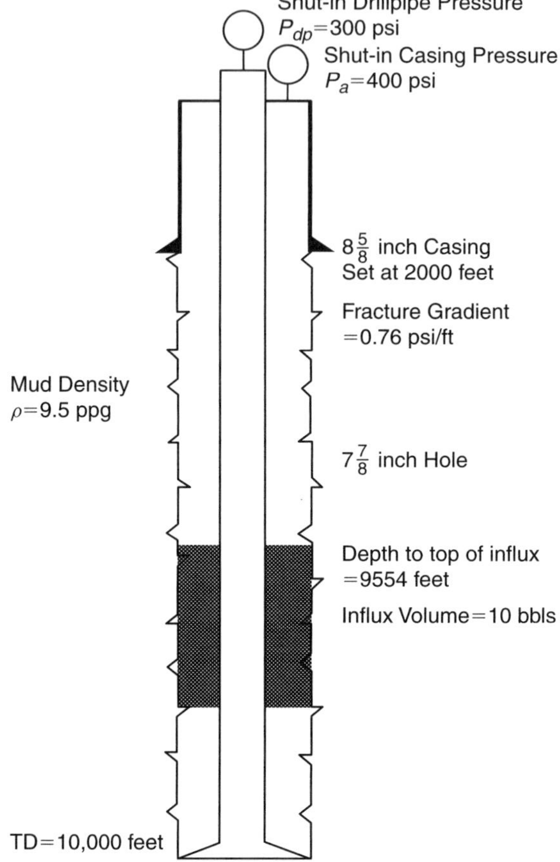

Figure 4.8   *Influx Migration after One Hour.*

by bleeding mud from the annulus. The casing pressure would have to be bled in small increments while noting the effect on the drillpipe pressure after a few seconds.

For example, in this instance it would not be proper to bleed 100 psi from the annulus and wait to observe the drillpipe pressure. The proper procedure would be to permit the surface pressure to build by 100-psi increments and then to bleed the casing pressure in 25-psi increments while observing the effect on the drillpipe pressure. The exact volumes of mud bled must be measured and recorded. The drillpipe pressure must be maintained at slightly over 200 psi. In that fashion the influx could be permitted to migrate to the surface. However, once the influx reaches the surface, the procedure generally must be terminated. Bleeding influx at the surface will usually result in additional influx at the bottom of the hole.

The procedure is illustrated in Example 4.3.

Example 4.3

Given:

Same conditions as Example 4.2.

Required:

Describe the procedure for permitting the influx to migrate to the surface.

Solution:

The effective hydrostatic of one barrel of mud, P_{hem}, in the annulus is given by Equation 4.10:

$$P_{hem} = \frac{0.052\rho}{C_{dpha}} \qquad (4.10)$$

Where:

$$\rho = \text{Mud weight, ppg}$$
$$C_{dpha} = \text{Annular capacity, bbl/ft}$$

$$P_{hem} = \frac{0.052(9.6)}{0.0406}$$

$$P_{hem} = \textbf{12.3 psi/bbl}$$

Therefore, for each barrel of mud bled from the annulus, the minimal acceptable annulus pressure must be increased by 12.3 psi.

The table on p. 139 summarizes the procedure.

As illustrated in Figure 4.9 and Table 4.1, the pressure on the surface is permitted to build to a predetermined value. This value should be calculated to consider the fracture gradient at the casing shoe in order that an underground blowout will not occur. In this first instance, the value is

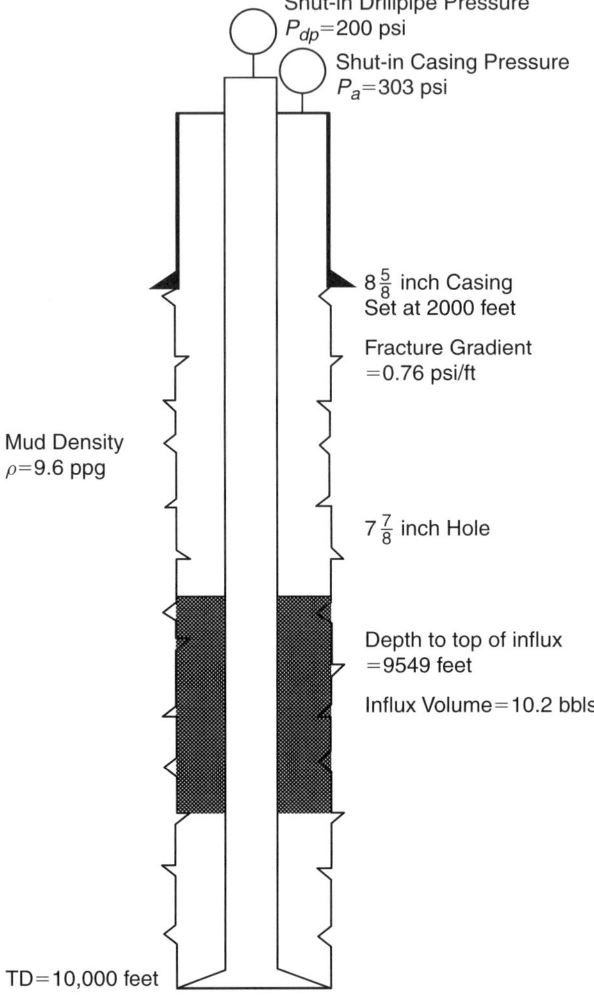

Figure 4.9   *Influx Migration after Bubble Expansion.*

a 100-psi increase in surface pressure. After the surface pressure has built 100 psi to 300 psi on the drillpipe and 400 psi on the casing, the influx is expanded and the surface pressure lowered by bleeding mud from the annulus. The pressure is bled in 25-psi increments.

Due to the expansion of the influx, the drillpipe pressure will return to 200 psi, but casing pressure will not return to 300 psi. Rather, the hydrostatic of the mud released from the annulus must be replaced

Table 4.1
Procedure for Influx Migration

Time	Drillpipe Pressure	Casing Pressure	Volume Bled	Minimum Casing Pressure
0900	200	300	0.00	300
1000	300	400	0.00	300
1005	275	375	0.05	301
1010	250	350	0.10	301
1015	225	325	0.15	302
1020	200	303	0.20	303
...	...	...	...	...

by the equivalent pressure at the surface. In this case, 0.20 barrels were bled from the annulus and the drillpipe pressure returned to 200 psi. However, the casing pressure could not be lowered below 303 psi. The 3-psi additional casing pressure replaces the 0.20 bbl of mud hydrostatic.

As further illustrated in Figure 4.9, the depth to the top of the influx after the mud has been bled from the annulus is 9549 feet. Therefore, the influx has expanded 5 feet, which was the volume formerly occupied by the 0.20 barrels of mud. Also, the influx volume has increased from 10 barrels to 10.2 barrels.

The calculations presented in this example are based on the actual theoretical calculations. In the field, the drillpipe pressure would probably be maintained at a value in excess of the original shut-in drillpipe pressure. However, the fracture gradient at the shoe must be considered in order to ensure that no underground blowout occurs.

Influx Migration—Volumetric Procedure

Influx migration without the ability to read the drillpipe pressure represents a much more difficult situation. The influx can safely be permitted to migrate to the surface if a volumetric procedure is used. Once again consider Equation 2.7:

$$P_b = \rho_f h + \rho_m (D - h) + P_a$$

Expanding Equation 2.7 gives

$$P_b = P_a + \rho_m D - \rho_m h + \rho_f h$$

The object of the procedure is to permit the influx to migrate while maintaining the bottomhole pressure constant. Therefore, the right side of Equation 2.7 must remain constant as the influx migrates. For any given conditions, $\rho_m D$ is constant. In addition, $\rho_f h$ is constant provided the geometry of the wellbore remains constant. To be pure theoretically, the geometry of the wellbore would have to be considered. However, to assume that the geometry is the same as on bottom is normally to err conservatively. That is, the cross-sectional area of the annulus might increase nearer the surface, thereby reducing the hydrostatic of the influx, but it almost never decreases nearer the surface. The one obvious exception is in floating drilling operations where the influx would have to migrate through a small choke line.

Therefore, in order to permit the influx to migrate while maintaining the bottomhole pressure constant, Equation 2.7 reduces to

$$\text{Constant} = P_a + \text{Constant} - \rho_m h + \text{Constant}$$

If the bottomhole pressure is to remain constant during the influx migration, any change in the mud hydrostatic due to the expansion of the influx must be offset by a corresponding increase in the annulus pressure. In this example, if one barrel of mud were released from the annulus, the shut-in casing pressure could not be reduced below 312 psi. If two barrels of mud were released, the shut-in casing pressure could not be reduced below 324 psi.

The procedure is illustrated in Example 4.4, which is the same as Example 4.3 with the exception that the drillpipe contains a float which does not permit the shut-in drillpipe pressure to be recorded.

Example 4.4

Given:

Same conditions as Example 4.3 except the drillpipe contains a float which does not permit the shut-in drillpipe pressure to be recorded. Further, the rig has lost all power and is unable to displace the influx. After one hour, the shut-in annulus pressure has increased to 400 psi.

Required:

Describe the procedure for permitting the influx to migrate to the surface.

Solution:

The effective hydrostatic of one barrel of mud, P_{hem}, in the annulus is given by Equation 4.10:

$$P_{hem} = \frac{0.052\rho}{C_{dpha}}$$

$$P_{hem} = \frac{0.052(9.6)}{0.0406}$$

$$P_{hem} = \textbf{12.3 psi/bbl}$$

Therefore, for each barrel of mud bled from the annulus, the minimum acceptable annulus pressure is increased by 12.3 psi.

Maximum acceptable increase in surface pressure prior to influx expansion is equal to 100 psi.

Bleed one barrel, but do not permit the surface pressure to fall below 312 psi.

After a cumulative volume of one barrel has been released, do not permit the surface pressure to fall below 324 psi.

After a cumulative volume of two barrels has been released, do not permit the surface pressure to fall below 336 psi.

After a cumulative volume of three barrels has been released, do not permit the surface pressure to fall below 348 psi. Continue in this manner until the influx reaches the surface.

When the influx reaches the surface, shut in the well.

The plan and instructions would be to permit the pressure to rise to a predetermined value considering the fracture gradient at the shoe. In this instance, a 100-psi increase is used. After the pressure had increased by 100 psi, mud would be released from the annulus. As much as one barrel would be released, provided that the shut-in casing pressure did not fall below 312 psi. Consider Table 4.2.

In this case, in the first step less than 0.20 barrels would have been bled and the casing pressure would be reduced to 312 psi. At that point, the well would be shut in and the influx permitted to migrate further up the hole.

Table 4.2
Volumetric Procedure for Influx Migration

Time	Surface Pressure	Volume Bled	Cumulative Volume Bled	Minimum Surface Pressure
0900	300	0.00	0.00	300
1000	400	0.00	0.00	300
1005	312	0.20	0.20	312
1100	412	0.00	0.20	312
1105	312	0.20	0.40	312
1150	412	0.00	0.40	312
1155	312	0.20	0.60	312
1230	412	0.00	0.60	312
1235	312	0.25	0.85	312
1300	412	0.00	0.85	312
1305	324	0.25	1.10	324
. . .	. . .	. . .	. . .	. . .

When the shut-in surface pressure reached 412 psi, the procedure would be repeated. After a total of one barrel was bled from the well, the minimum casing pressure would be increased to 324 psi and the instructions would be to bleed mud to a total volume of two barrels, but not to permit the casing pressure to fall below 324 psi.

When the shut-in surface pressure reached 424 psi, the procedure would be repeated. After a total of two barrels was bled from the well, the minimum casing pressure would be increased to 336 psi and the instructions would be to bleed mud to a total volume of three barrels, but not to permit the casing pressure to fall below 336 psi.

For each increment of one barrel of mud which was bled from the hole, the minimum shut-in surface pressure which would maintain the constant bottomhole pressure would be increased by the hydrostatic equivalent of one barrel of mud, which is 12.3 psi for this example. If the geometry changed as the influx migrated, the hydrostatic equivalent of one barrel of mud would be recalculated and the new value used.

Influx migration is a reality in well control operations and must be considered. Failure to consider the migration of the influx will usually result in unacceptable surface pressure, ruptured casing, or an underground blowout.

SAFETY FACTORS IN
CLASSICAL PRESSURE CONTROL PROCEDURES

It is well established that the Driller's Method and the Wait and Weight Method are based on the classical U-Tube Model as illustrated in Figure 4.1. The displacement concept for all classical procedures regardless of the name is to determine the bottomhole pressure from the mud density and the shut-in drillpipe pressure and to keep that bottomhole pressure constant while displacing the influx. For the conditions given in Figure 4.7, the shut-in bottomhole pressure would be 5200 psi. Therefore, as illustrated in Chapter 2, the goal of the control procedure would be to circulate the influx out of the wellbore while maintaining the bottomhole pressure constant at 5200 psi.

One of the most serious and frequent well control problems encountered in the industry is the inability to bring the influx to the surface without experiencing an additional influx or causing an underground blowout. In addition, in the field, difficulty is experienced starting and stopping displacement without permitting an additional influx. To address the latter problem, many have adopted "safety factor" methods.

The application of the "safety factors" arbitrarily alters the classical procedures and can result in potentially serious consequences. "Safety factors" are usually in three forms. The first is in the form of some arbitrary additional drillpipe pressure in excess of the calculated circulating pressure at the kill speed. The second is an arbitrary increase in mud density above that calculated to control the bottomhole pressure. The third is an arbitrary combination of the two. When the term "safety factor" is used, there is generally no question about the validity of the concept. Who could question "safety"? However, arbitrary "safety factors" can have serious effects on the well control procedure and can cause the very problems which they were intended to avoid! Consider Example 4.5:

Example 4.5

Given:

Wellbore schematic $=$ Figure 4.7

U-Tube schematic $=$ Figure 4.10

Well depth, $D = 10,000$ feet

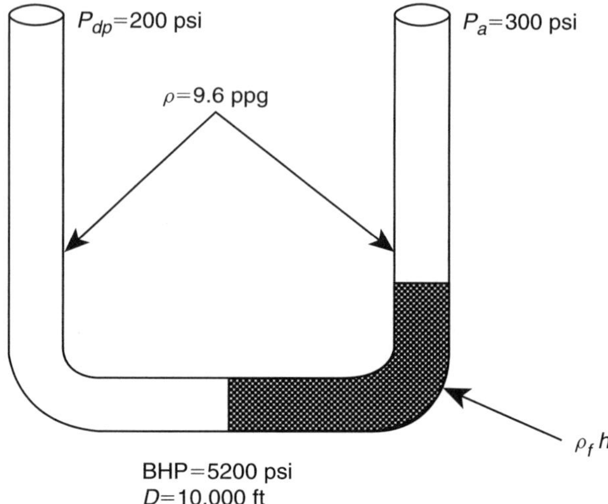

Figure 4.10   *U-Tube Schematic.*

Hole size, $D_h = 7\frac{7}{8}$ inches

Drillpipe size, $D_p = 4\frac{1}{2}$ inches

$8\frac{5}{8}$-inch surface casing = 2000 feet

Casing internal diameter, $D_{ci} = 8.017$ inches

Fracture gradient, $F_g = 0.76$ psi/ft

Fracture pressure = 1520 psi

Mud weight, $\rho = 9.6$ ppg

Mud gradient, $\rho_m = 0.50$ psi/ft

A kick is taken with the drill string on bottom and

Shut-in drillpipe pressure, $P_{dp} = 200$ psi

Shut-in annulus pressure, $P_a = 300$ psi

Pit level increase = 10 barrels

Normal circulation = 6 bpm at 60 spm

Kill rate $= 3$ bpm at 30 spm

Circulating pressure at kill rate, $P_{ks} = 500$ psi

Pump capacity, $C_p = 0.1$ bbl/stk

Capacity of the:

Drillpipe casing annulus, $C_{dpca} = 0.0428$ bbl/ft

Drillpipe hole annulus, $C_{dpha} = 0.0406$ bbl/ft

Initial displacement pressure, $P_c = 700$ psi @ 30 spm

Shut-in bottomhole pressure, $P_b = 5200$ psi

Maximum permissible surface pressure $= 520$ psi

Required:
1. The consequences of adding a 200-psi "safety factor" to the initial displacement pressure in the Driller's Method.

2. The consequences of adding a 0.5-ppg "safety factor" to the kill mud density in the Wait and Weight Method.

Solution:
1. The consequence of adding a 200-psi "safety factor" to the initial displacement pressure is that the pressure on the casing is increased by 200 psi to 500 psi.

    The pressure at the casing shoe is increased to 1500 psi, which is perilously close to the fracture gradient. See Figure 4.11.

2. With 10.5-ppg mud to the bit, the pressure on the left side of the U-Tube is:

    $$P_{10000} = \rho_{m1}D$$

    $$P_{10000} = 0.546(10,000)$$

    $$P_{10000} = \mathbf{5460\ psi}$$

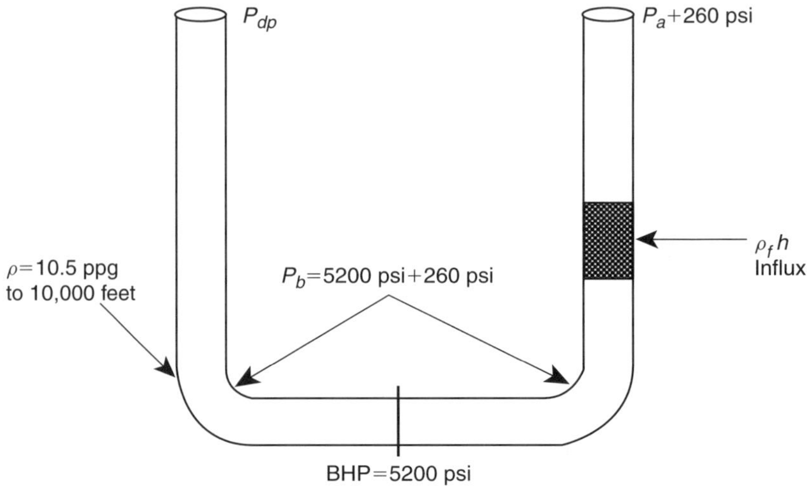

Figure 4.11   *U-Tube Schematic.*

Therefore, with the weighted mud at the bit, the pressure on the right side of the U-Tube is increased by **260 psi**, which would result in an underground blowout at the shoe.

In these examples, the "safety factors" were not safety factors after all. As illustrated in Figure 4.10, in the instance of a "safety factor" in the form of additional surface pressure, the additional pressure is added to the entire system. The bottomhole pressure is not being kept constant at the shut-in bottomhole pressure as intended. Rather, it is being held constant at the shut-in bottomhole pressure plus the "safety factor," which is 5400 psi in this example. To further aggravate the situation, the "safety factor" is applied to the casing shoe.

Thanks to the "safety factor," the pressure at the casing shoe is increased from 1300 psi to 1500 psi, which is within 20 psi of the pressure necessary to cause an underground blowout. As the influx is circulated to the casing shoe, the pressure at the casing shoe increases. Therefore, under the conditions in Example 4.5, with the 200 psi "safety factor," an underground blowout would be inevitable!

In Figure 4.11, it is illustrated that the increase in kill-mud weight to 10.5 ppg resulted in an additional 260 psi on the entire system. The bottomhole pressure was no longer being kept constant at 5200 psi as originally conceived. It was now being kept constant at 5460 psi. This additional burden was more than the fracture gradient was capable of withstanding. By the time that the kill mud reached the bit, the annulus pressure would be well above the maximum permissible 520 psi. Therefore, under these conditions, with the additional 0.5-ppg "safety factor," an underground blowout would be inevitable.

A kill-mud density higher than calculated by classical techniques can be used, provided that Equation 2.11 is strictly adhered to:

$$P_{cn} = P_{dp} - 0.052(\rho_1 - \rho)D + \left(\frac{\rho_1}{\rho}\right)P_{ks}$$

In Equation 2.11, any additional hydrostatic pressure resulting from the increased density is subtracted from the frictional pressure. Therefore, the bottomhole pressure can be maintained constant at the calculated bottomhole pressure, which is 5200 psi in this example. Following this approach, there would be no adverse effects as a result of using the 10.5-ppg mud as opposed to the 10.0-ppg mud. Further, there would be no "safety factor" in terms of pressure at the bottom of the hole greater than the calculated shut-in bottomhole pressure.

However, there would be a "safety factor" in that the pressure at the casing shoe with 10.5-ppg mud would be lower than the pressure at the casing shoe with 10.0-ppg mud. The annulus pressure profiles are further discussed in a following section. Another advantage would be that the circulating time would be less if 10.5-ppg mud were used because the trip margin would have been included in the original circulation.

A disadvantage is that the operation would fail if it became necessary to shut in the well any time after the 10.5-ppg mud reached the bit and before the influx reached the casing shoe. In that event, the pressure at the casing shoe would exceed the fracture gradient and an underground blowout would occur.

CIRCULATING A KICK OFF BOTTOM

All too often a drilling report reads like this, "Tripped out of hole. Well flowing. Tripped in with 10 stands and shut in well. Shut-in pressure 500 psi. Circulated heavy mud, keeping the drillpipe pressure constant. Shut-in pressure 5000 psi."

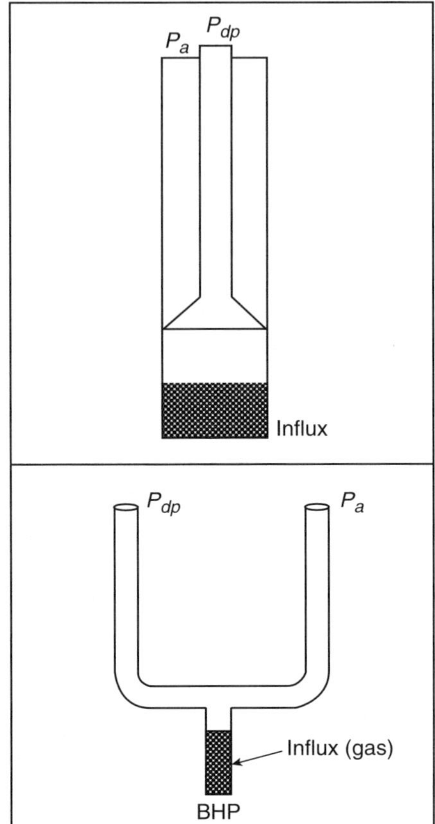

Figure 4.12   *Circulating Off Bottom Alters U-Tube Model.*

Attempting to circulate with the bit off bottom in a kick situation has caused as many well control situations to deteriorate as any other single operation. Simply put, **there is no classical well control procedure that applies to circulating with the bit off bottom with a formation influx in the wellbore!** The reason is that the classical U-Tube Model does not describe the wellbore condition and is not valid in this situation.

If the bit is off bottom as illustrated in Figure 4.12, the U-Tube Model becomes a Y-Tube Model. The drillpipe pressure can be influenced by the operations at the choke. However, the drillpipe pressure can also be affected by the wellbore condition and activity in the bottom of the Y-Tube. It is not possible to know the relative effect of each factor.

Therefore, the concepts, technology, and terminology of classical well control have no meaning or application under these circumstances. The Driller's Method is not valid. The Wait and Weight Method is not valid. Keeping the drillpipe pressure constant has no meaning. These are valid only if the U-Tube Model describes the wellbore conditions.

A well can be circulated safely off bottom with a kick in the hole provided that it exhibits all the characteristics of a dead well. That is, the drillpipe pressure must remain constant, the casing pressure must remain constant, the choke size must remain constant, the circulation rate must remain constant, and the pit volume must remain constant. Continuing to circulate with any of these factors changing usually results in more serious well control problems.

CLASSICAL PROCEDURES—PLUGGED NOZZLE EFFECT

While a kick is being circulated out utilizing the Driller's Method or the Wait and Weight Method, it is possible for a nozzle to plug. In the event that a nozzle does plug, the choke operator would observe a sudden rise in circulating drillpipe pressure with no corresponding increase in annulus pressure. The normal reaction would be for the choke operator to open the choke in an attempt to keep the drillpipe pressure constant. Of course, when the choke is opened, the well becomes underbalanced and additional influx is permitted. Unchecked, the well will eventually unload and a blowout will follow.

A plugged nozzle does not alter the U-Tube Model. The U-Tube Model and classical pressure control procedures are still applicable. What has been altered is the frictional pressure losses in the drill string. The circulating pressure at the kill speed has been increased as a result of the plugged nozzle. The best procedure to follow when the drillpipe pressure increases suddenly is to shut in the well and restart the procedure as outlined in Chapter 2 for either the Wait and Weight Method or the Driller's Method.

CLASSICAL PROCEDURES— DRILL STRING WASHOUT EFFECT

When a washout occurs in the drill string, a loss in drillpipe pressure will be observed with no corresponding loss in annulus pressure.

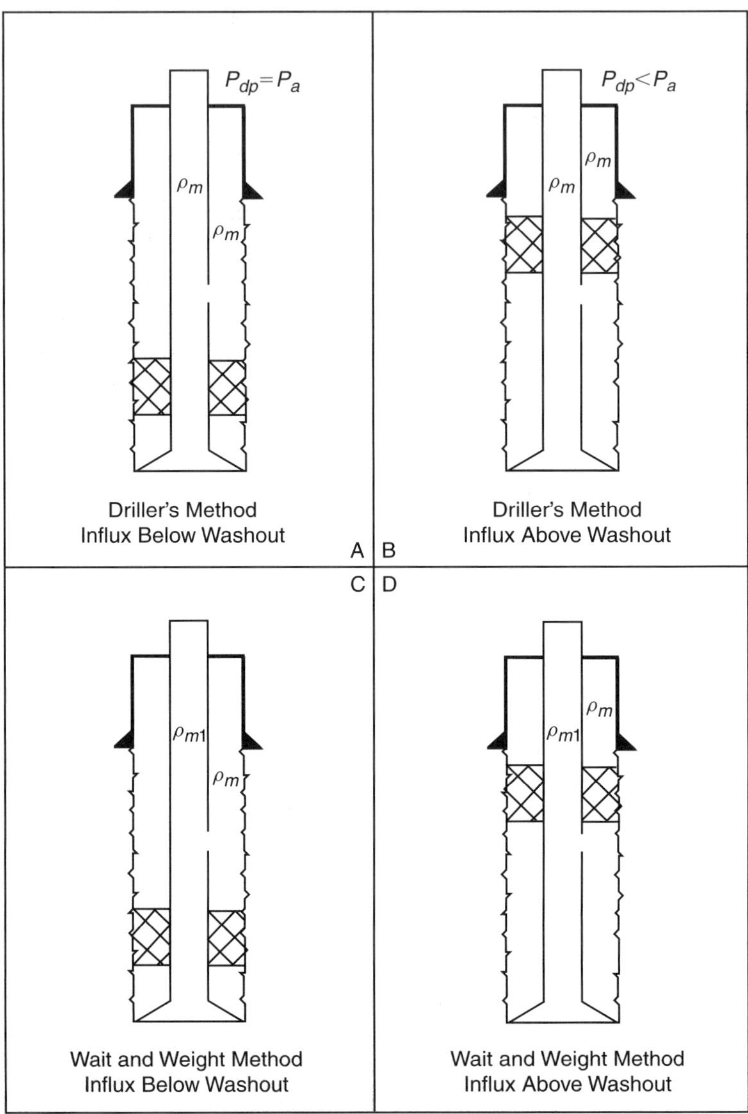

Figure 4.13

The only alternative is to shut in the well and analyze the problem. If the Driller's Method is being used, the analysis is simplified. As illustrated in Figure 4.13a, if the well is shut in and the influx is below the washout, the shut-in drillpipe pressure and the shut-in annulus pressure will be equal.

Under these conditions, the U-Tube Model is not applicable and no classical procedure is appropriate.

There are several alternatives. Probably the best general alternative is to permit the influx to migrate to the surface pursuant to the prior discussions and, once the influx has reached the surface, circulate it out. Another alternative is to locate the washout, strip out to the washout, repair the bad joint or connection, strip in the hole to bottom, and resume the well control procedure.

If the Driller's Method is being utilized when the washout occurs, the well is shut in, and the influx is above the washout as illustrated in Figure 4.13b, the shut-in drillpipe pressure will be less than the shut-in annulus pressure. Under these conditions, the U-Tube Model is applicable and the influx can be circulated out by continuing the classical Driller's Method as outlined in Chapter 2.

The frictional pressure losses in the drill string have been altered and the circulating pressure at the kill speed which was originally established is no longer applicable. A new circulating pressure at the kill speed must be established as outlined in Chapter 2. That is, hold the casing pressure constant while bringing the pump to speed. Read the new drillpipe pressure and keep that drillpipe pressure constant while the influx is circulated to the surface.

If the Wait and Weight Method is being used, the analysis is considerably more complicated because, as illustrated in Figure 4.13c and 4.13d, the kill-weight mud has been introduced to the system. Therefore, the differences in mud hydrostatic must be included in the analysis to determine the relationship between the shut-in drillpipe pressure and the shut-in casing pressure. Since the depth of the washout is not usually known, it may not be possible to determine a reliable relationship between the shut-in drillpipe pressure and the shut-in casing pressure. Once the analysis is performed, the alternatives are the same as those discussed for the Driller's Method.

DETERMINATION OF SHUT-IN DRILLPIPE PRESSURES

Generally, the drillpipe pressure will stabilize within minutes after shut-in and is easily determined. In some instances, the drillpipe pressure may never build to reflect the proper bottomhole pressure, particularly in cases of long open-hole intervals at or near the fracture gradient coupled with very low productivities. When water is used as the drilling fluid, gas migration can be rapid, thereby masking the shut-in drillpipe pressure. In these instances, a good knowledge of anticipated bottomhole pressures and anticipated drillpipe pressures is beneficial in recognizing and identifying problems and providing a base for pressure control procedures.

A float in the drill string complicates the determination of the drillpipe pressure; however, it can be readily determined by pumping slowly on the drillpipe and monitoring both the drillpipe and annulus pressure. When the annulus pressure first begins to increase, the drillpipe pressure at that instant is the shut-in drillpipe pressure.

Another popular procedure is to pump through the float for a brief moment, holding the casing pressure constant, and then shut in with the original annulus pressure, thereby trapping the drillpipe pressure on the stand pipe gauge. An additional technique is to bring the pump to kill speed and compare the circulating pressure with the pre-recorded circulating pressure at the kill rate with the difference being the drillpipe pressure. Still another alternative is to use a flapper-type float with a small hole drilled through the flapper that permits pressure reading but not significant flow.

DETERMINATION OF THE TYPE OF FLUID THAT ENTERED THE WELLBORE

Of primary interest in the determination of fluid types is whether gas has entered the wellbore. If only liquid is present, control is simplified. An accurate measurement of increase in pit level is mandatory if a reliable determination is to be made. Example 4.6 illustrates the calculation:

Example 4.6

Given:

Wellbore schematic = Figure 4.7

Well depth, $D = 10,000$ feet

Hole size, $D_h = 7\frac{7}{8}$ inches

Drillpipe size, $D_p = 4\frac{1}{2}$ inches

$8\frac{5}{8}$-inch surface casing $= 2000$ feet

Casing internal diameter, $D_{ci} = 8.017$ inches

Fracture gradient, $F_g = 0.76$ psi/ft

Mud weight, $\rho = 9.6$ ppg

Mud gradient, $\rho_m = 0.50$ psi/ft

A kick is taken with the drill string on bottom and

Shut-in drillpipe pressure, $P_{dp} = 200$ psi

Shut-in annulus pressure, $P_a = 300$ psi

Pit level increase $= 10$ barrels

Capacity of

Drillpipe hole annulus, $C_{dpha} = 0.0406$ bbl/ft

Required:
Determine density of the fluid entering the wellbore.

Solution:
From Equation 2.6:

$$P_b = \rho_m D + P_{dp}$$

From Equation 2.7:

$$P_b = \rho_f h + \rho_m (D - h) + P_a$$

The height of the bubble, h, is given by Equation 3.7:

$$h = \frac{\textit{Influx Volume}}{C_{dpha}}$$

$$h = \frac{10}{0.0406}$$

$$h = \textbf{246 feet}$$

Solving Equations 2.6 and 2.7 simultaneously gives

$$\rho_m D + P_{dp} = \rho_f h + \rho_m (D - h) + P_a$$

The only unknown is ρ_f; therefore, substituting and solving yield

$$\rho_f (246) = 0.5(10{,}000) + 200$$

$$- 0.5(10{,}000 - 246) - 300$$

$$\rho_f = \frac{23}{246}$$

$$\rho_f = \mathbf{0.094 \ psi/ft}$$

Confirmation can be attained utilizing Equation 3.5:

$$\rho_f = \frac{S_g P_b}{53.3 z_b T_b}$$

$$\rho_f = \frac{0.6(5200)}{53.3(1.1)(620)}$$

$$\rho_f = \mathbf{0.0858 \ psi/ft}$$

Therefore, since the calculated influx gradient is approximately the same as the fluid gradient determined using an assumed influx specific gravity of 0.6, the fluid in the wellbore is gas. Natural gas will usually have a fluid gradient of 0.15 psi/ft or less, while brine water has a density of approximately 0.45 psi/ft, and oil has a density of approximately 0.3 psi/ft. Obviously, combinations of gas, oil, and water can have a gradient anywhere between 0.1 and 0.45 psi/ft.

FRICTIONAL PRESSURE LOSSES

Frictional pressure losses inside the drill string are usually measured while frictional pressure losses in the annulus in conventional operations are usually ignored. However, in any kill operation, well-site personnel should have the means and ability to calculate the circulating pressure losses.

For laminar flow in the annulus, the frictional pressure loss is given by Equation 4.11 for the Power Law Fluid Model:

$$P_{fla} = \left[\left(\frac{2.4\bar{v}}{D_h - D_p}\right)\left(\frac{2n+1}{3n}\right)\right]^n \frac{Kl}{300(D_h - D_p)} \qquad (4.11)$$

For laminar flow inside pipe, the frictional pressure loss is given by Equation 4.12:

$$P_{fli} = \left[\left(\frac{1.6\bar{v}}{D}\right)\left(\frac{3n+1}{4n}\right)\right]^n \frac{Kl}{300} \qquad (4.12)$$

For turbulent flow in the annulus, the frictional pressure loss is given by Equation 4.13:

$$P_{fta} = \frac{7.7(10^{-5})\rho^{0.8}Q^{1.8}(PV)^{0.2}l}{(D_h - D_p)^3(D_h + D_p)^{1.8}} \qquad (4.13)$$

For turbulent flow inside pipe, the frictional pressure loss is given by Equation 4.14:

$$P_{fti} = \frac{7.7(10^{-5})\rho^{0.8}Q^{1.8}(PV)^{0.2}l}{D_i^{4.8}} \qquad (4.14)$$

For flow through the bit, the pressure loss is given by Equation 4.15:

$$P_{bit} = 9.14(10^{-5})\frac{\rho Q^2}{A_n^2} \qquad (4.15)$$

Where:

P_{bit} = Bit pressure losses, psi
P_{fla} = Laminar annular losses, psi
P_{fta} = Turbulent annular losses, psi
P_{fli} = Laminar losses inside pipe, psi
P_{fti} = Turbulent losses inside pipe, psi
$\bar{v}$ = Average velocity, fpm
D_h = Hole diameter, inches
D_p = Pipe outside diameter, inches
D_i = Pipe inside diameter, inches

$$\rho = \text{Mud weight, ppg}$$

$$\Theta_{600} = \text{Viscometer reading at 600 rpm,} \frac{lb_f}{100\,ft^2}$$

$$\Theta_{300} = \text{Viscometer reading at 300 rpm,} \frac{lb_f}{100\,ft^2}$$

$$n = 3.32 \log\left(\frac{\Theta_{600}}{\Theta_{300}}\right) \tag{4.16}$$

$$K = \frac{\Theta_{300}}{511^n} \tag{4.17}$$

$$l = \text{Length, feet}$$
$$Q = \text{Volume rate of flow, gpm}$$
$$PV = \text{Plastic viscosity, centipoise}$$
$$= \Theta_{600} - \Theta_{300} \tag{4.18}$$
$$A_n = \text{Total nozzle area, inches} \tag{4.19}$$

Flow inside the drill string is usually turbulent while flow in the annulus is normally laminar. When the flow regime is now known, make the calculations assuming both flow regimes. The calculation resulting in the greater value for frictional pressure loss is correct and defines the flow regime.

Example 4.7 illustrates a calculation.

Example 4.7

Given:

Wellbore schematic $=$ Figure 4.7

Well depth, $D = 10,000$ feet

Hole size, $D_h = 7\frac{7}{8}$ inches

Drillpipe size, $D_p = 4\frac{1}{2}$ inches

$8\frac{5}{8}$-inch surface casing $= 2000$ feet

Casing internal diameter, $D_{ci} = 8.017$ inches

Mud weight, $\rho = 9.6$ ppg

Mud gradient, $\rho_m = 0.50$ psi/ft

Normal circulation rate $= 6$ bpm at 60 spm

Kill circulation rate $= 3$ bpm at 30 spm

Capacity of

Drillpipe hole annulus, $C_{dpha} = 0.0406$ bbl/ft

$$\Theta_{600} = 25 \frac{lb_f}{100\, ft^2}$$

$$\Theta_{300} = 15 \frac{lb_f}{100\, ft^2}$$

Required:
The frictional pressure loss in the annulus assuming laminar flow.

Solution:

$$\bar{v} = \frac{Q}{Area}$$

$$\bar{v} = \frac{126 \left(\frac{gal}{min} \right)}{\frac{\pi}{4}(7.875^2 - 4.5^2)(in^2)} \left(\frac{144 \frac{in^2}{ft^2}}{7.48 \frac{gal}{ft^3}} \right)$$

$$\bar{v} = \mathbf{74\ fpm}$$

$$n = 3.32 \log \left(\frac{\Theta_{600}}{\Theta_{300}} \right)$$

$$n = 3.32 \log \left(\frac{25}{15} \right)$$

$$n = \mathbf{0.74}$$

$$K = \frac{\Theta_{300}}{511^n}$$

$$K = \frac{15}{511^{0.74}}$$

$$K = \mathbf{0.15}$$

$$P_{fla} = \left[\left(\frac{2.4\bar{v}}{D_h - D_p} \right) \left(\frac{2n + 1}{3n} \right) \right]^n \frac{Kl}{300(D_h - D_p)}$$

$$P_{fla} = \left[\left(\frac{2.4(74)}{7.875 - 4.5} \right) \left(\frac{2(0.74) + 1}{3(0.74)} \right) \right]^{0.74}$$

$$\times \frac{(0.15)(10,000)}{300(7.875 - 4.5)}$$

$$P_{fla} = \textbf{30 psi}$$

In this example, the frictional pressure loss in the annulus is only 30 psi. However, it is important to understand that the frictional pressure loss in the annulus is neglected in classical pressure control procedures. Therefore, the actual bottomhole pressure during a displacement procedure is greater than the calculated pressure by the *value* of the frictional pressure loss in the annulus. In this case, the bottomhole pressure would be held constant at 5230 psi during the Driller's Method and the Wait and Weight Method. In the final analysis, the frictional pressure loss in the annulus is a true "safety factor."

In deep wells with small annular areas, the frictional pressure loss in the annulus could be very significant and should be determined. Theoretically, if the fracture gradient at the shoe is a problem, the circulating pressure at the kill speed could be reduced by the frictional pressure loss in the annulus. For instance, using the Driller's Method in this example, the circulating pressure at the kill speed could be reduced from 700 psi at 30 spm to 670 psi at 30 spm. The bottomhole pressure would remain constant at 5200 psi, and there would be no additional influx during displacement.

ANNULUS PRESSURE PROFILES WITH CLASSICAL PROCEDURES

The annulus pressure profile as well as analysis of the pressures at the casing shoe during classic pressure control procedures provide essential insight into any well control operation. Further, the determination of the gas volume at the surface and the time sequence for events are essential to the understanding and execution of classical pressure control procedures. Those responsible for killing the well must be informed of what is to be expected and the appropriate sequence of events.

As will be illustrated in the following example, the annulus pressure will increase by almost three times during the displacement procedure and at the end dry gas will be vented for 20 minutes. Those with little experience may not expect or be mentally prepared for 20 minutes of dry gas and might be tempted to alter an otherwise sound and prudent procedure. Furthermore, confidence might be shaken by the reality of an additional 50-barrel increase in pit level. In any well control procedure, the more complete and thorough the plan, the better the chance of an expeditious and successful completion.

One of the primary problems in well control is that of circulating a kick to the surface after the well has been shut in without losing circulation and causing an underground blowout. Analysis of the annulus pressure behavior prior to initiating the displacement procedure would permit the evaluation and consideration of alternatives and probably prevent a disaster.

The annulus pressure profile during classical pressure control procedures for an influx of gas can be calculated for both the Driller's Method and the Wait and Weight Method. For the Driller's Method, consider Figure 4.14. The pressure, P_x, at the top of the influx at any point in the annulus X feet from the surface is given by Equation 4.19:

$$P_x = P_a + \rho_m X \tag{4.19}$$

With the influx X feet from the surface, the surface pressure on the annulus is given by Equation 4.20:

$$P_a = P_b - \rho_m(D - h_x) - P_f \tag{4.20}$$

Where:
$\quad$ P_x = Pressure at depth X, psi
$\quad$ P_a = Annulus pressure, psi
$\quad$ P_b = Bottomhole pressure, psi
$\quad$ ρ_m = Mud gradient, psi/ft
$\quad$ D = Well depth, feet
$\quad$ h_x = Height of the influx at depth X, feet
$\quad$ P_f = Pressure exerted by the influx at depth X, psi

and pursuant to the Ideal Gas Law:

$$h_x = \frac{P_b z_x T_x A_b}{P_x z_b T_b A_x} h_b \tag{4.21}$$

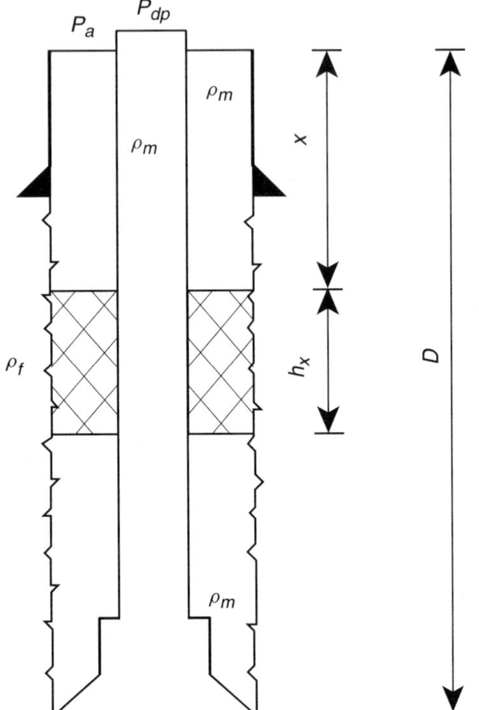

Figure 4.14 *Wellbore Schematic for Driller's Method.*

b—Denotes bottomhole conditions
x—Denotes conditions at depth X

The density of the influx is given by Equation 3.5:

$$\rho_f = \frac{S_g P_b}{53.3 z_b T_b}$$

Substituting and solving yield the Equation 4.22, which is an expression for the pressure at the top of the influx when it is any distance X from the surface when the Driller's Method is being used:

$$P_{xdm} = \frac{B}{2} + \left[\frac{B^2}{4} + \frac{P_b \rho_m z_x T_x h_b A_b}{z_b T_b A_x} \right]^{\frac{1}{2}} \tag{4.22}$$

$$B = P_b - \rho_m(D - X) - P_f \frac{A_b}{A_x} \qquad (4.23)$$

Where:

$$b = \text{Conditions at the bottom of the well}$$
$$x = \text{Conditions at } X$$
$$X = \text{Distance from surface to top of influx, feet}$$
$$D = \text{Depth of well, feet}$$
$$P_a = \text{Annular pressure, psi}$$
$$\rho_m = \text{Mud gradient, psi/ft}$$
$$P_b = \text{Bottomhole pressure, psi}$$
$$P_f = \text{Hydrostatic of influx, psi}$$
$$z = \text{Compressibility factor}$$
$$T = \text{Temperature, °Rankine}$$
$$A = \text{Annular area, in}^2$$
$$S_g = \text{Specific gravity of gas}$$

Pursuant to analysis of Figure 4.15, for the Wait and Weight Method, the pressure at X is also given by Equation 4.18. However, the expression for the pressure on the annulus becomes Equation 4.24:

$$P_a = P_b - \rho_{m1}(D - X - h_x - l_{vds}) - \rho_m(l_{vds} + X) - P_f \frac{A_b}{A_x} \qquad (4.24)$$

Where:

$$\rho_{m1} = \text{Kill mud gradient, psi/ft}$$
$$l_{vds} = \text{Length of drill string volume in annulus, feet}$$

Solving Equations 3.5, 4.19, 4.21, and 4.24 simultaneously results in Equation 4.25, which is an expression for the pressure at the top of the influx at any distance X from the surface when the Wait and Weight Method of displacement is being used:

$$P_{xww} = \frac{B_1}{2} + \left[\frac{B_1^2}{4} + \frac{P_b \rho_{m1} z_x T_x h_b A_b}{z_b T_b A_x} \right]^{\frac{1}{2}} \qquad (4.25)$$

$$B_1 = P_b - \rho_{m1}(D - X) - P_f \frac{A_b}{A_x} + l_{vds}(\rho_{m1} - \rho_m) \qquad (4.26)$$

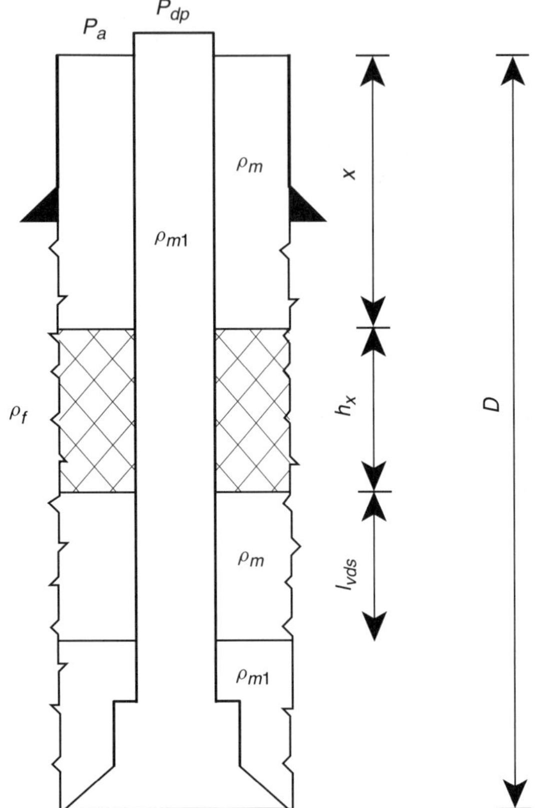

Figure 4.15   *Wellbore Schematic for Wait and Weight Method.*

Equation 4.22 can be used to calculate the pressure at the top of the gas bubble at any point in the annulus X distance from the surface, assuming there is no change in mud weight (Driller's Method). Similarly, Equation 4.25 can be used to calculate the pressure at the top of the gas bubble at any point in the annulus X distance from the surface, assuming that the gas bubble is displaced with weighted, ρ_{m1}, mud (Wait and Weight Method).

Depending on drill string geometry, the maximum pressure at any point in the annulus will generally occur when the bubble first reaches that point. The exception occurs when the drill collars are sufficiently larger than the drillpipe to cause a significant shortening of the influx as it passes from the drill collar annulus to the drillpipe annulus. In that instance, the pressure in the annulus will be lower than the initial shut-in annulus pressure until the influx has expanded to a length equal to its original length around the drill collars. From that point upward, the pressure in the annulus at the top of the influx will be greater than when the well was first shut in.

Example 4.8 illustrates the use of Equations 4.22 and 4.25 along with the significance and importance of the calculations:

Example 4.8

Given:

Wellbore schematic $=$ Figure 4.7

Well depth, $D = 10{,}000$ feet

Hole size, $D_h = 7\frac{7}{8}$ inches

Drillpipe size, $D_p = 4\frac{1}{2}$ inches

$8\frac{5}{8}$-inch surface casing $= 2000$ feet

Casing internal diameter, $D_{ci} = 8.017$ inches

Fracture gradient, $F_g = 0.76$ psi/ft

Fracture pressure $= 1520$ psi

Mud weight, $\rho = 9.6$ ppg

Mud gradient, $\rho_m = 0.50$ psi/ft

A kick is taken with the drill string on bottom and

Shut-in drillpipe pressure, $P_{dp} = 200$ psi

Shut-in annulus pressure, $P_a = 300$ psi

Pit level increase $= 10$ barrels

Kill-mud weight, $\rho_1 = 10$ ppg

Kill-mud gradient, $\rho_{m1} = 0.52$ psi/ft

Normal circulation = 6 bpm at 60 spm

Kill rate = 3 bpm at 30 spm

Circulating pressure at kill rate, $P_{ks} = 500$ psi

Pump capacity, $C_p = 0.1$ bbl/stk

Capacity of the:

Drillpipe (inside), $C_{dpi} = 0.0142$ bbl/ft

Drillpipe casing annulus, $C_{dpca} = 0.0428$ bbl/ft

Drillpipe hole annulus, $C_{dpha} = 0.0406$ bbl/ft

Initial displacement pressure, $P_c = 700$ psi @ 30 spm

Shut-in bottomhole pressure, $P_b = 5200$ psi

Maximum permissible surface pressure = 520 psi

Ambient temperature = 60°F

Geothermal gradient = 1.0°/100 feet

$\rho_f h = P_f = 23$ psi

$h_b = 246$ feet

Annular Area, $A_b = 32.80$ in^2

Required:

A.   Assuming the Driller's Method of displacement,

1.   The pressure at the casing seat when the well is first shut in.

2.   The pressure at the casing seat when the top of the gas bubble reached that point.

3.   The annulus pressure when the gas bubble first reaches the surface.

4.   The height of the gas bubble at the surface.

5.   The pressure at the casing seat when the gas bubble reaches the surface.

6.   The total pit volume increase with the influx at the surface.

7.   The surface annulus pressure profile during displacement using the Driller's Method.

B.   Using the Wait and Weight Method of displacement,

1.   The pressure at the casing seat when the well is first shut in.

2.   The pressure at the casing seat when the top of the gas bubble reached that point.

3.   The annulus pressure when the gas bubble first reaches the surface.

4.   The height of the gas bubble at the surface.

5.   The pressure at the casing seat when the gas bubble reaches the surface.

6.   The total pit volume increase with the influx at the surface.

7.   The surface annulus pressure profile during displacement using the Driller's Method.

8.   The rate at which barite must be mixed and the minimum barite required.

C.   Compare the two procedures.

D.   The significance of 600 feet of 6-inch drill collars on the annulus pressure profile.

Solution:

A.

1.   The pressure at the casing shoe at 2000 feet when the well is first shut in is given by Equation 4.19:

$$P_x = P_a + \rho_m X$$

$$P_{2000} = 300 + 0.5(2000)$$

$$P_{2000} = \mathbf{1300 \ psi}$$

2.   For the Driller's Method of displacement, the pressure at the casing seat at 2000 feet when the top of the influx reaches that point is given by Equation 4.22:

$$P_{xdm} = \frac{B}{2} + \left[\frac{B^2}{4} + \frac{P_b \rho_m z_x T_x h_b A_b}{z_b T_b A_x} \right]^{\frac{1}{2}}$$

$$B = P_b - \rho_m (D - X) - P_f \frac{A_b}{A_x}$$

$$B = 5200 - 0.5(10{,}000 - 2000) - 23 \left[\frac{32.80}{32.80} \right]$$

$$B = \mathbf{1177}$$

$$P_{2000dm} = \frac{1177}{2} + \frac{1177^2}{4}$$

$$+ \frac{5200(0.5)(0.811)(540)(246)(32.80)}{(620)(1.007)(32.80)} \Bigg]^{\frac{1}{2}}$$

$$P_{2000dm} = \mathbf{1480 \ psi}$$

3.   For the Driller's Method of displacement, the annulus pressure when the gas bubble first reaches

the surface may be calculated using Equation 4.22 as follows:

$$P_{xdm} = \frac{B}{2} + \left[\frac{B^2}{4} + \frac{P_b \rho_m z_x T_x h_b A_b}{z_b T_b A_x} \right]^{\frac{1}{2}}$$

$$B = P_b - \rho_m(D - X) - P_f \frac{A_b}{A_x}$$

$$B = 5200 - 0.5(10{,}000 - 0) - 23$$

$$B = \mathbf{177}$$

$$A_0 = \left(\frac{\pi}{4}\right)(8.017^2 - 4.5^2)$$

$$A_0 = 34.58 \text{ in}^2$$

$$P_{0dm} = \frac{177}{2} + \left[\frac{177^2}{4} \right.$$

$$\left. + \frac{5200(0.5)(0.875)(520)(246)(32.80)}{(620)(1.007)(34.58)} \right]^{\frac{1}{2}}$$

$$P_{0dm} = \mathbf{759 \ psi}$$

4.   The height of the gas bubble at the surface can be determined using Equation 4.21.

$$h_x = \frac{P_b z_x T_x A_b}{P_x z_b T_b A_x} h_b$$

$$h_0 = \frac{5200(0.875)(520)(32.80)}{759(1.007)(620)(34.58)}(246)$$

$$h_0 = \mathbf{1165 \ feet}$$

5.   The pressure at the casing seat when the influx reaches the surface may be calculated by adding

the annulus pressure to the mud and influx hydrostatics as follows:

$$P_{2000} = P_{0dm} + P_f \frac{A_b}{A_0} + \rho_m(2000 - h_0)$$

$$P_{2000} = 759 + 23\left[\frac{32.80}{34.58}\right] + 0.50(2000 - 1165)$$

$$P_{2000} = \mathbf{1198\ psi}$$

6. The total pit volume increase with the influx at the surface. From part 4, the length of the influx when it reaches the surface is 1165 feet.

$$\text{Total pit gain} = h_0 C_{dpca}$$

$$\text{Total pit gain} = (1165)(0.0428)$$

$$\text{Total pit gain} = \mathbf{50\ bbls}$$

7. The surface annulus pressure profile during displacement using the Driller's Method.

Example calculation, from part 2 with the top of the influx at 2000 feet, the pressure at 2000 feet is calculated to be

$$P_{2000dm} = \mathbf{1480\ psi}$$

With the top of the influx at 2000 feet, the surface annulus pressure is

$$P_x = P_a + \rho_m X$$

$$P_0 = 1480 - 0.50(2000)$$

$$P_0 = \mathbf{480\ psi}$$

The surface annulus pressure profile is summarized in Table 4.3 and illustrated as Figure 4.16.

Table 4.3
Surface Annulus Pressures—Driller's Method

Depth to Top of Bubble, feet	Volume Pumped, bbls	Annular Pressure, psi
9754	0	300
9500	10	301
9000	30	303
8500	50	306
8000	69	310
7500	89	313
7000	109	318
6500	128	323
6000	148	329
5500	167	336
5000	186	346
4500	205	357
4000	224	371
3500	242	389
3000	260	412
2500	277	442
2000	294	480
1500	311	517
1000	326	579
500	340	659
0	353	759

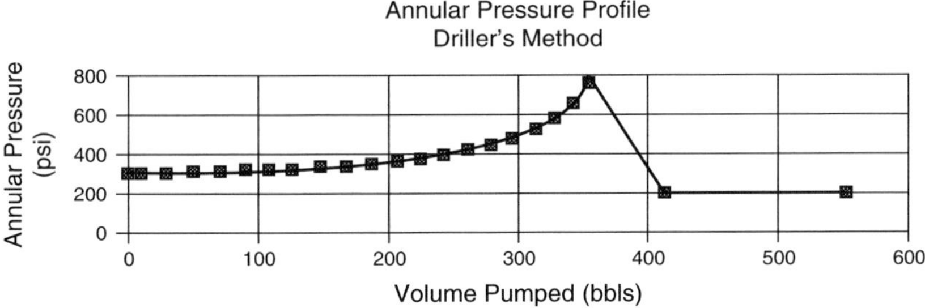

Figure 4.16

B.

1. The pressure at the casing shoe at 2000 feet when the well is first shut in is the same for both the Driller's Method and the Wait and Weight Method:

$$P_{2000} = \mathbf{1300\ psi}$$

2. For the Wait and Weight Method, the pressure at the casing seat at 2000 feet when the top of the influx reaches that point is given by Equation 4.25:

$$P_{xww} = \frac{B_1}{2} + \left[\frac{B_1^2}{4} + \frac{P_b \rho_{m1} z_x T_x h_b A_b}{z_b T_b A_x} \right]^{\frac{1}{2}}$$

$$B_1 = P_b - \rho_{m1}(D - X) - P_f \frac{A_b}{A_x}$$
$$+ l_{vds}(\rho_{m1} - \rho_m)$$

$$B_1 = 5200 - 0.52(10{,}000 - 2000)$$

$$- 23\left(\frac{32.80}{32.80}\right) + \left(\frac{142}{0.0406}\right)(0.52 - 0.50)$$

$$B_1 = \mathbf{1087}$$

$$P_{2000ww} = \frac{1087}{2} + \left[\frac{1087^2}{4} \right.$$

$$+ \frac{5200(0.52)(0.816)(540)(246)(32.80)}{(620)(1.007)(32.80)} \left. \right]^{\frac{1}{2}}$$

$$P_{2000ww} = \mathbf{1418\ psi}$$

3. For the Wait and Weight Method of displacement, the annulus pressure when the gas bubble

first reaches the surface may be calculated using Equation 4.25 as follows:

$$P_{xww} = \frac{B_1}{2} + \left[\frac{B_1^2}{4} + \frac{P_b \rho_{m1} z_x T_x h_b A_b}{z_b T_b A_x} \right]^{\frac{1}{2}}$$

$$B_1 = P_b - \rho_{m1}(D - X) - P_f \frac{A_b}{A_x}$$

$$+ l_{vds}(\rho_{m1} - \rho_m)$$

$$B_1 = 5200 - 0.52(10{,}000 - 0) - 23\left(\frac{32.80}{34.58}\right)$$

$$+ \left(\frac{142}{0.0406}\right)(0.52 - 0.50)$$

$$B_1 = \mathbf{48}$$

$$P_{0ww} = \frac{48}{2} + \left[\frac{48^2}{4} \right.$$

$$\left. + \frac{5200(0.52)(0.883)(520)(246)(32.80)}{(620)(1.007)(34.58)} \right]^{\frac{1}{2}}$$

$$P_{0ww} = \mathbf{706\ psi}$$

4. The height of the gas bubble at the surface can be determined using Equation 4.21:

$$h_x = \frac{P_b z_x T_x A_b}{P_x z_b T_b A_x} h_b$$

$$h_0 = \frac{5200(0.883)(520)(32.80)}{706(1.007)(620)(34.58)}(246)$$

$$h_0 = \mathbf{1264\ feet}$$

5. The pressure at the casing seat when the influx reaches the surface may be calculated by adding

the annulus pressure to the mud and influx hydrostatics as follows:

$$P_{2000} = P_{0ww} + P_f \frac{A_b}{A_0} + \rho_m (2000 - h_0)$$

$$P_{2000} = 706 + 23 \left[\frac{32.80}{34.58}\right] + 0.50(2000 - 1264)$$

$$P_{2000} = \mathbf{1096 \ psi}$$

6.   The total pit volume increase with the influx at the surface. From part 4, the length of the influx when it reaches the surface is 1264 feet.

$$\text{Total pit gain} = h_0 C_{dpca}$$

$$\text{Total pit gain} = (1264)(0.0428)$$

$$\text{Total pit gain} = \mathbf{54 \ bbls}$$

7.   The surface annulus pressure profile during displacement using the Wait and Weight Method.

Example calculation, from part 2 with the top of the influx at 2000 feet, the pressure at 2000 feet is calculated to be

$$P_{2000ww} = \mathbf{1418 \ psi}$$

With the top of the influx at 2000 feet, the surface annulus pressure is given by Equation 4.19:

$$P_x = P_a + \rho_m X$$

$$P_0 = 1418 - 0.50(2000)$$

$$P_0 = \mathbf{418 \ psi}$$

The surface annulus pressure profile is summarized in Table 4.4 and illustrated as Figure 4.17.

Table 4.4

Surface Annulus Pressures—Wait and Weight Method

Depth to Top of Bubble, feet	Volume Pumped, bbls	Annular Pressure, psi
9754	0	300
9500	10	301
9000	30	303
8500	50	306
8000	69	310
7500	89	313
7000	109	318
6500	128	323
6000	148	326
5500	167	323
5000	186	323
4500	205	325
4000	223	331
3500	241	341
3000	259	358
2500	276	382
2000	293	416
1500	309	450
1000	324	513
500	337	596

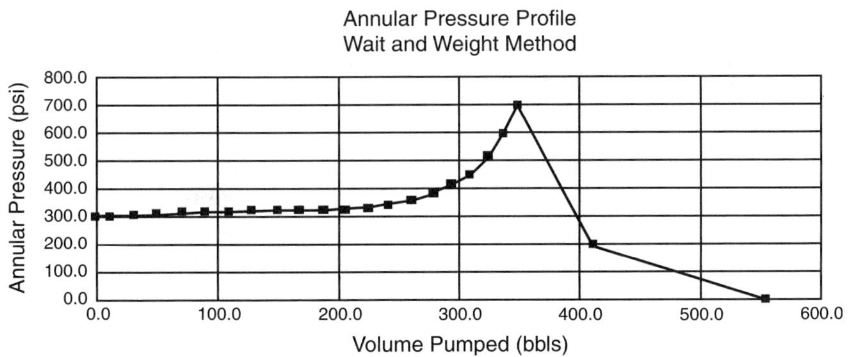

Figure 4.17

8.   The rate at which barite must be mixed and the minimum barite required:

$$X' = 350 \times S_m \left[\frac{W_2 - W_1}{(S_m \times 8.33) - W_2} \right] \quad (4.27)$$

Where:

$X' =$ Amount of barite, lbs barite/bbl mud
$W_2 =$ Final mud weight, ppg
$W_1 =$ Initial mud weight, ppg
$S_m =$ Specific gravity of the weight
material, water $= 1$

Note: S_m for barite $= 4.2$

$$X' = 350(4.2) \left[\frac{(10 - 9.6)}{(8.33 \times 4.2) - 10} \right]$$

$$X' = \textbf{23.5 lbs barite/bbl mud}$$

Minimum volume of mud $= 548$ bbls in the drillpipe and annulus:

Barite required $= (23.5)(548)$

Barite required $= \textbf{12,878 lbs of barite}$

Rate at which barite must be mixed:

Rate $= (23.5)(3)$

Rate $= \textbf{70.5 lbs barite/minute}$

C.   Compare the two procedures.

Table 4.5 and Figure 4.18 compare the two methods:

D.   The significance of 600 feet of 6-inch drill collars on the annulus pressure profile.

The significance of adding 600 feet of drill collars to each example is that the initial shut-in annulus pressure will

Table 4.5

Comparison of Driller's Method and Wait and Weight Method

	Driller's Method	Wait and Weight Method
Pressure at casing seat when well is first shut in	1300 psi	1300 psi
Pressure at casing seat when influx reaches that point	1480 psi	1418 psi
Fracture gradient at casing seat	1520 psi	1520 psi
Annulus pressure with gas to the surface	759 psi	706 psi
Height of the gas bubble at the surface	1165 feet	1264 feet
Total pit volume increase	50 bbls	54 bbls

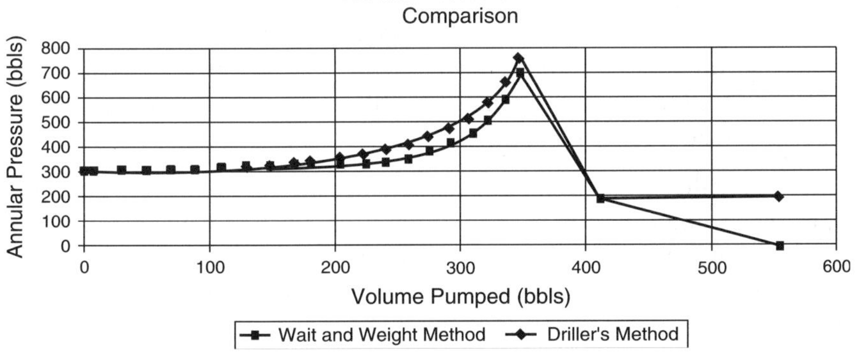

Figure 4.18

be higher because the bubble is longer. The new annular capacity due to the additional drill collars, C_{dcha}, is 0.0253 bbls/ft. Pursuant to Equation 3.7 the new influx height is:

$$h_b = \frac{Influx\ Volume}{C_{dcha}}$$

$$h_b = \frac{10}{0.0253}$$

$$h_b = \textbf{395 feet}$$

The annular pressure is given by Equation 4.20:

$$P_a = P_b - \rho_m(D - h_x) - P_f$$

$$P_a = 5200 - 0.5(10{,}000 - 395) - 0.094(395)$$

$$P_a = \textbf{360 psi}$$

As Figures 4.16, 4.17, and 4.18 and Table 4.5 illustrate, the annulus pressure profile is exactly the same for both the Driller's Method and the Wait and Weight Method until the weighted mud reaches the bit. After the weighted mud passes the bit in the Wait and Weight Method, the annulus pressures are lower than those experienced when the Driller's Method is used.

In this example, the Driller's Method may result in an underground blowout since the maximum pressure at the casing shoe, 1480 psi, is perilously close to the fracture gradient, 1520 psi. The well is more safely controlled using the Wait and Weight Method since the maximum pressure at the casing seat, 1418 psi, is almost 100 psi below the fracture gradient. Obviously, both displacement techniques would fail if a 200-psi "safety factor" was added to the circulating drillpipe pressure. This is another reason the Wait and Weight Method is preferred.

In reality, the difference between the annulus pressure profiles may not be as pronounced due to influx migration during displacement. In the final analysis, the true annular pressure profile for the Wait and Weight Method is probably somewhere between the profile for the Driller's Method and the profile for the Wait and Weight Method.

In part D, with 600 feet of drill collars in the hole, the annular area is smaller. Therefore, the length of the influx will be increased from 246 feet to 395 feet. As a result, the initial shut-in annulus pressure will be 360 psi is opposed to 300 psi. Using either the Driller's Method or the Wait and Weight Method, after 5 bbls are pumped, the influx will begin to shorten as it occupies the larger volume around the drillpipe. After 15 bbls of displacement are pumped, the influx is around the drillpipe and the annulus profiles for both techniques are unaltered from that point forward (see Figure 4.19).

The calculations of Example 4.8 are important so that rig personnel can be advised of the coming events. Being mentally prepared and forewarned may prevent a costly error. For example, if the Driller's Method

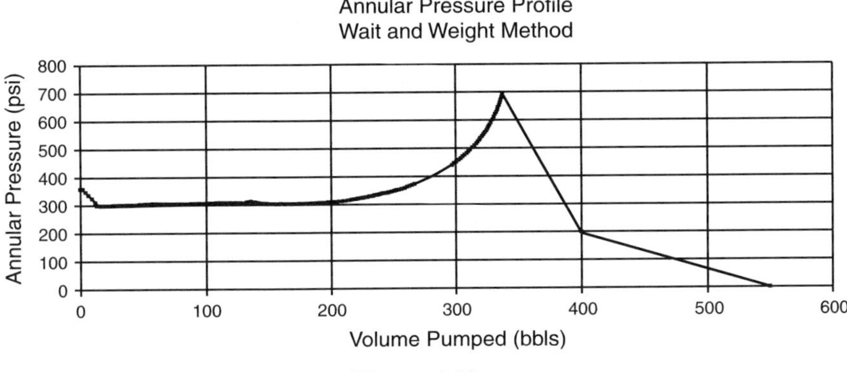

Figure 4.19

is used, rig personnel would expect that the maximum annulus pressure would be 759 psi, with gas to the surface; that the total pit level increase would be 50 barrels, which might cause the pits to run over; that gas would be at the surface in two hours [(406 − 50)/3] or less, depending on influx migration; and that dry gas might be vented at the surface for a period of 20 minutes, at an equivalent rate of 12 million cubic feet of gas per day. Certainly, these events are enough to challenge the confidence of unsuspecting and uninitiated rig personnel and cause a major disaster. Be prepared and prepare all involved for the events to come.

CONSTANT CASING PRESSURE, CONSTANT DRILLPIPE PRESSURE, AND MODIFICATION OF THE WAIT AND WEIGHT METHOD

After a well kick, the basic philosophy of control is the same, regardless of the procedure used. The total pressure against the kicking formation is maintained at a value sufficient to prevent further fluid entry and below the pressure that would fracture exposed formations. Simplicity is generally emphasized because decisions have to be made quickly when a well kicks, and many people are involved in the control procedure.

With the inception of current well control procedures, there was a considerable emphasis on the Driller's Method because the formation influx was immediately displaced. In recent years, it has become apparent that the Driller's Method, while simple, may result in an underground blowout that could have been prevented. Thus, many operators have adopted the Wait and Weight Method and accepted the necessity of increasing mud weight before displacement or during displacement of formation fluids. As illustrated in this chapter, the increase in mud weight requires an adjustment of drillpipe pressure during displacement of the weighted mud down the drill string. As a result, calculations are necessary to determine the pumping schedule required to maintain a constant bottomhole pressure.

To minimize the calculations, a modification of the Wait and Weight Method has become common. This procedure is known as the Constant Casing Pressure, Constant Drillpipe Pressure Wait and Weight Method. It is very simple. The only modification is that the casing pressure is kept constant at the initial shut-in casing pressure until the weighted mud reaches the bit. At that point, the drillpipe pressure is recorded and kept constant until the influx has been displaced.

The significance of this approach is illustrated by analyzing Table 4.4 and Figures 4.17 and 4.19. As illustrated, if there is no change in drill string geometry, as in the case of only one or two drill collars with heavy-weight drillpipe, the casing pressure would be kept constant at 300 psi for the first 142 barrels of displacement. At that point, the casing pressure should be 352 psi. Obviously, the equivalent hydrostatic is less than the formation pressure and will continue to be less than the formation pressure throughout the displacement of the influx. Additional influx of formation fluid will be permitted and the condition of the well will deteriorate into an underground blowout.

Pursuant to Figure 4.19, if 600 feet of drill collars are present and the casing pressure is kept constant at 360 psi while the drillpipe is displaced with 142 barrels, the well will probably be safely controlled since the casing pressure at 142 barrels should be approximately 325 psi or only 35 psi less than the 360 psi being rather arbitrarily held. After the weighted mud reaches the bit, the drillpipe pressure would be held 35 psi higher than necessary to maintain the bottomhole pressure constant at 5200 psi while the influx is displaced. In this instance, that additional 35 psi would have no detrimental or harmful effects. However, each situation is unique and should be considered. For example, if larger collars are being used, the margin would be even greater.

The obvious conclusion is that the Constant Casing, Constant Drillpipe Wait and Weight Method results in arbitrary pressure profiles, which can just as easily cause deterioration of the condition of the well or loss of the well. Therefore, use of this technique is not recommended without careful consideration of the consequences which could result in the simplification being more complicated than the conventional Wait and Weight technique.

THE LOW CHOKE PRESSURE METHOD

Basically, the Low Choke Pressure Method dictates that some pre-determined maximum permissible surface pressure will not be exceeded. In the event that pressure is reached, the casing pressure is maintained constant at that maximum permissible surface pressure by opening the choke, which obviously permits an additional influx of formation fluids. Once the choke size has to be reduced in order to maintain the maximum permissible annulus pressure, the drillpipe pressure is recorded and that drillpipe pressure is kept constant for the duration of the displacement procedure.

The Low Choke Pressure Method is considered by many to be a viable alternative in classic pressure control procedures when the casing pressure exceeds the maximum permissible casing pressure. However, close scrutiny dictates that this method is applicable only when the formations have low productivity.

It is understood and accepted that an additional influx will occur. However, it is assumed that the second kick will be smaller than the first. If the assumption that the second kick will be smaller than the first is correct and the second kick is in fact smaller than the first, the well might ultimately be controlled. However, if the second kick is larger than the first, the well will be lost.

In Example 4.8, the maximum permissible annulus pressure is 520 psi, which is that surface pressure which would cause fracturing at the casing seat at 2000 feet. If the Low Choke Pressure Method was used in conjunction with the Driller's Method in Example 4.8, the condition of the well would deteriorate to an underground blowout. Consider Table 4.3 and Figure 4.17. Assuming that the Driller's Method for displacement was being used, the drillpipe pressure would be held constant at 700 psi at a pump speed of 30 strokes per minute until 311 barrels had been pumped and the casing pressure reached 520 psi.

After 311 barrels of displacement, the choke would be opened in order to keep the casing pressure constant at 520 psi. As the choke was opened to maintain the casing pressure at 520 psi, the drillpipe pressure would drop below 700 psi and additional formation fluids would enter the wellbore. The well would remain underbalanced until the influx, which is approximately 406 barrels, was circulated to the surface. At a pump rate of three barrels per minute, the well would be underbalanced for approximately 32 minutes by as much as 186 psi!

Considering that during the original influx the underbalance was only 200 psi for much less than 32 minutes, the second influx is obviously excessive. Clearly, only a miracle will prevent an influx of formation fluids greater than the original influx of 10 barrels. That means that the well would ultimately be out of control since each successive bubble would be larger.

The Low Choke Pressure Method originated in West Texas, where formations are typically high pressure and low volume. In that environment, the well is circulated on a choke until the formation depletes. In any event, the productivity is seldom sustained at more than 1 mmscfpd. Occasionally, even in that environment, the productivity is high, resulting in a serious well control problem. Therefore, in general the Low Choke Pressure Method is an acceptable procedure only in areas of known low productivity and even in that environment can result in very serious well control problems and ultimate loss of the well. When inadvertently used in areas of high productivity, disasters of major proportions can result. The Low Choke Pressure Method is not generally recommended.

REVERSE THE BUBBLE OUT THROUGH THE DRILLPIPE

It is becoming increasingly common to reverse the influx out through the drillpipe. When the bubble is reversed out, the pressure profiles for the drillpipe and annulus are reversed, resulting in a reduction in annulus pressure when the Wait and Weight Method is used and a constant casing pressure when the Driller's Method is used. The potential hazards of bridging the annulus or plugging the bit or drillpipe are the primary objections to utilizing the reverse circulation technique; however, industry experience utilizing this technique has been successful and the industry has not experienced either drill string plugging, bit plugging, or bridging in the annulus.

Planning is essential if reverse circulation is anticipated since it must be convenient to tie the drillpipe into the choke manifold system. In addition, if time permits, it is recommended that the jets be blown out of the bit. A float in the drill string is not an insurmountable obstacle in that it can be blown out of the drill string along with the jets in the bit. As an alternative, a metal bar can be pumped through the float to hold it open during the reverse circulating operation. To date, reverse circulating the influx to the surface through the drillpipe has not become a common technique; however, the limited experience of the industry to date has been good and the technical literature is promising.

Operationally, reverse circulating is difficult due to the problems involved in commencing. As illustrated in Figure 4.20, the influx is in the annulus at first and within a few strokes it moves into the drillpipe. Therefore, the pressures on the drillpipe and annulus are changing rapidly during the first few strokes. The pressure on the drillpipe will increase to a value greater than the original shut-in annulus pressure while the pressure on the annulus will decline to the original shut-in drillpipe pressure. The U-Tube Model does not apply to the first few strokes; hence, there is no definitive and easy procedure for circulating the influx into the drillpipe.

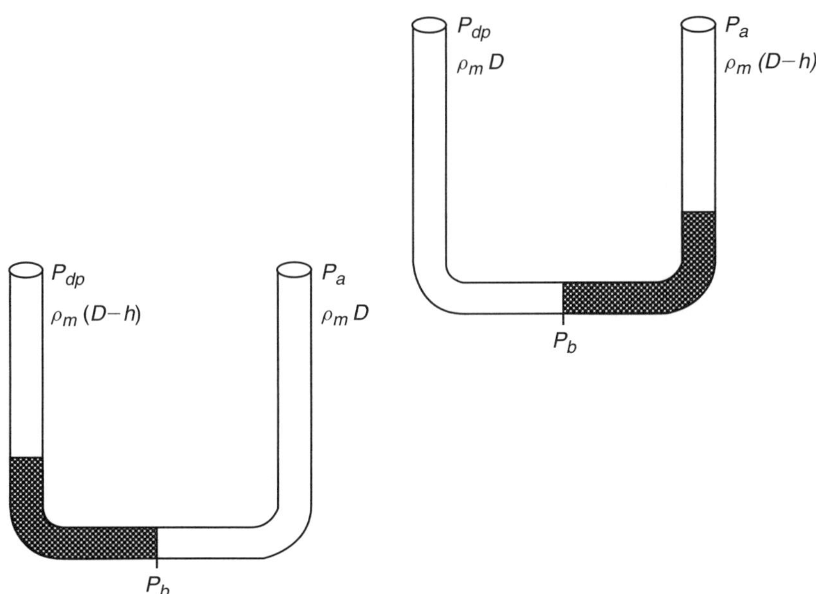

Figure 4.20   *Effect of Reverse Circulation on U-Tube Model.*

Another consideration is that in classical pressure control, most of the frictional pressure losses are in the drillpipe. Therefore, the circulating pressure at the casing shoe is not affected by the frictional pressure losses. However, with reverse circulation, any frictional pressure losses in the system are applied to the casing seat, which is the weak point in the circulating system. It is imperative that the procedure be designed with emphasis on the fracture pressure at the casing shoe. Consider Example 4.9:

Example 4.9

Given:

Wellbore schematic = Figure 4.7

Well depth, $D = 10{,}000$ feet

Hole size, $D_h = 7\frac{7}{8}$ inches

Drillpipe size, $D_p = 4\frac{1}{2}$ inches

$8\frac{5}{8}$-inch surface casing = 2000 feet

Casing internal diameter, $D_{ci} = 8.017$ inches

Fracture gradient, $F_g = 0.76\,\text{psi/ft}$

Fracture pressure = 1520 psi

Mud weight, $\rho = 9.6\,\text{ppg}$

Mud gradient, $\rho_m = 0.50\,\text{psi/ft}$

A kick is taken with the drill string on bottom and

Shut-in drillpipe pressure, $P_{dp} = 200\,\text{psi}$

Shut-in annulus pressure, $P_a = 300\,\text{psi}$

Pit level increase = 10 barrels

Kill mud weight, $\rho_1 = 10\,\text{ppg}$

Kill mud gradient, $\rho_{m1} = 0.52\,\text{psi/ft}$

Normal circulation $= 6$ bpm at 60 spm

Kill rate $= 3$ bpm at 30 spm

Circulating pressure at kill rate, $P_{ks} = 500$ psi

Pump capacity, $C_p = 0.1$ bbl/stk

Capacity of the:

Drillpipe (inside), $C_{dpi} = 0.0142$ bbl/ft

Drillpipe casing annulus, $C_{dpca} = 0.0428$ bbl/ft

Drillpipe hole annulus, $C_{dpha} = 0.0406$ bbl/ft

Initial displacement pressure, $P_c = 700$ psi @ 30 spm

Shut-in bottomhole pressure, $P_b = 5200$ psi

Maximum permissible surface pressure $= 520$ psi

Ambient temperature $= 60°$F

Geothermal gradient $= 1.0°/100$ feet

$\rho_f h = P_f = 23$ psi

$h_b = 246$ feet

Required:

Prepare a schedule to reverse circulate the influx to the surface.

Solution:

1.   The first problem is to get the gain into the drillpipe.

The maximum annulus pressure is given as 520 psi.

Therefore, the circulating annulus pressure must not exceed 520 psi.

2. Determine the volume rate of flow, Q, such that the frictional pressure is less than the difference between the maximum permissible annulus pressure and the shut-in annulus pressure.

   The maximum permissible annular friction pressure is

   $$520 - 300 \text{ psi} = 220 \text{ psi}$$

   From the relationship given in Equation 4.14:

   $$P \propto Q^{1.8}$$

   If at $Q = 30$ spm and $P = 500$ psi, Q may be determined for $P = 220$ psi as follows:

   $$\left(\frac{Q}{30}\right)^{1.8} = \left(\frac{220}{500}\right)$$

   $$Q = \textbf{19 spm}$$

   Therefore, the bubble must be reversed at rates less than 19 spm (1.9 bpm) to prevent fracturing the shoe.

   Choose 1 barrel per minute $= 10$ spm.

3. Conventionally, pump the influx up the annulus 100 strokes (an arbitrary amount) using the Driller's Method, keeping the drillpipe pressure constant.

4. Shut in the well.

5. Begin reverse circulation by keeping the drillpipe pressure constant at 200 psi while bringing the pump to speed approximately equal to 10 spm, with the annulus pressure not to exceed 500 psi. Total volume pumped in this step must not exceed that pumped in step 3 (100 strokes in this example).

6. Once the rate is established, read the annulus pressure and keep that pressure constant until the influx is completely displaced.

7.   Once the influx is out, shut in and read that the drillpipe pressure equals the annulus pressure equals 200 psi.

8.   Circulating conventionally (the long way), circulate the kill-weight mud to the bit keeping the annulus pressure equal to 200 psi.

   The choke should not change during this step. The only change in pit level should be that caused by adding barite. If not, shut in.

9.   With kill mud at the bit, read the drillpipe pressure and circulate kill mud to the surface, keeping the drillpipe pressure constant.

10.   Circulate and weight up to provide desired trip margin.

Once the influx is in the drillpipe, the procedure is the same as that for the Driller's Method. Calculations such as presented in Example 4.8 can be used. With the influx inside the drillpipe, the drillpipe pressure would be 487 psi. The maximum pressure on the drillpipe would be 1308 psi and would occur when the influx reached the surface. The influx would be 2347 feet long at the surface and the total pit gain would be 33 barrels. The time required to bring the gas to the surface would be 109 minutes and displacement of the influx would require 33 minutes. The distinct advantage is that the casing shoe would be protected from excessive pressure and an underground blowout would be avoided.

Once the influx is definitely in the drillpipe, the annulus pressure could be reduced by the difference between the original shut-in drillpipe pressure and the original shut-in annulus pressure, which is 100 psi in this example. The pressure on the annulus necessary to keep the bottomhole pressure constant at the original shut-in bottomhole pressure is the equivalent of the original shut-in drillpipe pressure plus the frictional pressure loss in the circulating system.

THE OVERKILL WAIT AND WEIGHT METHOD

The Overkill Wait and Weight Method is the Wait and Weight Method using a mud density greater than the calculated density for the kill-weight mud. Analysis of Example 4.8 indicates that the use of kill-weight mud to displace the gas bubble reduced the pressures in the annulus. It follows that, if increasing the mud weight from 9.6 pounds per gallon to 10 pounds per gallon reduced the pressure at the casing seat by 62 psi as shown in Table 4.5, then additional increases in mud weight would further reduce the pressure at the casing seat. The maximum practical density would be that which would result in a vacuum on the drillpipe. The effect on the pressure at the casing seat of utilizing such a technique is calculable.

Consider the utilization under the conditions described in Example 4.10:

Example 4.10

Given:

Example 4.6

Required:

1. The effect on the pressure at the casing seat when the density of the kill-weight mud is increased to 11 ppg.

2. The appropriate pumping schedule for displacing the influx with 11-ppg mud.

Solution:

1. For the Wait and Weight Method, the pressure at the casing seat at 2000 feet when the top of the influx reaches that point is given by Equation 4.25:

$$P_{xww} = \frac{B_1}{2} + \left[\frac{B_1^2}{4} + \frac{P_b \rho_{m1} z_x T_x h_b A_b}{z_b T_b A_x} \right]^{\frac{1}{2}}$$

$$B_1 = P_b - \rho_{m1}(D - X) - P_f \frac{A_b}{A_x} + l_{vds}(\rho_{m1} - \rho_m)$$

$$B_1 = 5200 - 0.572(10{,}000 - 2000)$$

$$- 23 \left[\frac{32.80}{32.80} \right] + \left(\frac{142}{0.0406} \right)(0.572 - 0.50)$$

$$B_1 = \mathbf{852}$$

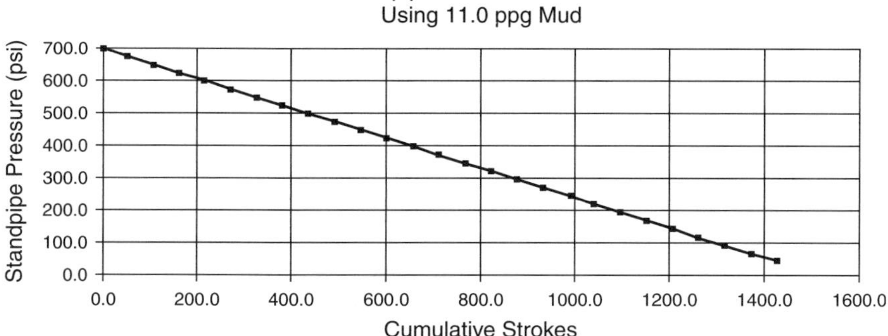

Figure 4.21

$$P_{2000ww} = \frac{852}{2} + \left[\frac{852^2}{4} \right.$$

$$+ \frac{5200(0.572)(0.828)(540)(246)(32.80)}{(620)(1.007)(32.80)} \left. \right]^{\frac{1}{2}}$$

$$P_{2000ww} = \textbf{1267 psi}$$

2. The appropriate pumping schedule may be determined using Equations 2.11 and 2.14 and is illustrated in Figure 4.21:

$$P_{cn} = P_{dp} - 0.052(\rho_1 - \rho)D + \left(\frac{\rho_1}{\rho} \right) P_{ks}$$

$$P_{cn} = 200 - 0.052(11.0 - 9.6)10{,}000$$

$$+ \left(\frac{11.0}{9.6} \right) 500$$

$$P_{cn} = \textbf{45 psi}$$

From Equation 2.14:

$$\frac{STKS}{25 \text{ psi}} = \frac{25(STB)}{P_c - P_{cn}}$$

$$\frac{STKS}{25} = \frac{25(1420)}{700 - 45}$$

$$\frac{STKS}{25} = 55$$

Obviously, the Driller's Method and the Wait and Weight Method result in the same annulus profile while the drillpipe is being displaced since it contains unweighted mud of density ρ. In this instance, the drillpipe capacity is 142 barrels, which means that both techniques have the same effect on annulus pressure for that period regardless of the density of the kill-weight mud being used in the Wait and Weight Method.

As illustrated in Table 4.4, the annulus pressure at the surface after 142 barrels is approximately 325 psi. Therefore, the pressure at the shoe at 2000 feet after 142 barrels is approximately 1325 psi. Pursuant to Example 4.10, the pressure at the casing seat when the influx reaches the casing seat is 1267 psi. Therefore, the maximum pressure at the casing seat occurred when the weighted mud reached the bit. In this example, with the 11-ppg kill-weight mud, the maximum pressure at the casing seat was almost 200 psi less than the fracture pressure of 1520 psi. The annulus pressure profile is illustrated as Figure 4.22.

In order to maintain the bottomhole pressure constant at 5200 psi, the drillpipe pressure must be reduced systematically to 45 psi and held constant throughout the remainder of the displacement procedure. Failure to consider properly reduction in drillpipe pressure resulting from the increased density is the most common cause of failure of the Overkill Wait and Weight Method. Properly utilized as illustrated in Example 4.10, the Overkill Wait and Weight Method can be a good alternative well control displacement procedure when casing shoe pressures approach fracture pressures and an underground blowout threatens.

SLIM HOLE DRILLING— CONTINUOUS CORING CONSIDERATIONS

Continuous coring utilizing conventional hard-rock mining equipment to conduct slim hole drilling operations offers unique considerations in pressure control. A typical slim hole wellbore schematic is illustrated

Table 4.6

Standpipe Pressure Schedule

Cumulative Strokes	Standpipe Pressure
0	700
55	675
110	649
165	624
220	599
275	573
330	548
385	523
440	497
495	472
550	446
605	421
660	396
715	370
770	345
825	320
880	294
935	269
990	244
1045	218
1100	193
1155	168
1210	142
1265	117
1320	91
1375	66
1430	45

in Figure 4.23. As previously discussed, the classical pressure control displacement procedures assume that the only significant frictional pressure losses are in the drill string and that the frictional pressure losses in the annulus are negligible.

As can be seen from an analysis of Figure 4.23, in slim hole drilling the conditions are reversed. That is, the frictional pressure losses in the drill string are negligible and the frictional pressure losses in the annulus are considerable. In addition, in slim hole drilling the volume of the influx

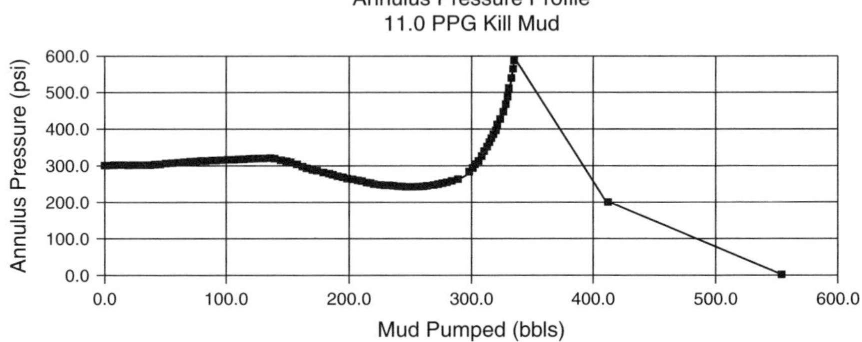

Figure 4.22

is much more critical because the annulus area is small. Under normal conditions, a one-barrel influx would not result in a significant surface annulus pressure. However, in slim hole drilling a one-barrel influx may result in excessive annular pressures.

The best and most extensive work in this area has been done by Amoco Production Company.[1] Sensitive flow meters have been developed which are capable of detecting extremely small influxes. Provided that the casing seats are properly selected and the influx volume is limited, classical pressure control procedures can be used. However, consideration must be given to frictional pressure losses in the annulus.

The first step is to measure the surface pressure as a function of the circulation rate as discussed in Chapter 2. Modeling must then be performed to match the measured values for frictional pressure losses in order that the frictional pressure losses in the annulus may be accurately determined. Figure 4.24 illustrates a typical pressure determination for the wellbore schematic in Figure 4.23. If, for example, the influx was to be circulated out at 40 gallons per minute, the drillpipe pressure during displacement would have to be reduced by the frictional pressure loss in the annulus, which is approximately 1000 psi in Figure 4.24. The circulating pressure at the kill speed is given by Equation 4.28:

$$P_{kssl} = P_c + P_{dp} - P_{fa} \qquad (4.28)$$

Where:

P_{kssl} = Circulating pressure at kill speed, psi
P_c = Initial circulating pressure loss, psi

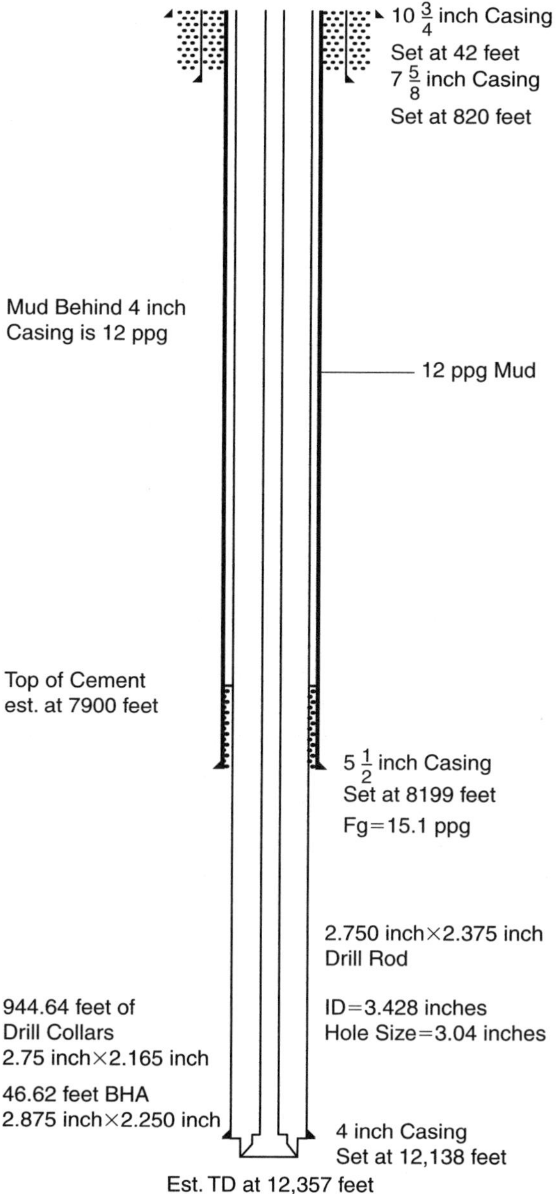

Figure 4.23 *Typical Slim Hole Wellbore.*

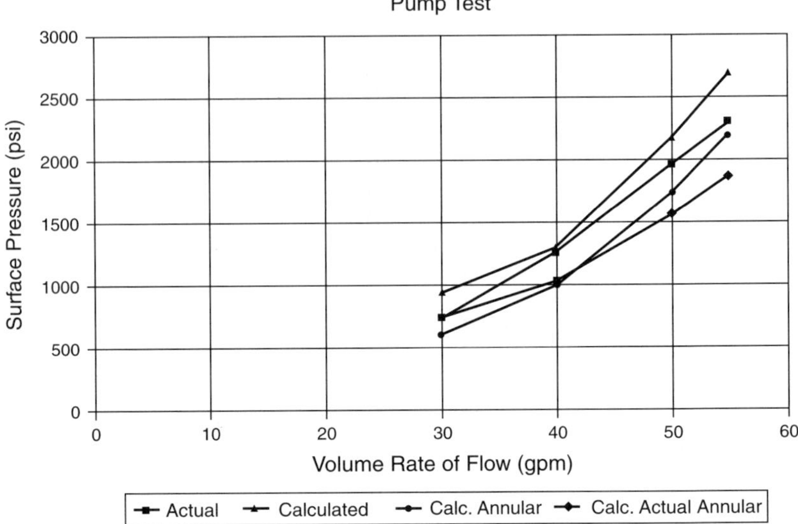

Figure 4.24

P_{dp} = Shut-in drillpipe pressure, psi

P_{fa} = Frictional pressure loss in annulus, psi

Since the influx will be displaced from the annulus before the weighted mud can reach the bit, the only classical displacement procedure applicable to most slim hole drilling operations is the Driller's Method. Therefore, pursuant to Figure 4.24 and Equation 4.28, if the shut-in drillpipe pressure was 1500 psi, the shut-in casing pressure was 1700 psi and the kill rate was 40 gpm, then the displacement pressure at the kill speed would be given by Equation 4.28 as follows:

$$P_{kssl} = P_c + P_{dp} - P_{fa}$$

$$P_{kssl} = 1300 + 1500 - 1000$$

$$P_{kssl} = \textbf{1800 psi}$$

Bringing the pump to speed for the displacement of the influx is not as straightforward as in conventional drilling operations. That is, keeping the casing pressure constant while bringing the pump to speed results in excessive pressure at the casing shoe. Therefore, the pump must be brought to speed while the casing pressure is being permitted to decline by the amount of the frictional pressure in the annulus. For example, if the shut-in casing pressure is 1700 psi, the pump would be brought to a speed of 40 gpm, permitting the casing pressure to decline by 1000 psi to 700 psi. At the same time, the drillpipe pressure would be increasing to 1800 psi. Once the appropriate drillpipe pressure is obtained, it would be held constant while the influx is circulated to the surface.

Since the annular frictional pressure losses are high and can result in an equivalent circulating density which is several pounds per gallon greater than the mud density, it is anticipated that influxes will occur when circulation is stopped for a connection. In that event, the influx can be circulated to the surface dynamically using the same circulating rate and drillpipe pressure as was used during the drilling operation.

Slim hole drilling offers the advantage of continuous coring. However, continuous coring requires the wire line retrieval of a core barrel after every 20 to 40 feet. The operation is particularly vulnerable to a kick during the retrieval of the core due to the potential swabbing action of the core barrel. The standard practice is to pump the drill rod down slowly during all core retrieval operations to insure that full hydrostatic is maintained.

The equipment required in slim hole drilling operations must be as substantial as in normal drilling operations. The choking effect of the annulus on a well flowing out of control is not as significant as might be expected. Therefore, the full complement of well control equipment is needed. Since it is more likely that the well will flow to the surface, particular attention must be focused on the choke manifold and flare lines.

The systems outlined in Chapter 1 are applicable to slim hole drilling with two exceptions. Since a top drive is used and the drill rods have a constant outside diameter, an annular preventer is not required. In addition, since the drill rods do not have conventional upset tool joints, slip rams should be included in the blowout preventer stack to prevent the drill rods from being blown out of the hole.

STRIPPING WITH INFLUX MIGRATION

The proper procedure for stripping was discussed in Chapter 3. However, the procedure presented in Chapter 3 does not consider influx migration. To perform a stripping operation accurately with influx migration, the classical procedure in Chapter 3 must be combined with the volumetric procedures outlined in this chapter. The combined procedure is not difficult. The stripping operation is carried out normally as presented in Chapter 3.

As the influx migrates, the surface pressure will slowly increase. When the surface pressure becomes unacceptable or reaches a predetermined maximum, the stripping operation is discontinued and the influx is expanded pursuant to the volumetric procedures presented in this chapter. Once the influx is expanded, the stripping operation is resumed.

Shell Oil has developed and reported a rather simple and technically correct procedure for stripping into the hole with influx migration.[2] Equipment is required which is not normally available on the rig; therefore, preplanning is a necessity. The required equipment consists of a calibrated trip tank and a calibrated stripping tank and is illustrated in Figure 4.25.

Basically, the surface pressure is held constant while a stand is stripped into the hole. The mud is displaced into the calibrated trip tank.

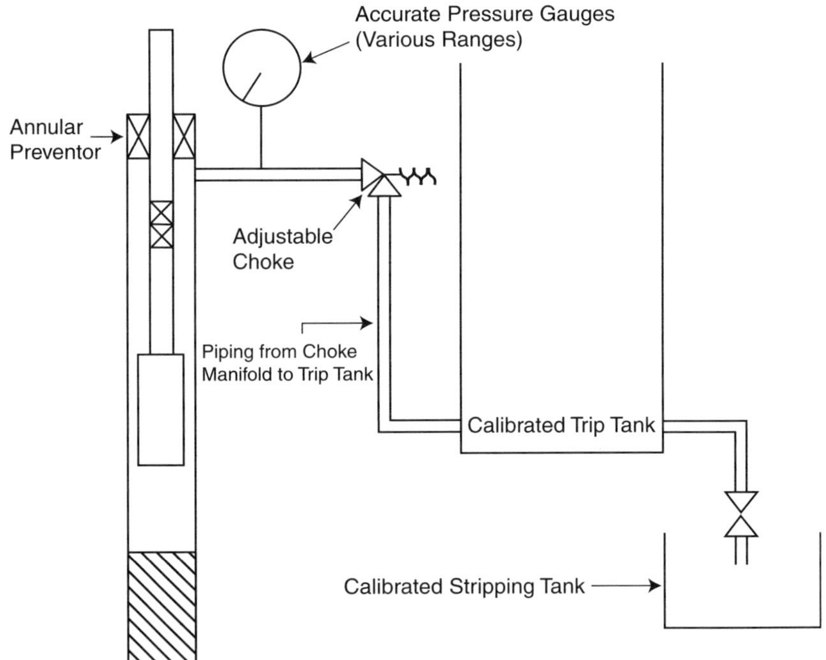

Figure 4.25   *Stripping Equipment for Shell Method.*

The theoretical displacement of the stand is then drained into the calibrated stripping tank. Any increase in the volume of the mud in the trip tank is recorded. The hydrostatic equivalent of the increase in volume is then added to the choke pressure. As a safety factor, the minimum annular areas are used in determining the equivalent hydrostatic of mud displaced from the hole. The procedure is as follows:

Step 1

After closing in the well, determine the influx volume and record the surface pressure.

Step 2

Determine the volume of drilling mud in the open-hole, drill collar annulus equivalent to 1 psi of mud hydrostatic.

Step 3

Adopt a convenient working pressure increment, P_{wpi}. The working pressure increment is arbitrary, but the fracture gradient should be considered.

Step 4

Determine the volume increase in the trip tank represented by the equivalent hydrostatic of the working pressure increment.

Step 5

Determine the additional back pressure, P_{hydl}, required when the influx is in the minimum annular area, which is usually the drill collar, open-hole annulus.

Step 6

While stripping the first stand into the hole, permit the surface pressure to increase to P_{choke} where

$$P_{choke} = P_a + P_{hydl} + P_{wpi} \qquad (4.29)$$

Step 7

Strip pipe into the hole maintaining P_{choke} constant.

Step 8

After each stand is stripped into the hole, drain the theoretical displacement into the calibrated tripping tank. Record any change in the volume of mud in the trip tank.

Step 9

When the increase in volume of mud in the trip tank equals the volume determined in Step 4, increase P_{choke} by P_{wpi} and continue.

Step 10

Repeat steps 7 through 9 until the pipe is returned to bottom.

Consider Example 4.11:

Example 4.11

Given:

Wellbore schematic $=$ Figure 3.2

Well depth, $D = 10,000$ feet

Number of stands pulled $= 10$ stands

Length per stand, $L_{std} = 93$ ft/std

Stands to be stripped $= 10$ stands

Drillpipe to be stripped $= 4\frac{1}{2}$ inches

Drill string displacement, $DSP_{ds} = 2$ bbl/std

Mud density, $\rho = 9.6$ ppg

Influx $= 10$ bbls of gas

Capacity of drillpipe annulus, $C_{dpha} = 0.0406$ bbl/ft

Hole diameter, $D_h = 7\frac{7}{8}$ inches

Hole capacity, $C_h = 0.0603$ bbl/ft

Bottomhole pressure, $P_b = 5000$ psi

Bottomhole temperature, $T_b = 620°$ Rankine

Gas specific gravity, $S_g = 0.6$

Shut-in casing pressure, $P_a = 75$ psi

$P_{wpi} = 50$ psi

Trip tank volume $= 2$ inches per barrel

Required:
Describe the procedure for stripping the 10 stands back to bottom using the Shell Method.

Solution:
Equivalent hydrostatic of one barrel of mud in the drillpipe annulus is given by Equation 4.10:

$$P_{hem} = \frac{0.052\rho}{C_{dpha}}$$

$$P_{hem} = \frac{0.052(9.6)}{0.0406}$$

$$P_{hem} = \textbf{12.3 psi/bbl}$$

Therefore, for a 10-barrel influx:

$$P_{hydl} = 10(12.3)$$

$$P_{hydl} = \textbf{123 psi}$$

From Equation 4.29:

$$P_{choke} = 75 + 123 + 50$$

$$P_{choke} = \textbf{248 psi}$$

Therefore, strip the first stand into the hole and permit the choke pressure to increase to 250 psi. Bleed mud proportional to the amount of drillpipe stripped after reaching 250 psi.

Continue to strip pipe into the hole, keeping the surface pressure constant at 250 psi.

After each stand, bleed two barrels of mud into the stripping tank and record the volume in the trip tank.

When the volume in the trip tank has increased by eight inches, increase the surface pressure to 300 psi and continue. Eight inches represents the volume of the equivalent hydrostatic of the 50 psi working pressure increment (50 psi divided by 12.3 psi/bbl multiplied by 2 inches per barrel).

OIL-BASE MUD IN PRESSURE
AND WELL CONTROL OPERATIONS

The widespread use of oil-base drilling muds in deep drilling operations is relatively new. However, from the beginning, well control problems with unusual circumstances associated will oil-base muds have been observed. Typically, field personnel reports of well control problems with oil muds are as follows:

"Nothing adds up with an oil mud."

"It happened all at once! We had a 200-barrel gain in 2 minutes! There was nothing we could do! It was on us before we could do anything!"

"We saw nothing—no pit gain ... nothing—until the well was flowing wildly out of control! We shut it in as fast as we could, but the pressures were too high! We lost the well!"

"We started out of the hole and it just didn't act right. It was filling OK, but it just didn't seem right. We shut it in and didn't see anything—no pressure, nothing! We still weren't satisfied; so we circulated all night. Still we saw

nothing—no pit gain—no gas-cut mud—nothing. We circulated several hole volumes and just watched it. It looked OK! We pulled 10 stands, and the well began to flow. It flowed 100 barrels before we could get it shut in! I saw the pressure on the manifold go over 6000 psi before the line blew. It was all over then! We lost the rig!"

"We had just finished a trip from below 16,000 feet. Everything went great! The hole filled like it was supposed to, and the pipe went right to bottom. We had gone back to drilling. We had already drilled our sand. We didn't drill anything new but shale and had not circulated bottoms up. Everything was going good. Then, I looked around and there was mud all over the location! I looked back toward the floor and there was mud going over the bushings! Before I could close the Hydril, the mud was going to the board! I got it shut in okay, but somehow the oil mud caught on fire in the derrick and on the rotary hose! The fire burned the rotary hose off and the well blew out up the drillpipe! It took about 30 minutes for the derrick to go! It was a terrible mess! We were paying attention! It just got us before we could do anything."

FIRE

The most obvious problem is that oil-base muds will burn. The flash point of a liquid hydrocarbon is the temperature to which it must be heated to emit sufficient flammable vapor to flash when brought into contact with a flame. The fire point of a hydrocarbon liquid is the higher temperature at which the oil vapors will continue to burn when ignited. In general, the open flash point is 50 to 70 degrees Fahrenheit less than the fire point.

Most oil-base muds are made with # 2 diesel oil. The flash point for diesel is generally accepted to be about 140 degrees Fahrenheit. On that basis, the fire point would be about 200 degrees Fahrenheit. Mixing the oil-base mud with hydrocarbons from the reservoir will only increase the tendency to burn. The exposure of gas with the proper concentrations of air to any open flame or a source capable of raising the temperature

of the air-gas mixture to about 1200 degrees Fahrenheit will result in
a fire.

SOLUBILITY OF NATURAL GAS IN OIL-BASE MUD

It is well known in reservoir engineering that such hydrocarbons
as methane, hydrogen sulfide, and carbon dioxide are extremely soluble
in oil. With the popularity of oil muds used in routine drilling operations
in recent years, considerable research has been performed relative to the
solubility of hydrocarbons in oil muds.[3] As illustrated in Figure 4.26,
the solubility of methane increases virtually linearly to approximately
6000 psi at 250 degrees Fahrenheit. The methane solubility becomes
asymptotic at 7000 psi, which basically means that the solubility of methane
is infinite at pressures of 7000 psi or greater and temperatures of 250 degrees
Fahrenheit or greater. The pressure at which the solubility becomes infinite
is defined as the Miscibility Pressure.

Note in Figure 4.26 that the solubility of carbon dioxide and hydrogen sulfide is higher than that of methane. The Miscibility Pressure of

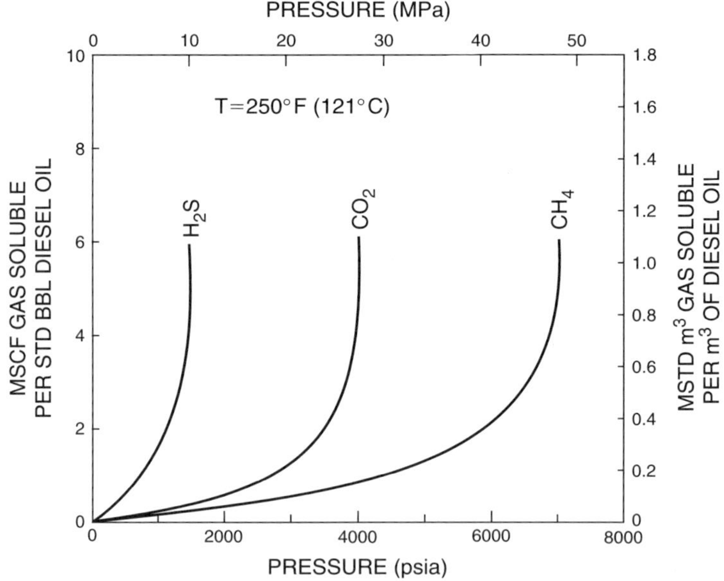

Figure 4.26   *Gas Solubility.*

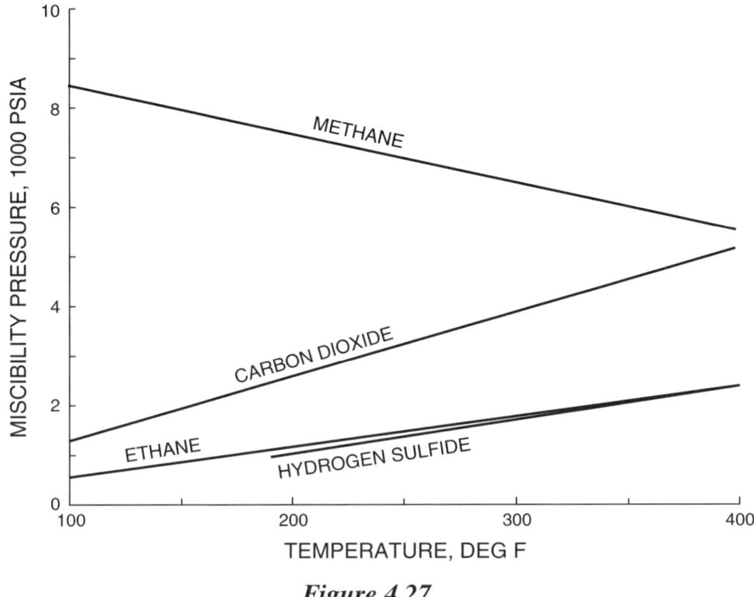

Figure 4.27

methane decreases with temperature, as illustrated in Figure 4.27, while the Miscibility Pressure of carbon dioxide and hydrogen sulfide increases with temperature. As further illustrated in Figure 4.28, the Miscibility Pressure of methane decreases from about 8000 psi at 100 degrees to approximately 3000 psi at 600 degrees.

The significance of this research to field operations is that in deep, high-pressure gas wells an influx of reservoir hydrocarbons can, in part, dissolve in an oil-base mud. Another variable which will be discussed is the manner in which the influx occurs. However, in the most simple illustration, when a kick is taken while drilling with an oil-base mud and the influx is primarily methane, the influx will dissolve into the mud system and effectively mask the presence of the influx. That is not to say that one barrel of oil-base mud plus one barrel of reservoir hydrocarbon results in one barrel of a combination of the two.

However, it is certain that under the aforementioned conditions one barrel of oil-base mud plus one barrel of reservoir hydrocarbon in the gaseous phase will yield something less than two barrels. Therefore, the danger signals that the man in the field normally observes are more subtle.

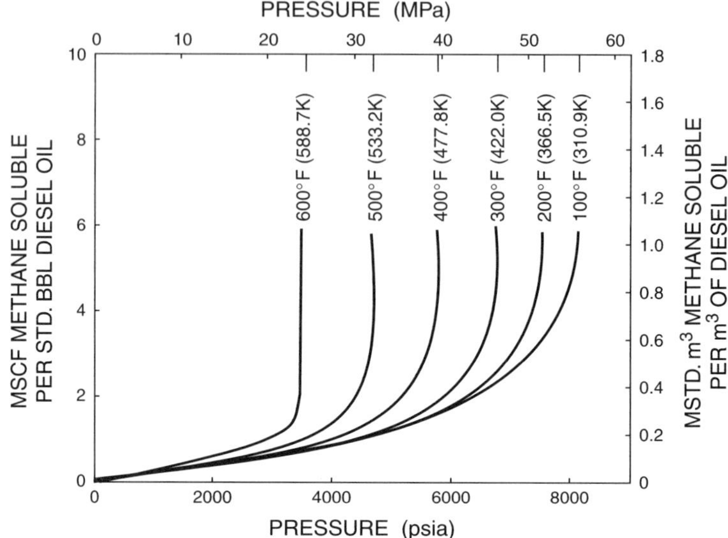

Figure 4.28 *Methane Solubility.*

The rate of gain in the pit level when using an oil-mud system will be much less than the rate of gain when using a water-base system.

The exact behavior of a particular system is unpredictable. The phase behavior of hydrocarbons is very complex and individual to the precise composition of the system. Furthermore, the phase behavior changes as the phases change. That is, when the gas does begin to break out of solution, the phase behavior of the remainder of the liquid phase shifts and changes. Therefore, only generalized observations can be made.

Again, assuming the most simple example of taking a kick while drilling on bottom, the influx is partially dissolved into the oil phase of the mud system. A typical phase diagram is illustrated in Figure 4.29. Under such conditions, the drilling fluid is represented by point "A." Point A represents a hydrocarbon system above the bubble point with all gas in solution. As the influx is circulated up the hole, the gas will remain in solution until the bubble point is reached.

The hydrocarbon system then enters the two-phase region. As the hydrocarbon continues up the hole, more and more gas breaks out. As the gas breaks out, the liquid hydrostatic is replaced by the gas hydrostatic and the effective hydrostatic on bottom will decrease, permitting additional

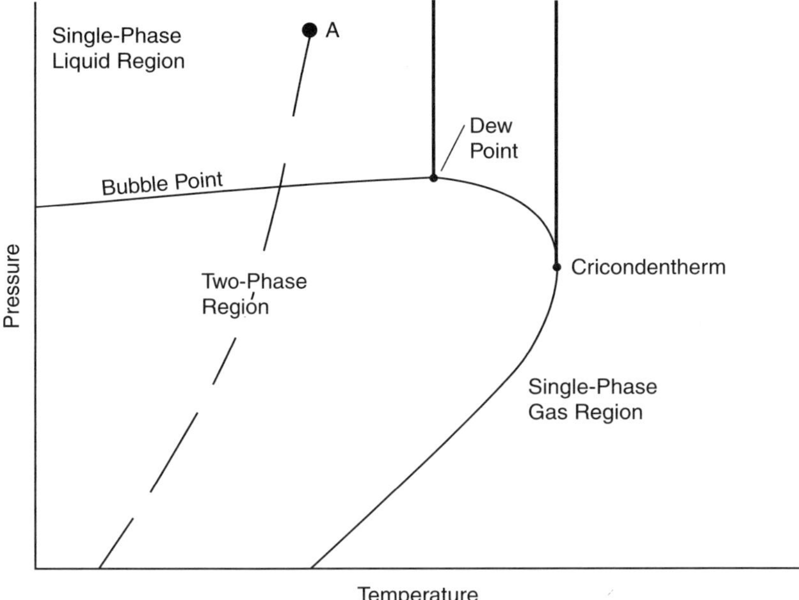

Figure 4.29   *Typical Hydrocarbon Phase Diagram.*

influx at an exponentially increasing rate. This can account for the field observation of high flow rates and rapidly developing events.

The pressure behavior of kicks in oil-base muds can be confusing. Using the previous illustration, if a well is shut in after observing a 10-barrel gain at the surface, it is probable that a much larger gain has been taken. However, due to the compressibility of the system, the surface pressure on the annulus may be less than that observed with a 10-barrel surface gain in a water-base mud.

As the gas is circulated to the surface and begins to break out of solution, the kick will begin to behave more like a kick in a water-base mud and the annular pressure will respond accordingly. As illustrated in Figure 4.28, the solubility of methane in diesel oil is very low at low pressure and almost any reasonable temperature. Therefore, when the gas reaches the surface, the annular pressure will be much higher than expected and almost that anticipated with the same kick in a water-base mud. This scenario is schematically illustrated in Figure 4.30.

Research has shown that the solubility of the influx into the mud system is a more significant problem when the influx is widely distributed.

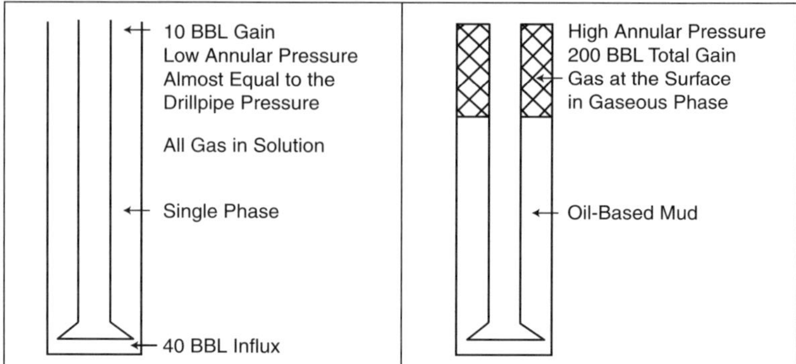

Figure 4.30 *Complexities of Oil-Based Muds.*

A kick taken while on bottom drilling is an example. An influx which is in a bubble and not mixed behaves more like an influx in a water-base system. A kick which is taken while tripping or while making a connection with the pump off is an illustration.

To further complicate matters, the density of an oil-base mud is affected by temperature and pressure. As a result, an oil-base mud may have a density of 17 ppg at the surface but a different density under bottom hole conditions. These concepts are very difficult to quantitize, visualize, and verbalize. However, they are worthy of consideration. The experience of the industry along with a technical analysis dictates that much more caution must be exercised when drilling with an oil-base mud. For example, if a well is observed for flow for 15 minutes using a water-base mud, it should be observed longer when an oil-base mud is being used. Unfortunately, precise field calculations cannot be made due to the complexity of the phase behavior of the hydrocarbons involved. Well control problems are significantly complicated when oil-base muds are used and the consequences are more severe.

FLOATING DRILLING AND SUBSEA OPERATION CONSIDERATIONS

SUBSEA STACK

Recently, in some explanatory operations, operators have utilized high pressure risers and surface blowout preventers. However, in most

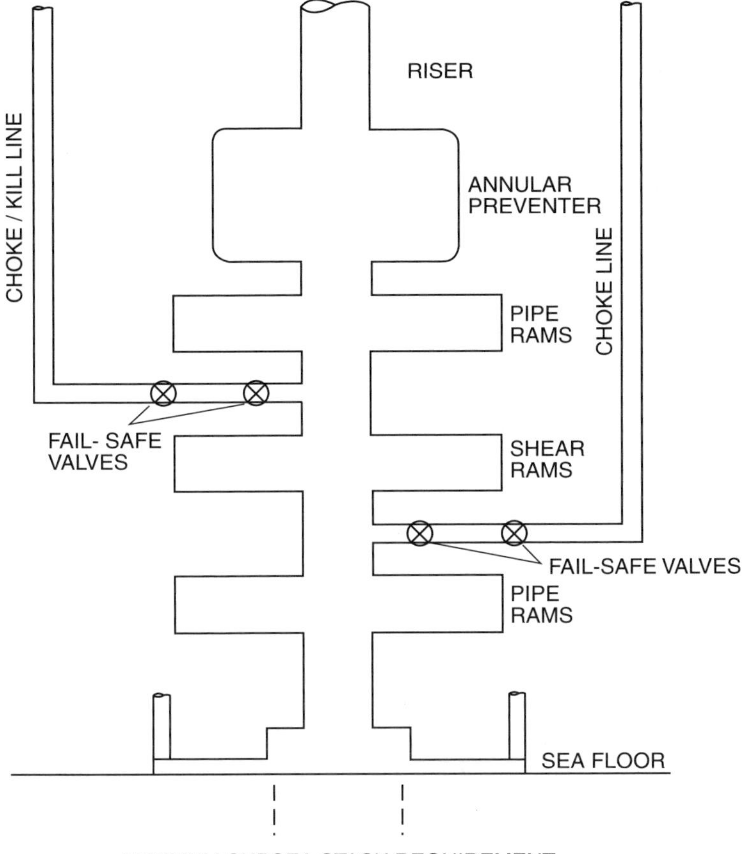

Figure 4.31

floating drilling operations, the blowout preventers are located on the sea floor, necessitating redundancy on much of the equipment. Figure 4.31 demonstrates the minimum subsea stack requirement. Some operators use a double annular preventer configuration with a connector in between. The connector allows for the top package to be pulled and the top annular preventer to be repaired. The lowermost annular preventer is used as a backup when the top annular preventer fails.

Shear rams, as opposed to blind rams, are a necessity in the event that conditions dictate that the drill string must be sheared and the drilling vessel moved off location.

The list of alternate stack arrangements is endless and it would be pointless to list the advantages and disadvantages of each. In many cases, the operator has no choice but to take the stack that comes with the rig. When a choice is available, the operator should review the potential problems in his area to determine the stack arrangement that best suits his needs, with the underlying requirement being redundancy on all critical components.

Accumulators are sometimes mounted on the subsea stack to improve response time or to activate the BOP stack acoustically in the event of an emergency when a ship is moved off location. In these instances, the precharge pressure requires adjusting to account for the hydrostatic pressure in the control lines or

$$P_{pc} = P_{inipc} + 14.7 + (l_{contline})(S_g)(0.433) \qquad (4.30)$$

Where:

P_{pc} = Precharge pressure, psi
P_{inipc} = Initial precharge pressure, psi
$l_{contline}$ = Length of control line, feet
S_g = Specific gravity

The use of higher precharge pressure also results in a reduction in the usable volume of an accumulator bottle. It therefore follows that more bottles than normal are required. Using the Ideal Gas Law and equating volumes at precharge, minimum, and maximum operating pressure:

$$\frac{P_{pc} V_{pc}}{z_{pc} T_{sc}} = \frac{P_{min} V_{min}}{Z_{min} T_{ss}} = \frac{P_{max} V_{max}}{z_{min} T_{ss}}$$

Where:
ss = Subsea
pc = Precharge
min = Minimum
max = Maximum
sc = Standard conditions
P = Pressure, psi
V = Volume, bbls
z = Compressibility factor
T = Temperature, °Absolute

The usable volume per bottle is then $V_{min} - V_{max}$.

SPACING OUT

Due to the problems associated with pressure surge in the system and wear on the preventers associated with heave on semi-submersibles and drillships, the drill string is usually hung off on the upper pipe rams when the heave becomes excessive. The distance to the rams should be calculated after the riser has been run to ensure that the rams do not close on a tool joint. To do this, run in hole, position the tool joint 15 feet above the upper rams, close the upper rams, and slowly lower the drill string until the rams take the weight of the string. Record the tide and distance to the tool joint below the rotary table for future reference.

SHUT-IN PROCEDURES

Classical shut-in procedures are modified because of the special considerations of floating drilling operations with a subsea stack. Some of the modifications are as follows:

A.   While drilling

    1.   At the first indication of a kick, stop the mud pumps. Close the BOPs and, if conditions dictate, hang off as described in the previous section.

    2.   Notify the toolpusher and company representative.

    3.   Read and record the pit gain, SIDPP and SICP.

B.   While tripping

    1.   At the first indication of a kick, set the slips; install and close the drillpipe safety valve.

    2.   Pick up the kelly and close the BOPs. If conditions dictate, hang off as described in the previous section.

    3.   Notify the toolpusher and company representative.

4.   Open the safety valve; read and record the pit gain, SIDPP and SICP.

FLOATING DRILLING WELL CONTROL PROBLEMS

There are four well control problems peculiar to floating drilling operations. They are:

1.   Fluctuations in flow rate and pit volume due to the motion of the vessel.
2.   Friction loss in the choke line.
3.   Reduced fracture gradient.
4.   Gas trapped in BOP stack after circulating out a kick.

Fluctuations in Flow Rate and Pit Volume

Due to the heaving of the vessel and the related change in volume of the riser on floaters, the flow rate and pit volume fluctuate, making these primary indications of a kick difficult to interpret. Monitoring the pit volume for indications of a kick is further complicated by the pitch and roll of the vessel, as this will cause the fluid in the pits to "slosh" with the motion of the vessel—even if there is not fluid flow in or out of the pits.

Many techniques have been proposed to decrease the effect of vessel movement. The pit volume totalizer, as opposed to the mechanical float, is a step in the right direction, but it requires infinite sensors to compensate totally for the entire range of vessel motion. An electronic sea-floor flow rate indicator has been devised to alleviate the problem of vessel movement. The idea is sound, but experience with this equipment is limited. For now, the industry will have to continue to monitor the surface equipment for changes in the trend. Naturally, this causes a delay in the reaction time and allows a greater influx. Knowing this, the rig personnel must be particularly alert to other kick indicators.

Frictional Loss in the Choke Line

Frictional pressure losses in the small internal-diameter (ID) choke line are negligible on land rigs but can be significant in deep-water subsea stack operations. The degree is proportional to the length and ID of the choke line. For the land rig U-Tube Dynamic Model, the bottomhole pressure, P_b, is equal to the hydrostatic of the annulus fluids, ρ_m, plus the choke back pressure, P_{ch1}:

$$P_b = \rho_m D + P_{ch1}$$

Normally, the equivalent circulating density (ECD) resulting from frictional pressure losses in the annulus is not considered since it is difficult to calculate, positive, and minimal. However, in the case of a long choke line, the effects are dramatic, particularly during a start-up and shut-down operation.

With the long, small choke line the dynamic equation becomes bottomhole pressure equals the hydrostatic of the annulus fluids, ρ_m, plus the choke back pressure, P_{ch2}, plus the friction loss in the choke line, P_{fcl}:

$$P_b = \rho_m D + P_{ch2} + P_{fcl}$$

Solving the equations simultaneously results in

$$P_{ch1} = P_{ch2} + P_{fcl}$$

or

$$P_{ch2} = P_{ch1} - P_{fcl}$$

Simply said, the pressure on the choke at the surface must be reduced by the frictional pressure in the choke line. The need to understand this concept is paramount because, if the choke operator controls the back pressure to equal SICP during start up, an additional and unnecessary pressure will be imposed on the open hole formations equal to the choke line friction pressure, P_{fcl}, often with catastrophic results.

On the other hand, if the operator maintains choke pressure constant during a shut-in operation, the choke line friction pressure is reduced to 0, reducing the bottomhole pressure by the choke line friction pressure, P_{fcl}, and allowing an additional influx.

There is a very simple solution, one that has gained acceptance in the industry. During any operation when the pump rate is changed, including start-up and shut-in operations, monitor the secondary choke line pressure and operate the choke to maintain this pressure constant. Since the choke line pressure loss is above the stack, then

$$P_b = \rho_m D + P_{ch2}$$

Therefore, monitor the secondary choke line pressure in the same manner that the primary choke back pressure (also called casing or annulus) on a land rig is monitored. The choke gauge must still be monitored for evidence of plugging.

Reduced Fracture Gradient

The classic works done in fracture gradient determination were developed for land operations and cannot be applied directly to offshore operations. The fracture gradient offshore will normally be less than an onshore gradient at equivalent depth as a result of the reduction in total overburden stress due to the air gap and sea-water gradient.

Various charts and procedures have been developed for specific areas to modify the classic methods (Eaton, Kelly and Matthews, etc.) for offshore use. The basic premise is to reduce the water depth to an equivalent section of formation by the ratio of the sea-water gradient to the overburden stress gradient at the point of interest. When using these charts, it is important to realize that some are referenced by the ratio of the subsea depth to the depth rotary table (or RKB). Consider Example 4.12:

Example 4.12

Given:

Depth at shoe $= 2600$ feet rotary table

Water depth $= 500$ feet

Air gap $= 100$ feet

Required:

Determine the fracture gradient at the shoe.

Solution:

Sediment thickness $= 2600 - 500 - 100 = 2000$ feet

Fracture pressure from Figure 4.32 $= 1500$ psi

$$F_g = \frac{1500}{(2600)(0.052)}$$

$$F_g = \mathbf{11.1\ ppg}$$

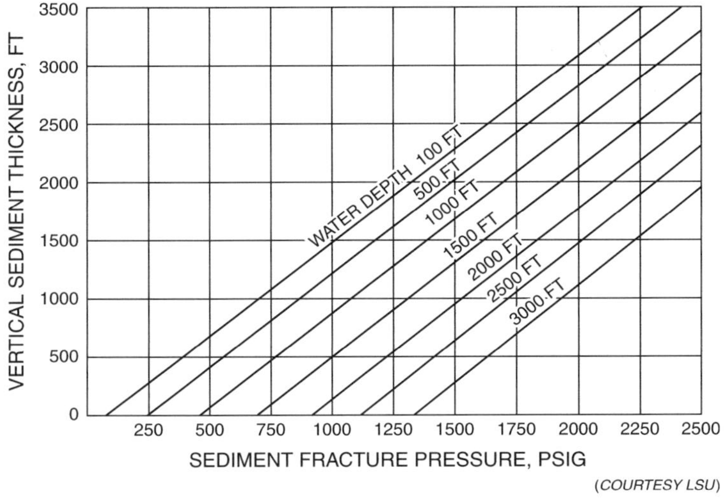

(*COURTESY LSU*)

ESTIMATED FRACTURE PRESSURE FOR MARINE SEDIMENTS

Figure 4.32

Trapped Gas after Circulating Out a Kick

Most BOP stack arrangements do not provide for the removal of the gas bubble remaining between the annular preventer and the choke line after circulating out a kick. This presents no problem in surface operations, as the pressure of this gas is minimal. However, with the use of subsea stacks, the pressure of this gas bubble is equal to the hydrostatic weight of the kill mud existing in the choke line. If improperly handled, the expansion of this bubble after opening the BOP could result in an extremely hazardous situation. Closing the lowest pipe ram to isolate the hole while displacing kill-weight mud into the riser via the kill line with the diverter closed is an acceptable method in more shallow situations. A more conservative method for deeper situations is recommended as follows:

After circulating out all kicks

1.   Close the lower pipe rams and calculate the difference in pressure between the hydrostatic head of the kill mud versus a column of water in the choke line.

2.   Displace the kill-weight mud in the BOP stack and choke line with water pumped down the kill line while holding the back pressure calculated in step 1.

3.   Close the kill-line valve and bleed the pressure off through the gas buster.

4.   Close the diverter and open the annular preventer, allowing the remaining gas to U-tube up the choke line to the gas buster.

5.   Displace the mud in riser with kill-weight mud via kill line.

6.   Open lower pipe rams.

DEEP-WATER FLOATING DRILLING

Well control principles are no different for operations in deep water. A well is deep and dumb and doesn't have any idea how much water, if any, is between the wellhead and the producing formation. However, with the increase in deep-water operations, a kind of mystic veil has evolved.

Some aspects of well control operations in deep water discourage and/or compromise classic well control procedures. However, there is no mystery. The laws of physics and chemistry still apply. The general philosophy among most operators seems to be to keep the problems below the casing seat.

As the water depth increases, the margin between fracture gradient and pore pressure generally decreases. Therefore, more drilling liners are required to maintain any reasonable kick tolerance.

As the water depth increases, the length of the choke line and, of course, the frictional pressure loss in the choke line increase proportionally. These conditions make circulating out a kick without losing circulation a slow, difficult, and delicate (and often impossible) process.

The length of the choke line introduces other complications in addition to added friction. Because the frictional pressure loss is a function of the fluid properties, the frictional pressure loss with gas or gas cut mud in the choke line can be significantly different than that determined with only mud. In addition, the annulus pressure has been known to fluctuate erratically with gas cut mud in the choke line, making keeping the drillpipe pressure constant very difficult.

As the choke line lengthens, the length of the gas column increases, resulting in higher surface pressures. Consequently, there is a reluctance to bring gas into the choke line and to the surface. Consider the following example.

Example 4.13

The maximum annular pressure will occur when the gas bubble first reaches the surface and is given by the following expression:

$$P_{s\,max} = \frac{B_s + \left[B_s^2 + 4A_sX_s\right]^{\frac{1}{2}}}{2A_s} \tag{4.31}$$

Where:

$$A_s = 1 + \rho_m l_{cl}\left(1 - \frac{C_{cl}}{C_h}\right)\left(\frac{S}{53.3\,Z_s\,T_s}\right)$$

$$B_s = P_b - \rho_m D + \rho_m l_{cl}\left(1 - \frac{C_{cl}}{C_h}\right) + \frac{S\rho_m\,P_b h_b}{53.3\,Z_b\,T_b}$$

$$X_s = \frac{\rho_m\,P_b\,Z_s\,T_s\,h_b}{Z_b\,T_b}$$

s = at surface
b = at bottom of the hole
ρ_m = Mud density, psi/ft
l_{cl} = Length of choke line, feet
C_{cl} = Capacity of choke line, bbls/ft
C_h = Capacity of annulus beneath BOPs, bbls/ft
S = Specific gravity of the gas
P = Pressure, psi
Z = Compressibility
T = Temperature, °Rankine
h_b = Height of initial influx with geometry beneath BOPs

Assumptions:

This equation assumes that the Driller's Method is used for displacement of the influx. Therefore, it is a "worst case" scenario. In order to eliminate the complexities introduced by different geometries between the bottom of the hole and the blowout preventers, the height of the influx, h_b, is determined using the geometry of the annulus immediately below the blowout preventers.

Consider Figure 4.30 and the following:

Gas specific gravity $= 0.6$

Influx height $= 246$ feet

Bottomhole temperature $= 640°$ Rankine

Surface temperature $= 535°$ Rankine

Compressibility at bottomhole and surface conditions $= 1.0$

Bottomhole pressure $= 5200$ psi

Mud density $= 9.6$ ppg

Length of choke line $= 1500$ feet

Capacity of choke line $= 0.0087$ bbls/ft

Capacity of annulus $= 0.0406$

Depth $= 10,000$ feet

Substituting these values into Equation 4.31, the pressure when the gas reaches the surface is determined to be:

$$P_{s\,\text{max}} = 1233 \text{ psi}$$

This same problem was solved in Example 4.8 on page 163 using the Driller's Method for a constant annular area to the surface. In that example, the maximum surface pressure was determined to be 759 psi. Therefore, the 1500 feet of small-diameter choke line would add 474 psi to the surface pressure (Figure 4.33).

As deep-water floating drilling operations become more routine, the comfort level in well control operations will rise. The BOP equipment is generally very reliable. The general procedure should include a "pain threshold pressure." When that pressure is reached, the influx can still be bullheaded to the casing or liner shoe.

As of this writing, most influxes are "sandwiched." The hole is displaced simultaneously through both the drillpipe and the annulus,

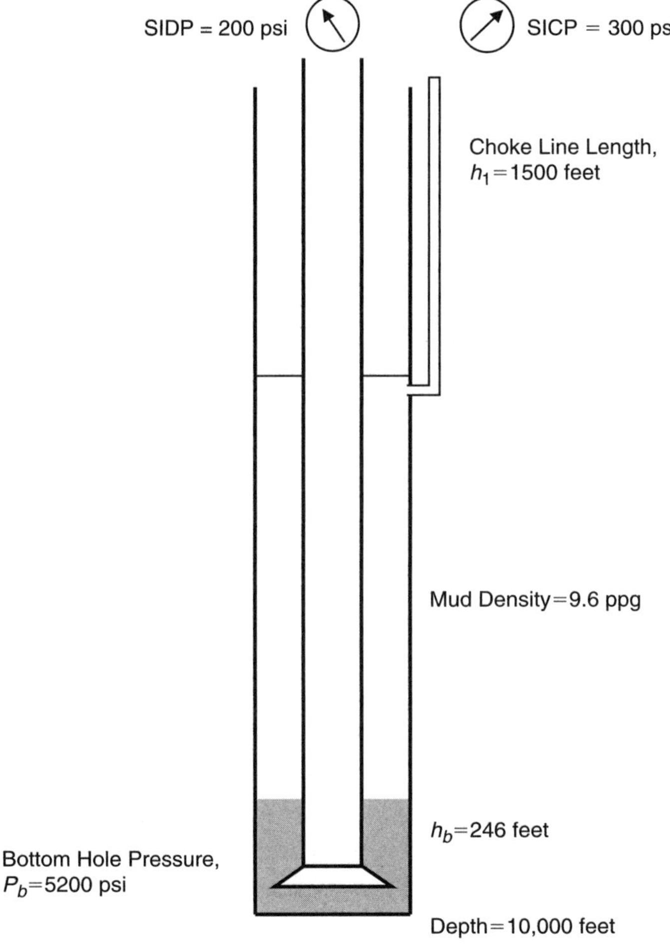

Figure 4.33

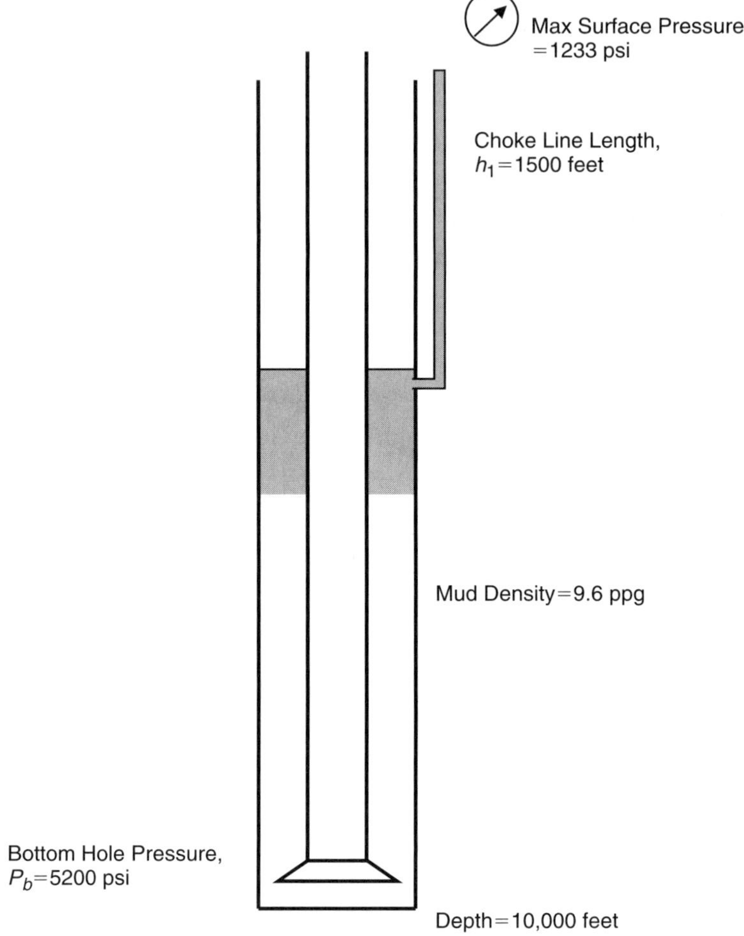

Figure 4.33   *cont'd*

thereby "sandwiching" the influx into the lost zone at the shoe. Although this procedure is arbitrary, it has enjoyed a fair amount of success.

In addition, the sediments exposed in deep water are usually very young. Therefore, there is a very good chance that the well will bridge around the bottom hole assembly and kill itself.

Deep water can be an asset in the event of a catastrophic failure. Consider the consequences of a dramatic failure that results in a discharge at the sea floor. In shallow water, the boil will be in the immediate proximity of the vessel. Alternatives for further operations are compromised or eliminated. In deep water, a similar situation would most likely not have such alarming results. If the water depth is sufficient, the current will move the boil a substantial distance away from the vessel and alternatives should be more numerous.

SHALLOW GAS KICKS

Shallow gas blowouts can be catastrophic. There is not a perfect technique for handling shallow gas blowouts. There are basically two accepted methods for handling a shallow gas kick, and the difference centers on whether the gas is diverted at the sea floor or at the surface. Operators are evenly divided between the two methods and equally dedicated to their favorite.

The disadvantage to diverting the gas at the sea floor, particularly when operating in deep water and controlling it via the choke line(s), is the additional back pressure exerted on the casing seat due to the frictional pressure drop in the choke line system.

The disadvantage of diverting the gas at the surface, particularly when operating in deep water, is the possibility of collapsing the marine riser when rapid gas expansion evacuates the riser. In addition, the gas is brought to the rig floor. Surface sands are usually unconsolidated and severe erosion is certain. Plugging and bridging are probable in some areas and can result in flow outside the casing and in cratering. Advances in diverters and diverter system design have improved surface diverting operations substantially.

References

1. Grace, Robert, D., "Further Discussion of Application of Oil Muds," *SPE Drilling Engineering*," September 1987, page 286.

2. O'Bryan, P.L. and Bourgoyne, A.T., "Methods for Handling Drilled Gas in Oil Muds," SPE/IADC #16159, SPE/IADC Drilling Conference held in New Orleans, LA, March, 15–18, 1987.

3. O'Bryan, P.L. and Bourgoyne, A.T., "Swelling of Oil-based Drilling Fluids Due to Dissolved Gas," SPE 16676, presented at the 62nd Annual Technical Conference and Exhibition of the Society of Petroleum Engineers held in Dallas, TX, September 27–30, 1987.

4. O'Bryan, P.L., et al., "An Experimental Study of Gas Solubility in Oil-base Drilling Fluids," SPE #15414 presented at the 61st Annual Technical Conference and Exhibition of the Society of Petroleum Engineers held in New Orleans, LA, October, 5–8, 1986.

5. Thomas, David, C., et al., "Gas Solubility in Oil-based Drilling Fluids: Effects on Kick Detection," *Journal of Petroleum Technology*, June 1984, page 959.

CHAPTER FIVE

FLUID DYNAMICS IN WELL CONTROL

The use of kill fluids in well control operations is not new. However, one of the newest technical developments in well control is the engineered application of fluid dynamics. The technology of fluid dynamics is not fully utilized because most personnel involved in well control operations do not understand the engineering applications and do not have the capabilities to apply the technology at the rig in the field. The best well control procedure is the one that has predictable results from a technical as well as a mechanical perspective.

Fluid dynamics have an application in virtually every well control operation. Appropriately applied, fluids can be used cleverly to compensate for unreliable tubulars or inaccessibility. Often, when blowouts occur, the tubulars are damaged beyond expectation. For example, at a blowout in Wyoming an intermediate casing string subjected to excessive pressure was found by survey to have failed in NINE places.[1] One failure is understandable. Two failures are imaginable, but NINE failures in one string of casing?

After a blowout in South Texas, the $9\frac{5}{8}$-inch surface casing was found to have parted at 3200 feet and again at 1600 feet. Combine conditions such as those just described with the intense heat resulting from an oil-well fire or damage resulting from a falling derrick or collapsing substructure, and it is easy to convince the average engineer that after a blowout the wellhead and tubulars could be expected to have little integrity. Properly applied, fluid dynamics can offer solutions that do not challenge the integrity of the tubulars in the blowout.

The applications of fluid dynamics to be considered are:

1. Kill-Fluid Bullheading
2. Kill-Fluid Lubrication
3. Dynamic Kill
4. Momentum Kill

KILL-FLUID BULLHEADING

"Bullheading" is the pumping of the kill fluid into the well against any pressure and regardless of any resistance the well may offer. Kill-fluid bullheading is one of the most common misapplications of fluid dynamics. Because bullheading challenges the integrity of the wellhead and tubulars, the result can cause further deterioration of the condition of the blowout. Many times wells have been lost, control delayed, or options eliminated by the inappropriate bullheading of kill fluids.

Consider the following for an example of a proper application of the bullheading technique. During the development of the Ahwaz Field in Iran in the early 1970s, classic pressure control procedures were not possible. The producing horizon in the Ahwaz Field is so prolific that the difference between circulating and losing circulation is a few psi. The typical wellbore schematic is presented as Figure 5.1.

Drilling in the pay zone was possible by delicately balancing the hydrostatic with the formation pore pressure. The slightest underbalance resulted in a significant kick. Any classic attempt to control the well was unsuccessful because even the slightest back pressure at the surface caused lost circulation at the $9\frac{5}{8}$-inch casing shoe. Routinely, control was regained by increasing the weight of two hole volumes of mud at the surface by .1 to .2 ppg and pumping down the annulus to displace the influx and several hundred barrels of mud into the productive formation. Once the influx was displaced, routine drilling operations were resumed.

After the blowout at the Shell Cox in the Piney Woods of Mississippi, a similar procedure was adopted in the deep Smackover tests. In these operations, bringing the formation fluids to the surface was hazardous due to the high pressures and high concentrations of hydrogen sulfide. In response to the challenge, casing was set in the top of the Smackover. When a kick was taken, the influx was overdisplaced back into the Smackover. When a kick was taken, the influx was overdisplaced back into the Smackover by bullheading kill-weight mud down the annulus.

The common ingredients of success in these two examples are pressure, casing seat, and kick size. The surface pressures required to pump into the formation were low because the kick sizes were always small. In addition, it was of no consequence that the formation was fractured in the process and damaged by the mud pumped. The most important aspect

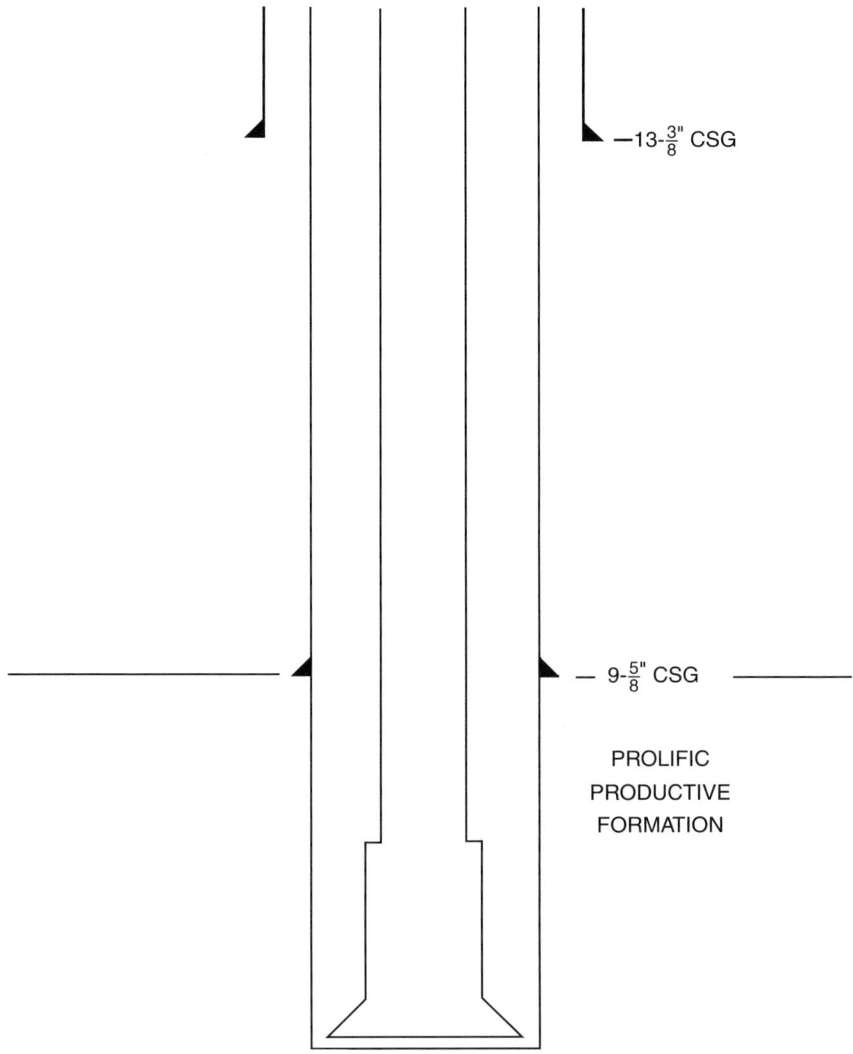

Figure 5.1   *Ahwaz Field.*

was the casing seat. The casing that sat at the top of the productive interval in each example ensured that the kill mud as well as the influx would be forced back into the interval from which the kick occurred. It is most important to understand that, when bullheading, the kill fluid will almost always exit the wellbore at the casing seat.

Consider an example of misapplication from the Middle East:

Example 5.1

Given:

Figure 5.2

Depth, $D = 10{,}000$ feet

Mud weight, $\rho = 10$ ppg

Mud gradient, $\rho_m = 0.052$ psi/ft

Gain $= 25$ bbl

Hole size, $D_h = 8\frac{1}{2}$ inches

Intermediate casing $(9\frac{5}{8}$-inch$) = 5000$ feet

Fracture gradient, $F_g = 0.65$ psi/ft

Fracture pressure:

    @ 5000 ft, $P_{frac5000} = 3250$ psi

    @ 10,000 ft, $P_{frac10000} = 6500$ psi

Shut-in annulus pressure, $P_a = 400$ psi

Drillpipe $= 4\frac{1}{2}$ inch

Internal diameter, $D_i = 3.826$ inch

Drill collars: $= 800$ feet

        $= 6$ inches $\times 2.25$ inches

Gas specific gravity, $S_g = 0.60$

Temperature gradient $= 1.2\,°/100$ ft

Ambient temperature $= 60\,°\text{F}$

Compressibility factor, $z = 1.00$

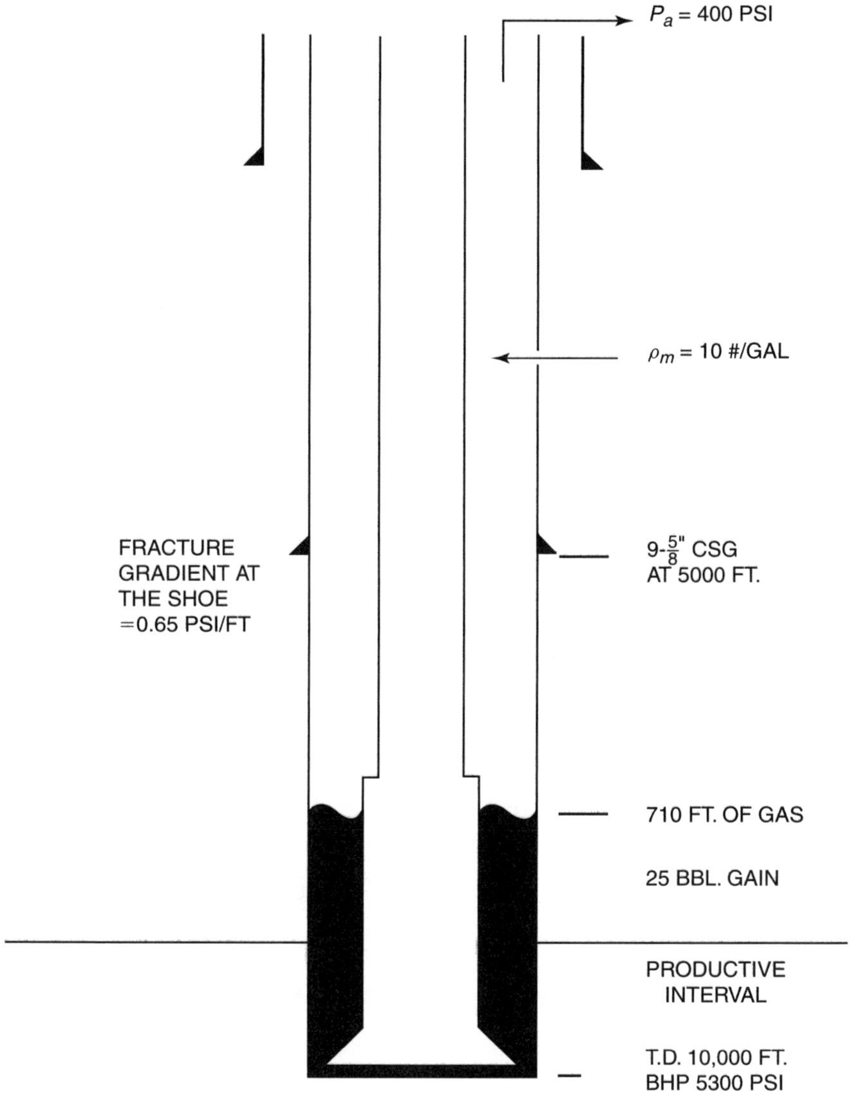

Figure 5.2

Capacity of:

Drill collar annulus, $C_{dcha} = 0.0352$ bbl/ft

Drillpipe, $C_{dpha} = 0.0506$ bbl/ft

The decision was made to weight up the mud to the kill weight and displace down the annulus since the drillpipe was plugged.

From Figure 5.2 and Equation 2.7:

$$P_b = \rho_f h + \rho_m (D - h) + P_a$$

and from Equation 3.7:

$$h_b = \frac{Influx\ Volume}{C_{dcha}}$$

$$h_b = \frac{25}{0.0352}$$

$$h_b = \textbf{710 feet}$$

and from Equation 3.5:

$$\rho_f = \frac{S_g P_b}{53.3 z_b T_b}$$

$$\rho_f = \frac{(0.6)(5200)}{53.3(1.0)(640)}$$

$$\rho_f = \textbf{0.091 psi/ft}$$

Therefore:

$$P_b = 400 + 0.52(10,000 - 710) + (0.091)(710)$$

$$P_b = \textbf{5295 psi}$$

The kill-mud weight would then be given from Equation 2.9:

$$\rho_1 = \frac{P_b}{0.052 D}$$

$$\rho_1 = \frac{5295}{(10,000)(0.052)}$$

$$\rho_1 = \textbf{10.2 ppg}$$

The annular capacity:

$$Capacity = (9200)(0.0506) + (800)(0.0352)$$

$$Capacity = \textbf{494 bbls}$$

As a result of these calculations, 500 barrels of mud was weighted up at the surface to the kill weight of 10.2 ppg and pumped down the annulus. After pumping the 500 barrels of 10.2-ppg mud, the well was shut in and surface pressure was observed to be 500 psi, or 100 psi more than the 400 psi originally observed!

Required:
1. Explain the cause of the failure of the bullhead operation.

2. Explain the increase in the surface pressure.

Solution:
1. The pressure at 5000 feet is given by:

$$P_{5000} = (0.052)(10.0)(5000) + 400$$

$$P_{5000} = \textbf{3000 psi}$$

The fracture gradient at 5000 feet is 3250 psi. Therefore, the difference is

$$\Delta P_{5000} = 3250 - 3000$$

$$\Delta P_{5000} = \textbf{250 psi}$$

The pressure at 10,000 feet is 5296 psi. The fracture gradient at 10,000 feet is 6500 psi. Therefore, the difference is

$$\Delta P_{10000} = 6500 - 5296$$

$$\Delta P_{10000} = \textbf{1204 psi}$$

The pressure required to pump into the zone at the casing shoe was only 250 psi above the shut-in pressure and

954 psi less than the pressure required to pump into the zone at 10,000 feet. Therefore, when pumping operations commenced, the zone at the casing seat fractured and the mud was pumped into that zone.

2.   The surface pressure after the bullheading operation was 100 psi more than the surface pressure before the pumping job (500 psi versus 400 psi) because the 25 bbl influx had risen during the pumping operation. (See Chapter 4 for a further discussion of bubble rise.)

There are other reasons that a bullheading operation can fail. For example, after a well has been shut in, the influx often migrates to the surface, leaving drilling mud opposite the kick zone. Once pumping begins, the surface pressure must be increased until the zone to be bull-headed into is fractured by the drilling mud. The fracture pressure may be several hundred to several thousand pounds per square inch above the shut-in pressure. This additional pressure may be enough to rupture the casing in the well and cause an underground blowout.

Sometimes bullheading operations are unsuccessful when an annulus in a well is completely filled with gas that is to be pumped back into the formation. The reason for the failure in an instance such as this is that the kill mud bypasses the gas in the annulus during the pumping operation. Therefore, after the kill mud is pumped and the well is shut in to observe the surface pressure, there is pressure at the surface and gas throughout the system. The result is that the well unloads and blows out again.

Another consideration is the rate at which the mud being bull-headed is pumped. In the discussion concerning influx migration, it was noted that the influx usually migrates up one side of the annulus while the mud falls down the other side of the annulus. Further, when the influx nears the surface, the velocity of migration can be very high, as evidenced by the rate of surface pressure increase. Under those conditions, bullheading at $\frac{1}{4}$ barrel per minute will not be successful because the mud will simply bypass the migrating influx. This is particularly problematic when the annulus area is large. The bullheading rate may have to be increased to more than 10 barrels per minute in order to be successful. In any event, the bullheading rate will have to be increased until the shut-in surface pressure is observed to be decreasing as the mud is bullheaded into the annulus.

Bullheading is often used in deep, high-pressure well control situations to maintain acceptable surface pressures. Consider Example 5.2 during underground blowouts:

Example 5.2

Given:

Depth, $D = 20,000$ feet

Bottomhole pressure, $P_b = 20,000$ psi

Casing shoe at, $D_{shoe} = 10,000$ feet

Fracture gradient at shoe, $F_g = 0.9$ psi/ft

The production is gas and the well is blowing out underground.

Required:

Approximate the surface pressure if the gas is permitted to migrate to the surface.

Determine the surface pressure if 15-ppg mud is continuously bullheaded into the annulus.

Solution:

If the gas is permitted to migrate to the surface, the surface pressure will be approximately as follows:

$$P_{surf} = (F_g - 0.10)D_{shoe}$$

$$P_{surf} = (0.90 - 0.10)(10,000)$$

$$P_{surf} = \textbf{8000 psi}$$

With 15-ppg mud bullheaded from the surface to the casing shoe at 10,000 feet, the surface pressure would be

$$P_{surf} = 8000 - (0.052)(15)(10,000)$$

$$P_{surf} = \textbf{200 psi}$$

Therefore, as illustrated in Example 5.2, without the bullheading operation, the surface pressure would build to 8000 psi. At that pressure surface operations are very difficult at best. If 15-ppg mud is bullheaded into the lost circulation zone at the casing shoe, the surface pressure can be reduced to 200 psi. With 200 psi surface pressure, all operations such as snubbing or wire line are considerably easier and faster.

A deep well in the southern United States got out of control while tripping. The conditions and wellbore schematic are presented in Figure 5.3.

Since there were only 173 feet of open hole, it was concluded that the well would probably not bridge. However, when the cast iron bridge plug was set inside the drillpipe in preparation for stripping in the hole, the flow began to diminish. Within the hour, the flow declined to a small, lazy flare. The well had obviously bridged.

The primary question at that point was the location of the bridge. Normally, bridging occurs in the bottom hole assembly. The velocity is low deep in the well adjacent to the bottom hole assembly. The changes in diameter provide opportunity for the accumulation of formation particulate. In this case, drillpipe protector rubbers had been used. It is not unusual for protector rubbers to be problematic in the presence of high temperature gas. Due to the presence of the rubbers, the bridge could be high in the well.

A deep bridge would be acceptable since sufficient hydrostatic could be placed above the bridge to compensate for the high formation pressure. A shallow bridge would be very dangerous since full shut-in pressure could easily be immediately below the surface and held only by a few feet of packed debris.

Since the well was continuing to produce a small quantity of gas, it was considered that a temperature survey could identify the bridge. It was anticipated that the expansion of the gas across the bridge would result in a significant temperature anomaly. Accordingly, a temperature survey was run and is presented as Figure 5.4. As illustrated in Figure 5.4, interpretation of the temperature indicated that the bridge was around the bottom hole assembly. Accordingly, the annulus was successfully loaded with 432 barrels (the calculated volume was 431 barrels) of 20 ppg mud.

In summary, bullheading operations can have unpleasant results and should be thoroughly evaluated prior to commencing the operation. Too often, crews react to well control problems without analyzing the problem and get into worse condition than when the operation began. Remember,

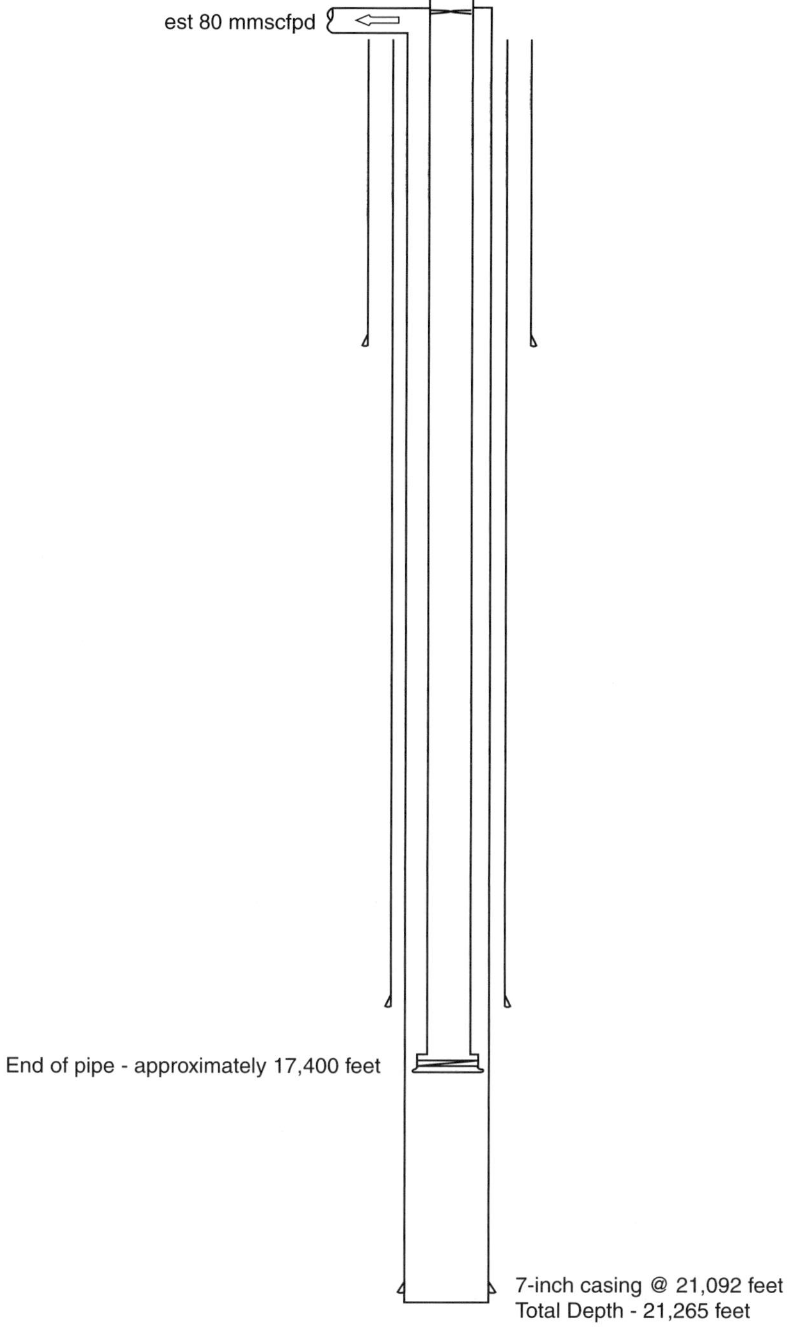

Figure 5.3 *Wellbore Schematic.*

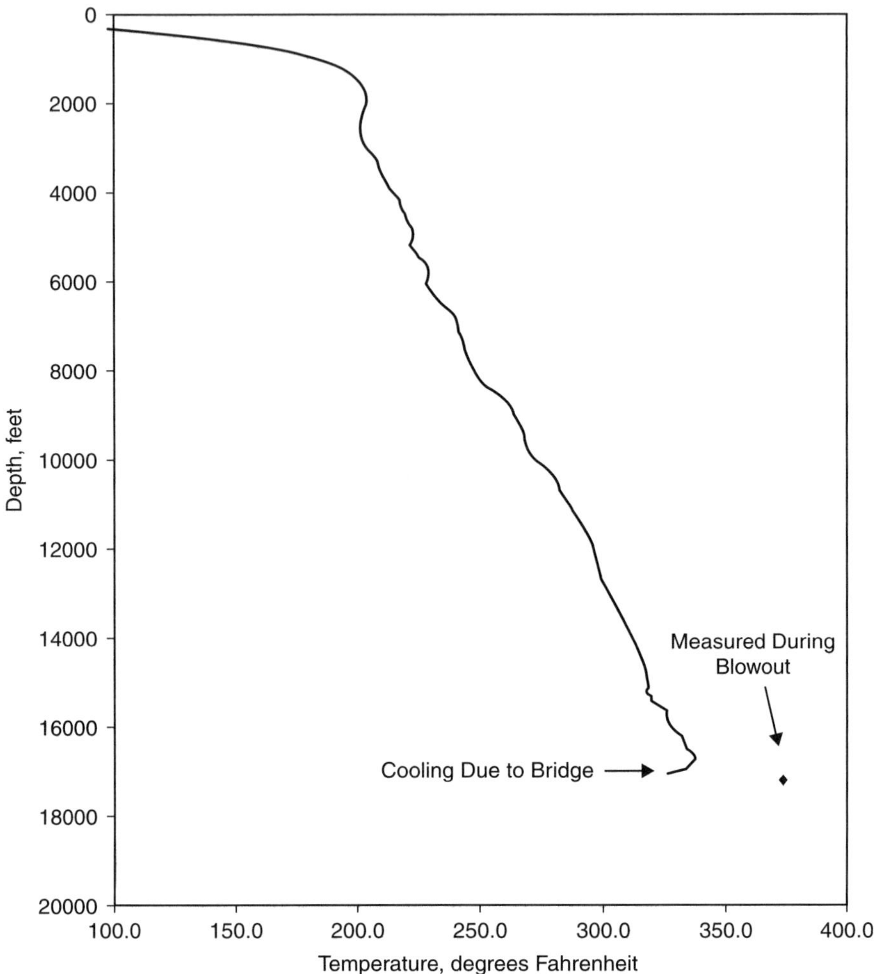

Figure 5.4   *Temperature Analysis Well Bridged, August 13, 1998.*

the best well control procedure is one that has predictable results from the technical as well as the mechanical perspective.

KILL-FLUID LUBRICATION— VOLUMETRIC KILL PROCEDURE

Kill-Fluid Lubrication, also sometimes called the Volumetric Kill Procedure, is the most overlooked well control technique. Lubricating the kill fluid into the wellbore involves an understanding of only the most fundamental aspects of physics. Basically, Kill-Fluid Lubrication is a technique whereby the influx is replaced by the kill fluid while the bottom-hole pressure is maintained at or above the formation pressure. The result of the proper Kill-Fluid Lubrication operation is that the influx is removed from the wellbore, the bottomhole pressure is controlled by the kill-mud hydrostatic and no additional influx is permitted during the operation.

Kill-Fluid Lubrication has application in a wide variety of operations. The only requirement is that the influx has migrated to the surface or, as in some instances, the wellbore is completely void of drilling mud. Kill-Fluid Lubrication has application in instances when the pipe is on bottom, the pipe is part-way out of the hole, or the pipe is completely out of the hole. The technique is applicable in floating drilling operations where the rig is often required, due to weather or some other emergency, to hang off, shut in, and move off the hole. In each instance, the principles are fundamentally the same.

Consider the following from a recent well control problem at a deep, high-pressure operation in southeastern New Mexico. A kick was taken while on a routine trip at 14,080 feet. The pipe was out of the hole when the crew observed that the well was flowing. The crew ran 1500 feet of drill string back into the hole. By that time the well was flowing too hard for the crew to continue the trip into the hole and the well was shut in. A sizable kick had been taken. Subsequently, the drillpipe was stripped into the hole in preparation for a conventional kill operation.

However, the back-pressure value, placed in the drill string 1500 feet above the bit to enable the drillpipe to be stripped into the hole, had become plugged during the stripping operation. In addition, during the time the snubbing unit was being rigged up and the drillpipe was being stripped to bottom, the gas migrated to the surface. The influx came from a prolific interval at 13,913 feet. The zone had been drill-stem tested in this wellbore and had flowed gas at a rate of 10 mmscfpd with a flowing surface pressure of 5100 psi and a shut-in bottomhole pressure of 8442 psi.

Since it was not possible to circulate the influx out of the wellbore in a classic manner, kill fluid was lubricated into the wellbore while efforts were being made to remove the obstruction in the drillpipe. The conditions as they existed at this location on that pleasant November afternoon are

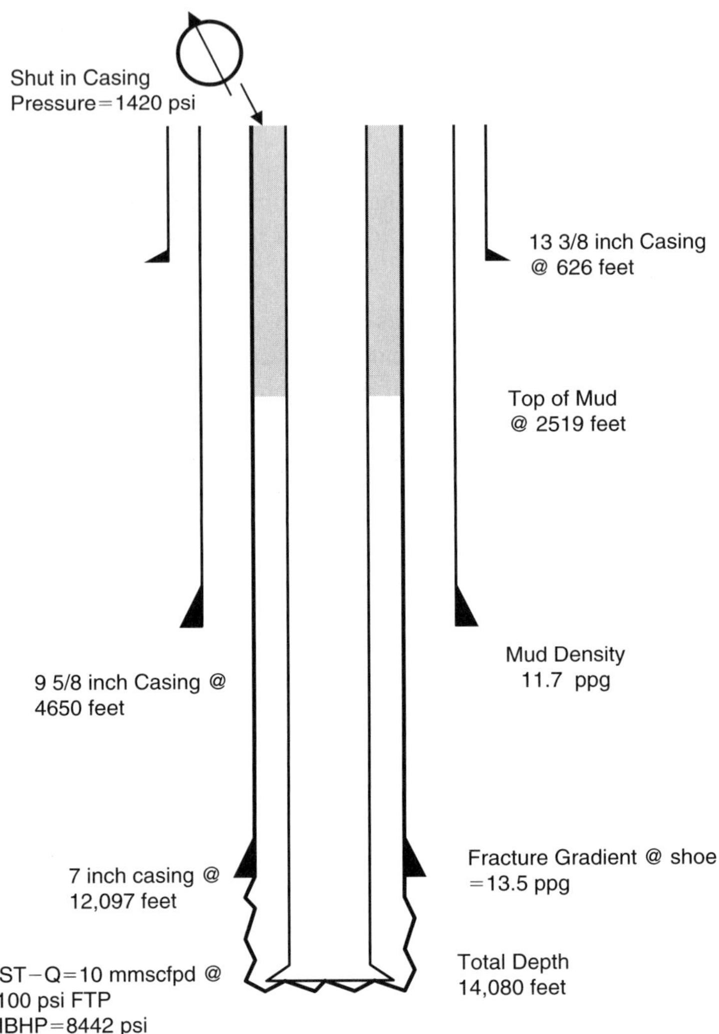

Figure 5.5

schematically illustrated in Figure 5.5. The following example illustrates the proper procedure for lubricating kill fluid into a wellbore:

Example 5.3

Given:

Figure 5.5

Depth, $D = 14,080$ feet

Surface pressure, $P_a = 1420$ psi

Mud weight, $\rho = 11.7$ ppg

Fracture gradient at shoe, $F_g = 0.702$ psi/ft

Intermediate casing:

7-inch casing @ $D_{shoe} = 12,097$ feet

29 #/ft P-110   82 feet

26 #/ft S-95   7800 feet

29 #/ft P-110   4200 feet

Gas gravity, $S_g = 0.6$

Bottomhole pressure, $P_b = 8442$ psi

Temperature, $T_s = 540$ °Rankine

Kill-mud weight, $\rho_1 = 12.8$ ppg

Compressibility factor, $z_s = 0.82$

Capacity of drillpipe annulus, $C_{dpca} = 0.0264$ bbl/ft

Drill-stem test at 13,913 feet

Volume rate of flow, $Q = 10$ mmscfpd @ 5100 psi

Plugged drillpipe at 12,513 feet

Required:

Design a procedure to lubricate kill mud into and the gas influx out of the annulus.

Solution:

Determine the height of the gas bubble, h, as follows from Equations 2.7 and 3.5:

$$P_b = \rho_f h + \rho_m (D - h) + P_a$$

$$\rho_f = \frac{S_g P_s}{53.3 z_s T_s}$$

$$\rho_f = \frac{(0.6)(1420)}{53.3(0.82)(540)}$$

$$\rho_f = \mathbf{0.035\ psi/ft}$$

Solve for h using Equation 2.7:

$$8442 = 1420 + (0.052)(11.7)(13913 - h) + 0.035h$$

$$h = \mathbf{2{,}520\ feet}$$

Gas volume at the surface, V_s:

$$V_s = (2520)(0.0264)$$

$$V_s = \mathbf{66.5\ bbls}$$

Determine the margin for pressure increase at the casing shoe using Equation 5.1:

$$P_{shoe} = \rho_f h + P_a + \rho_m (D_{shoe} - h) \tag{5.1}$$

Where:

P_{shoe} = Pressure at the casing shoe, psi
ρ_f = Influx gradient, psi/ft
h = Height of the influx, feet
P_a = Annulus pressure, psi
ρ_m = Original mud gradient, psi/ft
D_{shoe} = Depth to the casing shoe, feet

$$P_{shoe} = 0.036(2520) + 1420 + 0.6087(12{,}097 - 2520)$$

$$P_{shoe} = \mathbf{7340\ psi}$$

Determine maximum permissible pressure at shoe, P_{frac}:

$$P_{frac} = F_g D_{shoe} \tag{5.2}$$

$$P_{frac} = (0.052)(13.5)(12{,}097)$$

$$P_{frac} = \mathbf{8492\ psi}$$

Where:

F_g = Fracture gradient, psi/ft
D_{shoe} = Depth of the casing shoe, feet

Maximum increase in surface pressure and hydrostatic, ΔP_t, that will not result in fracturing at the shoe is given by Equation 5.3:

$$\Delta P_t = P_{frac} - P_{shoe} \tag{5.3}$$

$$\Delta P_t = 8492 - 7340$$

$$\Delta P_t = \mathbf{1152\ psi}$$

Where:

P_{frac} = Fracture pressure at casing shoe, psi
P_{shoe} = Calculated pressure at casing shoe, psi

The volume of the kill-weight mud, V_1, with density, ρ_1, to achieve ΔP_t is given by Equation 5.4:

$$V_1 = X_1 - \left[X_1^2 - \frac{\Delta P_t C_{dpca} V_s}{\rho_{m1}} \right]^{\frac{1}{2}} \tag{5.4}$$

$$X_1 = \frac{\rho_{m1} V_s + C_{dpca}(P_a - \Delta P_t)}{2(\rho_{m1})} \tag{5.5}$$

$$X_1 = \frac{0.667(66.5) + 0.0264(1420 + 1152)}{2(0.667)}$$

$$X_1 = \mathbf{84.150}$$

$$V_1 = 84.150 - \left[84.150^2 - \frac{1152(0.0264)(66.5)}{0.667} \right]^{\frac{1}{2}}$$

$$V_1 = \mathbf{20.5\ bbls}$$

Where:

$\Delta P_t =$ Maximum surface pressure, psi
$C_{dpca} =$ Annular capacity, bbl/ft
$V_s =$ Gas volume at the surface, bbl
$\rho_{m_1} =$ Kill mud gradient, psi/ft
$X_1 =$ Intermediate calculation

Determine the effect of pumping 20 bbls of kill mud with density $\rho_1 = 12.8$ ppg. The resulting additional hydrostatic, ΔHyd, is calculated with Equation 5.6:

$$\Delta Hyd = 0.052\rho_1 \left(\frac{V_1}{C_{dpca}} \right) \tag{5.6}$$

$$\Delta Hyd = (0.052)(12.8) \left(\frac{20}{0.0264} \right)$$

$$\Delta Hyd = \mathbf{504\ psi}$$

Additional surface pressure resulting from compressing the bubble at the surface with 20 bbls of kill mud is given by Equation 2.3:

$$\frac{P_1 V_1}{z_1 T_1} = \frac{P_2 V_2}{z_2 T_2}$$

1 = Prior to pumping kill mud
2 = After pumping kill mud

Therefore, by modifying Equation 2.2:

$$P_2 = \frac{P_1 V_1}{V_2}$$

$$P_2 = \frac{(1420)(66.5)}{(66.5 - 20)}$$

$$P_2 = \textbf{2031 psi}$$

Additional surface pressure, ΔP_s, is given as:

$$\Delta P_s = P_2 - P_a \tag{5.7}$$

$$\Delta P_s = 2031 - 1420$$

$$\Delta P_s = \textbf{611 psi}$$

Total pressure increase, ΔP_{total}, is given as:

$$\Delta P_{total} = \Delta Hyd + \Delta P_s \tag{5.8}$$

$$\Delta P_{total} = 504 + 611$$

$$\Delta P_{total} = \textbf{1115 psi}$$

Since ΔP_{total} is less than the maximum permissible pressure increase, ΔP_t, calculated using Equation 5.3, pump 20 bbls of 12.8-ppg mud at 1 bpm and shut in to permit the gas to migrate to the surface.

Observe initial P_2 after pumping $= 1950$ psi.

Observe 2-hour shut-in $P_2 = 2031$ psi.

The surface pressure, P_a, may now be reduced by bleeding **ONLY GAS** from P_2 to

$$P_{newa} = P_a - \Delta Hyd$$

$$P_{newa} = 1420 - 504$$

$$P_{newa} = \textbf{916 psi}$$

However, the new surface pressure must be used to determine the effective hydrostatic pressure at 13,913 feet to ensure no additional influx.

Equation 2.7 expands to Equation 5.9:

$$P_b = P_a + \rho_f h + \rho_m (D - h - h_1) + \rho_{m1} h_1 \qquad (5.9)$$

$$h = \frac{V_s}{C_{dpca}}$$

Where:

$$V_s = \text{Remaining volume of influx, bbls}$$

$$h = \frac{66.5 - 20}{0.0264}$$

$$h = \textbf{1762 feet}$$

$$h_1 = \frac{V_1}{C_{dpca}}$$

$$h_1 = \frac{20}{0.0264}$$

$$h_1 = \textbf{758 feet}$$

$$\rho_f = \frac{S_g P_s}{53.3 z_s T_s}$$

$$\rho_f = \frac{(0.6)(916)}{(53.3)(0.866)(540)}$$

$$\rho_f = \textbf{0.022 psi/ft}$$

$$P_b = 916 + (0.6084)(13{,}913 - 1762 - 758)$$

$$+ (0.022)(1762) + (0.667)(758)$$

$$P_b = \textbf{8392 psi}$$

However, the shut-in pressure is 8442 psi. Therefore, since the effective hydrostatic cannot be less than the reservoir pressure, the surface pressure can only be bled to

$$P_a = 916 + (8442 - 8392)$$

$$P_a = \textbf{966 psi}$$

***Important*: The pressure cannot be reduced by bleeding mud. If mud is bled from the annulus, the well must be shut in for a longer period to allow the gas to migrate to the surface.**

Now the procedure must be repeated until the influx is lubricated from the annulus and replaced by mud:

$$\rho_f = \frac{S_g P_s}{53.3 z_s T_s}$$

$$\rho_f = \frac{(0.6)(966)}{53.3(0.866)(540)}$$

$$\rho_f = \textbf{0.023 psi/ft}$$

Now, solving for h, use

$$h = \frac{V_s}{C_{dpca}}$$

$$h = \textbf{1762 feet}$$

Similarly, solving for h_1:

$$h_1 = \frac{V_1}{C_{dpa}}$$

$$h_1 = \frac{20}{0.0264}$$

$$h_1 = \textbf{758 feet}$$

Determine the margin for pressure increase at the casing shoe using Equation 5.10, which is Equation 5.1 modified for inclusion of kill mud, ρ_1:

$$P_{shoe} = \rho_f h + P_a + \rho_m(D_{shoe} - h - h_1) + \rho_1 h_1 \quad (5.10)$$

$$P_{shoe} = 0.023(1762) + 966 + 0.6087$$

$$\times (12{,}097 - 1762 - 758) + (0.052)(12.8)(758)$$

$$P_{shoe} = \mathbf{7339\ psi}$$

The maximum permissible pressure at shoe, P_{frac}, from Equation 5.2 is equal to 8492 psi.

The maximum increase in surface pressure and hydrostatic, ΔP_t, that will not result in fracturing at the shoe is given by Equation 5.3:

$$\Delta P_t = P_{frac} - P_{shoe}$$

$$\Delta P_t = 8492 - 7339$$

$$\Delta P_t = \mathbf{1153\ psi}$$

The volume of the kill-weight mud, V_1, with density, ρ_1, to achieve ΔP_t is given by Equations 5.4 and 5.5:

$$V_1 = X_1 - \left[X_1^2 - \frac{\Delta P_t C_{dpca} V_s}{\rho_{m_1}} \right]^{\frac{1}{2}}$$

$$X_1 = \frac{\rho_{m1} V_s + C_{dpca}(P_a - \Delta P_t)}{2(\rho_{m1})}$$

$$X_1 = \frac{0.667(46.5) + 0.0264(966 + 1153)}{2(0.667)}$$

$$X_1 = \mathbf{65.185}$$

$$V_1 = 65.185 - \left[65.185^2 - \frac{1153(0.0264)(46.5)}{0.667} \right]^{\frac{1}{2}}$$

$$V_1 = \mathbf{19.1\ bbls}$$

Determine the effect of pumping 19 bbls of kill mud with density $\rho_1 = 12.8$ ppg. The resulting additional hydrostatic, ΔHyd, is

$$\Delta Hyd = (0.052)(\rho_1)\left(\frac{V_1}{C_{dpca}}\right)$$

$$\Delta Hyd = (0.052)(12.8)\left(\frac{19}{0.0264}\right)$$

$$\Delta Hyd = \textbf{480 psi}$$

Additional surface pressure resulting from compressing the bubble at the surface with 19 bbls of kill mud is

$$P_2 = \frac{P_1 V_1}{V_2}$$

$$P_2 = \frac{(966)(46.5)}{(46.5 - 19)}$$

$$P_2 = \textbf{1633 psi}$$

Additional surface pressure, ΔP_s, is given as

$$\Delta P_s = P_2 - P_a$$

$$\Delta P_s = 1633 - 966$$

$$\Delta P_s = \textbf{677 psi}$$

Total pressure increase, ΔP_{total}, is given by Equation 5.8:

$$\Delta P_{total} = \Delta Hyd + \Delta P_s$$

$$\Delta P_{total} = 480 + 667$$

$$\Delta P_{total} = \textbf{1147 psi}$$

Since ΔP_{total} is less than the maximum permissible pressure increase, ΔP_t, calculated using Equation 5.3, pump 19 bbls of 12.8-ppg mud at 1 bpm and shut in to permit the gas to migrate to the surface.

Observe initial P_2 after pumping $= 1550$ psi.

Observe 2-hour shut-in $P_2 = 1633$ psi.

The surface pressure, P_a, may now be reduced by bleeding **ONLY GAS** from P_2 to

$$P_{newa} = P_a - \Delta Hyd$$

$$P_{newa} = 966 - 480$$

$$P_{newa} = \textbf{486 psi}$$

However, the new surface pressure must be used to determine the effective hydrostatic pressure at 13,913 feet to ensure no additional influx.

The bottomhole pressure is given by Equation 5.9:

$$P_b = P_a + \rho_f h + \rho_m (D - h - h_1) + \rho_{m1} h_1$$

$$h = \frac{V_s}{C_{dpca}}$$

$$h = \frac{46.5 - 19}{0.0264}$$

$$h = \textbf{1042 feet}$$

$$h_1 = \frac{V_1}{C_{dpca}}$$

$$h_1 = \frac{39}{0.0264}$$

$$h_1 = \textbf{1477 feet}$$

$$\rho_f = \frac{S_g P_s}{53.3 z_s T_s}$$

$$\rho_f = \frac{(0.6)(486)}{(53.3)(0.930)(540)}$$

$$\rho_f = \textbf{0.011 psi/ft}$$

$$P_b = 486 + (0.6084)(13,913 - 1042 - 1477)$$

$$+ (0.011)(1042) + (0.667)(1477)$$

$$P_b = \textbf{8415 psi}$$

However, the shut-in pressure is 8442 psi. Therefore, since the effective hydrostatic cannot be less than the reservoir pressure, the surface pressure can only be bled to

$$P_a = 486 + (8442 - 8415)$$

$$P_a = \textbf{513 psi}$$

The surface pressure must be bled to only 513 psi to ensure no additional influx by bleeding only dry gas through the choke manifold. *Important*: No mud can be bled from the annulus. If the well begins to flow mud from the annulus, it must be shut in until the gas and mud separate.

Again, the procedure must be repeated:

$$\rho_f = \frac{S_g P_s}{53.3\, z_s T_s}$$

$$\rho_f = \frac{(0.6)(513)}{53.3(0.926)(540)}$$

$$\rho_f = \textbf{0.012 psi/ft}$$

Now, solving for h, use

$$h = \frac{V_s}{C_{dpca}}$$

$$h = \frac{46.5 - 19.0}{0.0264}$$

$$h = \textbf{1042 feet}$$

Similarly, solving for h_1:

$$h_1 = \frac{V_1}{C_{dpca}}$$

$$h_1 = \frac{39}{0.0264}$$

$$h_1 = \mathbf{1477\ feet}$$

Determine the margin for pressure increase at the casing shoe using Equation 5.10:

$$P_{shoe} = \rho_f h + P_a + \rho_m (D_{shoe} - h - h_1) + \rho_1 h_1$$

$$P_{shoe} = 0.012(1042) + 513 + 0.6084$$

$$\times (12{,}097 - 1042 - 1477) + (0.052)(12.8)(1477)$$

$$P_{shoe} = \mathbf{7336\ psi}$$

The maximum permissible pressure at shoe, P_{frac}, from Equation 5.2 is equal to 8492 psi.

The maximum increase in surface pressure and hydrostatic, ΔP_t, that will not result in fracturing at the shoe is given by Equation 5.3:

$$\Delta P_t = P_{frac} - P_{shoe}$$

$$\Delta P_t = 8492 - 7336$$

$$\Delta P_t = \mathbf{1156\ psi}$$

The volume of the kill-weight mud, V_1, with density, ρ_1, to achieve ΔP_t is given by Equations 5.4 and 5.5:

$$V_1 = X_1 - \left[X_1^2 - \frac{\Delta P_t C_{dpca} V_s}{\rho_{m1}} \right]^{\frac{1}{2}}$$

$$X_1 = \frac{\rho_{m1} V_s + C_{dpca}(P_a - \Delta P_t)}{2(\rho_{m1})}$$

$$X_1 = \frac{0.667(27.5) + 0.0264(513 + 1156)}{2(0.667)}$$

$$X_1 = \mathbf{46.780}$$

$$V_1 = 46.780 - \left[46.780^2 - \frac{1156(0.0264)(27.5)}{0.667}\right]^{\frac{1}{2}}$$

$$V_1 = \mathbf{16.3\ bbls}$$

Determine the effect of pumping 16 bbls of kill mud with density $\rho_1 = 12.8$ ppg. The resulting additional hydrostatic, ΔHyd, is given by Equation 5.6:

$$\Delta Hyd = 0.052\rho_1 \left(\frac{V_1}{C_{dpca}}\right)$$

$$\Delta Hyd = (0.052)(12.8)\left(\frac{16}{0.0264}\right)$$

$$\Delta Hyd = \mathbf{404\ psi}$$

Additional surface pressure resulting from compressing the bubble at the surface with 16 bbls of kill mud is

$$P_2 = \frac{P_1 V_1}{V_2}$$

$$P_2 = \frac{(513)(27.5)}{(27.5 - 16)}$$

$$P_2 = \mathbf{1227\ psi}$$

Additional surface pressure, ΔP_s is given by Equation 5.7:

$$\Delta P_s = P_2 - P_a$$

$$\Delta P_s = 1227 - 513$$

$$\Delta P_s = \mathbf{714\ psi}$$

Total pressure increase, ΔP_{total}, is given by Equation 5.8:

$$\Delta P_{total} = \Delta Hyd + \Delta P_s$$

$$\Delta P_{total} = 404 + 714$$

$$\Delta P_{total} = \mathbf{1118\ psi}$$

Since ΔP_{total} is less than the maximum permissible pressure increase, ΔP_t, calculated using Equation 5.3, pump 16 bbls of 12.8-ppg mud at 1 bpm and shut in to permit the gas to migrate to the surface.

In 2 to 4 hours shut-in $P_2 = 1227$ psi.

The surface pressure, P_a, may now be reduced by bleeding **ONLY GAS** from P_2 to

$$P_{newa} = P_a - \Delta Hyd$$

$$P_{newa} = 513 - 404$$

$$P_{newa} = \mathbf{109\ psi}$$

However, the new surface pressure must be used to determine the effective hydrostatic pressure at 13,913 feet to insure no additional influx.

Pursuant to Equation 5.9:

$$P_b = P_a + \rho_f h + \rho_m (D - h - h_1) + \rho_{m1} h_1$$

$$h = \frac{V_s}{C_{dpca}}$$

$$h = \frac{27.5 - 16}{0.0264}$$

$$h = \mathbf{436\ feet}$$

$$h_1 = \frac{V_1}{C_{dpca}}$$

$$h_1 = \frac{55}{0.0264}$$

$$h_1 = \textbf{2083 feet}$$

$$\rho_f = \frac{S_g P_s}{53.3 z_s T_s}$$

$$\rho_f = \frac{(0.6)(109)}{(53.3)(0.984)(540)}$$

$$\rho_f = \textbf{0.0023 psi/ft}$$

$$P_b = 109 + (0.6084)(13{,}913 - 436 - 2083)$$

$$+ (0.011)(436) + (0.667)(2083)$$

$$P_b = \textbf{8431 psi}$$

However, the shut-in pressure is 8442 psi. Therefore, since the effective hydrostatic cannot be less than the reservoir pressure, the surface pressure can only be bled to

$$P_a = 109 + (8442 - 8431)$$

$$P_a = \textbf{120 psi}$$

The surface pressure must be bled to only 120 psi to ensure no additional influx by bleeding only dry gas through the choke manifold. *Important*: **No mud can be bled from the annulus. If the well begins to flow mud from the annulus, it must be shut in until the gas and mud separate.**

Now the procedure is repeated for the final increment:

$$\rho_f = \frac{S_g P_s}{53.3 z_s T_s}$$

$$\rho_f = \frac{(0.6)(120)}{53.3(0.983)(540)}$$

$$\rho_f = \textbf{0.0025 psi/ft}$$

Now, solving for h, use

$$h = \frac{V_s}{C_{dpca}}$$

$$h = \frac{27.5 - 16.0}{0.0264}$$

$$h = \textbf{436 feet}$$

Similarly, solving for h_1:

$$h_1 = \frac{V_1}{C_{dpca}}$$

$$h_1 = \frac{55}{0.0264}$$

$$h_1 = \textbf{2083 feet}$$

Determine the margin for pressure increase at the casing shoe using Equation 5.10:

$$P_{shoe} = \rho_f h + P_a + \rho_m(D_{shoe} - h - h_1) + \rho_1 h_1$$

$$P_{shoe} = 0.0025(436) + 120 + 0.6084$$

$$\times\ (12{,}097 - 436 - 2083) + (0.052)(12.8)(2083)$$

$$P_{shoe} = \textbf{7338 psi}$$

The maximum permissible pressure at shoe, P_{frac}, from Equation 5.2 is equal to 8489 psi.

The maximum increase in surface pressure and hydrostatic, ΔP_t, that will not result in fracturing at the shoe is given by Equation 5.3:

$$\Delta P_t = P_{frac} - P_{shoe}$$

$$\Delta P_t = 8492 - 7338$$

$$\Delta P_t = \textbf{1154 psi}$$

The volume of the kill-weight mud, V_1, with density, ρ_1, to achieve ΔP_t is given by Equations 5.4 and 5.5:

$$V_1 = X_1 - \left[X_1^2 - \frac{\Delta P_t C_{dpca} V_s}{\rho_{m1}} \right]^{\frac{1}{2}}$$

$$X_1 = \frac{\rho_{m1} V_s + C_{dpca}(P_a - \Delta P_t)}{2(\rho_{m1})}$$

$$X_1 = \frac{0.667(11.5) + 0.0264(120 + 1154)}{2(0.667)}$$

$$X_1 = \mathbf{30.963}$$

$$V_1 = 30.963 - \left[30.963^2 - \frac{1154(0.0264)(11.5)}{0.667} \right]^{\frac{1}{2}}$$

$$V_1 = \mathbf{10.14\ bbls}$$

Determine the effect of pumping 10 bbls of kill mud with density $\rho_1 = 12.8$ ppg. The resulting additional hydrostatic, ΔHyd, is given by Equation 5.6:

$$\Delta Hyd = 0.052 \rho_1 \left(\frac{V_1}{C_{dpca}} \right)$$

$$\Delta Hyd = (0.052)(12.8) \left(\frac{10}{0.0264} \right)$$

$$\Delta Hyd = \mathbf{252\ psi}$$

Additional surface pressure resulting from compressing the bubble at the surface with 10 bbls of kill mud is

$$P_2 = \frac{P_1 V_1}{V_2}$$

$$P_2 = \frac{(120)(11.5)}{(11.5 - 10)}$$

$$P_2 = \textbf{920 psi}$$

Additional surface pressure, ΔP_s, is given as

$$\Delta P_s = P_2 - P_a$$

$$\Delta P_s = 920 - 120$$

$$\Delta P_s = \textbf{800 psi}$$

Total pressure increase, ΔP_{total}, is given as

$$\Delta P_{total} = \Delta Hyd + \Delta P_s$$

$$\Delta P_{total} = 252 + 800$$

$$\Delta P_{total} = \textbf{1052 psi}$$

Since ΔP_{total} is less than the maximum permissible pressure increase, ΔP_t, calculated using Equation 5.3, pump 10 bbls of 12.8-ppg mud at 1 bpm and shut in to permit the gas to migrate to the surface.

In 2 to 4 hours shut-in $P_2 = 920$ psi.

The surface pressure, P_a, may now be reduced by bleeding **ONLY GAS** from P_2 to

$$P_{newa} = P_a - \Delta Hyd$$

$$P_{newa} = 120 - 252$$

$$P_{newa} = \textbf{0 psi}$$

and the hole filled and observed.

The well is killed.

Summary

	Volume Pumped	Old Surface Pressure	Surface Pressure after Adding Kill Mud	Surface Pressure after Bleeding
Initial Conditions	-0-	1420		
1st Stage	20	1420	2031	966
2nd Stage	19	966	1633	513
3rd Stage	16	513	1227	120
4th Stage	10	120	920	-0-

By following this schedule, the well can be killed safely without violating the casing seat and losing more mud and without permitting any additional influx of formation gas into the wellbore.

In summary, the well is controlled by pumping kill-weight mud into the annulus and bleeding dry gas out of the annulus. The volume of kill fluid that can be pumped without fracturing the casing seat is determined from Equation 5.3. Logically, the surface pressure should be reduced from the value prior to lubricating kill mud by the additional hydrostatic contributed by the kill-weight mud. However, for obvious reasons, the effective hydrostatic cannot be less than the reservoir pressure. At the lower surface pressures, the gas hydrostatic gradient is less and the final surface pressure must be higher to reflect the difference in gas gradient and prevent additional influx. The analysis is continued until the well is dead. Rudimentary analysis suggests higher mud weights for well control.

DYNAMIC KILL OPERATIONS

Simply put, to kill a well dynamically is to use frictional pressure losses to control the flowing bottomhole pressure and, ultimately, the static bottomhole pressure of the blowout.

Dynamic Kill implies the use of a kill fluid whose density results in a hydrostatic column which is less than the static reservoir pressure. Therefore, the frictional pressure loss of the kill fluid is required to stop the flow of reservoir fluids. Dynamic Kill in the purest sense was intended as an intermediate step in the well control procedure. Procedurally, after the blowout was dynamically controlled with a fluid of lesser density, it was ultimately controlled with a fluid of greater density, which resulted in a hydrostatic greater than the reservoir pressure.

Generically and in this text, Dynamic Kill includes control procedures utilizing fluids with densities less than and much greater than that required to balance the static reservoir pressure. In reality, when the density of the kill fluid is equal to or greater than that required to balance the static reservoir pressure, the fluid dynamics are more properly described as a Multiphase Kill Procedure.

Dynamically controlling a well using a Multiphase Kill Procedure is one of the oldest and most widely used fluid control operations. In the past, it has been a "seat-of-the-pants" operation with little or no technical evaluation. Most of the time, well control specialists had some arbitrary rules of thumb, for example, the kill fluid had to be 2 pounds per gallon heavier than the mud used to drill the zone, or the kill fluid had to achieve some particular annular velocity. Usually, that translated into manifolding together all the pumps in captivity, weighting the mud up as high as possible, pumping like hell, and hoping for the best. Sometimes it worked and sometimes it didn't.

There are many applications of the Dynamic and Multiphase Kill procedures and any well control operation should be studied from that perspective. The most common application is when the well is flowing out of control and the drillpipe is on bottom. The kill fluid is pumped to the bottom of the hole through the drillpipe, and the additional hydrostatic of the kill fluid along with the increased friction pressure resulting from the kill fluid controls the well.

Multiphase Kill operations were routine in the Arkoma Basin in the early 1960s. There, air drilling was popular. In air drilling operations, every productive well is a blowout. Under different circumstances, some of them would have made the front page of the *New York Times*. At one location just east of McCurtain, Oklahoma, the fire was coming out of an 8-inch blooie line and was almost as high as the crown of the 141-foot derrick. The blooie line was 300 feet long, and you could toast marshmallows on the rig floor. The usual procedure was the seat-of-the-pants Multiphase Kill and it killed the well.

The most definitive work on the pure Dynamic Kill was done by Mobil Oil Corporation and reported by Elmo Blount and Edy Soeiinah.[2] The biggest gas field in the world is Mobil's Arun Field in North Sumatra, Indonesia. On June 4, 1978, Well No. C-II-2 blew out while drilling and caught fire. The rig was immediately consumed. The well burned for 89 days at an approximate rate of 400 million standard cubic feet per day.

Due to the well's high deliverability and potential, it was expected to be extremely difficult to kill. The engineering was so precise that only one relief well was required. That a blowout of this magnitude was completely dead one hour and 50 minutes after pumping operations commenced is a tribute to all involved. One of the most significant contributions resulting from this job was the insight into the fluid dynamics of a Dynamic Kill.

The engineering concepts of a Dynamic Kill are best understood by considering the familiar U-tube of Figure 5.6. The left side of the U-tube may represent a relief well and the right side a blowout, as in the case just discussed, or the left side may represent drillpipe while the right side would correspond to the annulus, as is the case in many well control situations. The connecting interval may be the formation in the case of a relief well with the valve representing the resistance due to the flow of fluids through the formation. In the case of the drillpipe–annulus scenario, the valve may represent the friction in the drill string or the nozzles in the bit, or the situation may be completely different. Whatever the situation, the technical concepts are basically the same.

The formation is represented as flowing up the right side of the U-tube with a flowing bottomhole pressure, P_{flow}, which is given by the following equation:

$$P_{flow} = P_a + \Delta P_{fr} + \Delta P_h \tag{5.11}$$

Where:

P_a = Surface pressure, psi
ΔP_{fr} = Frictional pressure, psi
ΔP_h = Hydrostatic pressure, psi

The capability of the formation to deliver hydrocarbons to the wellbore is governed by the familiar back-pressure curve illustrated in

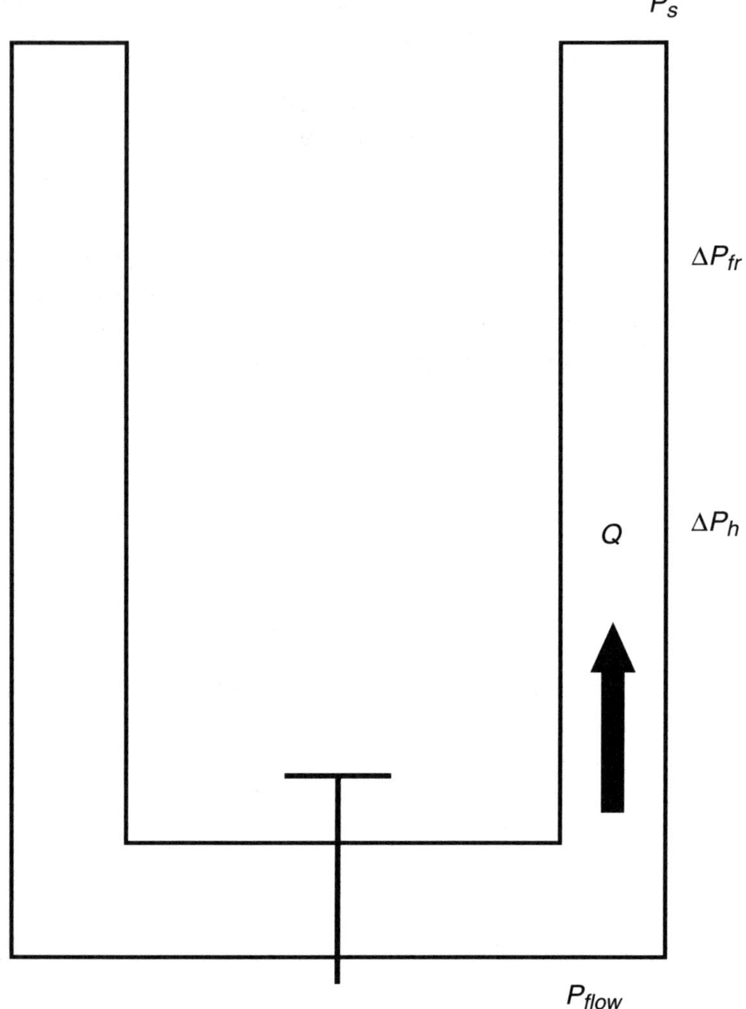

Figure 5.6 *Dynamic Kill Schematic.*

Figure 5.7 and described by the Equation 5.12:

$$Q = C(P_b^2 - P_{flow}^2)^n \tag{5.12}$$

Where:

Q = Flow rate, mmscfpd

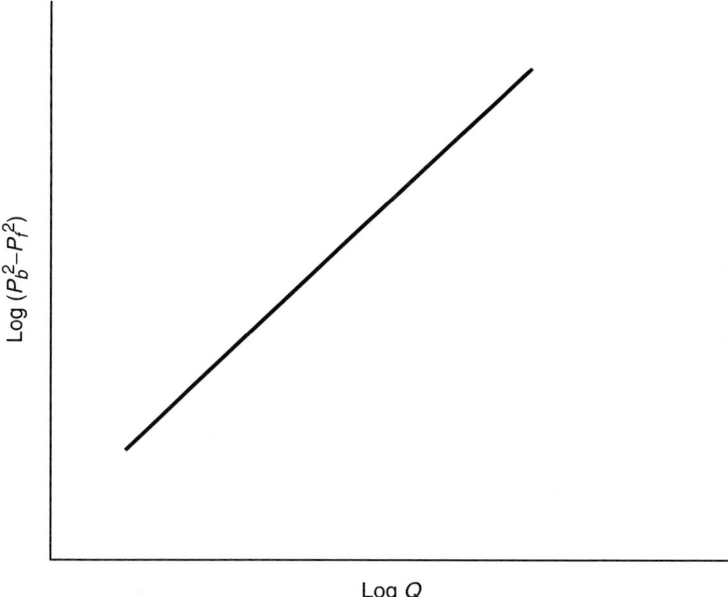

Figure 5.7   *Typical Open Flow Potential Curve.*

P_b = Formation pore pressure, psi
P_{flow} = Flowing bottmhole pressure, psi
C = Constant
n = Slope of back-pressure curve
  = 0.5 for turbulence
  = 1.0 for laminar flow

Finally, the reaction of the formation to an increase in flowing bottomhole pressure, P_{flow}, is depicted by the classic Horner Plot, illustrated in Figure 5.8. The problem is to model the blowout considering these variables.

In the past, shortcuts have been taken for the sake of simplicity. For example, the most simple approach is to design a kill fluid and rate such that the frictional pressure loss plus the hydrostatic is greater than the shut-in bottomhole pressure, P_b. This rate would be that which is sufficient to maintain control. Equations 4.13 through 4.14 for frictional pressure losses in turbulent flow can be used in such an analysis.

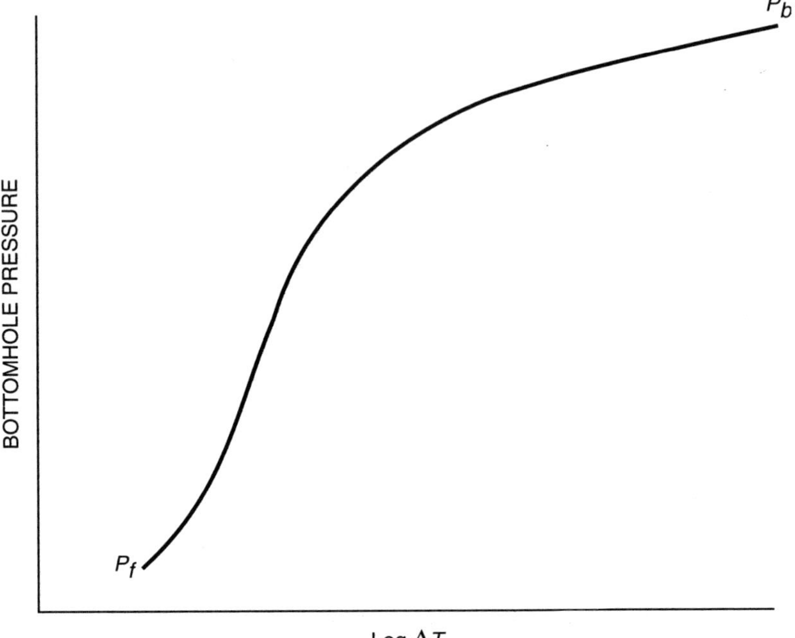

Figure 5.8   *Typical Pressure Build Up Curve.*

Consider the Example 5.4:

Example 5.4

Given:

Depth, $D = 10,000$ feet

Kill fluid, $\rho_1 = 8.33$ ppg

Kill-fluid gradient, $\rho_{m1} = 0.433$ psi/ft

Bottomhole pressure, $P_b = 5200$ psi

Inside diameter of pipe, $D_i = 4.408$ inches

Surface pressure, $P_a = $ atmospheric

Figure 5.6

Required:

Determine the rate required to kill the well dynamically with water.

Solution:

$$P_{flow} = P_a + \Delta P_{fr} + \Delta P_h$$

$$\Delta P_h = 0.433(10{,}000)$$

$$\Delta P_h = \textbf{4330 psi}$$

$$\Delta P_{fr} = 5200 - 4330$$

$$\Delta P_{fr} = \textbf{870 psi}$$

Rearranging Equation 4.14 where $\Delta P_{fr} = P_{fti}$:

$$Q = \left[\frac{P_{fti} D_i^{4.8}}{7.7(10^{-5})\rho_i^{.8} P V^{.2} l} \right]^{\frac{1}{1.8}}$$

$$Q = \left[\frac{(870)(4.408)^{4.8}}{7.7(10^{-5})(8.33)^{.8}(1)^{.2}(10{,}000)} \right]^{\frac{1}{1.8}}$$

$$Q = \textbf{1011 gpm}$$

$$Q = \textbf{24.1 bpm}$$

Therefore, as illustrated in Example 5.4, the well would be dynamically controlled by pumping fresh water through the pipe at 24 bpm. The Dynamic Kill Procedure would be complete when the water was followed with kill mud of sufficient density (10 ppg in this example) to control the bottomhole pressure.

The rate required to maintain control is insufficient in most instances to achieve control and is considered to be the minimum rate for a Dynamic Kill operation. Kouba, et al. have suggested that the rate sufficient to maintain control is the minimum and that the maximum rate is approximated by Equation 5.13:[3]

$$Q_{k\,\text{max}} = A \left(\frac{2 g_c D_{tvd} D_h}{f D_{md}} \right)^{\frac{1}{2}} \tag{5.13}$$

Where:

g_c = Gravitational constant, $\dfrac{lb_m \cdot ft}{lb_f \cdot \sec^2}$

A = Cross sectional area, ft^2
D_h = Hydraulic diameter, feet
D_{tvd} = Vertical well height, feet
D_{md} = Measured well length, feet
f = Moody friction factor, dimensionless

Consider Example 5.5:

Example 5.5

Given:

Same conditions as Example 5.4

Required:

Calculate the maximum kill rate using Equation 5.13.

Solution:

Solving Equation 5.13 gives

$$Q_{k\,\text{max}} = A \left(\frac{2 g_c D_{tvd} D_h}{f D_{md}} \right)^{\frac{1}{2}}$$

$$Q_{k\,\text{max}} = 0.106 \left(\frac{2(32.2)(10,000)(.3673)}{.019(10,000)} \right)^{\frac{1}{2}}$$

$$Q_{k\,\text{max}} = 3.74 \frac{ft^3}{\sec}$$

$$Q_{k\,\text{max}} = \textbf{40 bpm}$$

As illustrated in Examples 5.4 and 5.5, the minimum kill rate is 24 bpm and the maximum kill rate is 40 bpm. The rate required to kill the well is somewhere between these values and is very difficult to determine. Multiphase flow analysis is required. The methods of multiphase flow analysis are very complex and based upon empirical correlations obtained from laboratory research. The available correlations and research are based upon gas-lift models describing the flow of gas, oil, and water inside small pipes. Precious little research has been done to describe annular flow, much less the multiphase relationship between gas, oil, drilling mud, and water flowing up a very large, inclined annulus. The conditions and boundaries describing most blowouts are very complex to be described by currently available multiphase models. It is beyond the scope of this work to offer an in-depth discussion of multiphase models.

Further complicating the problem is the fact that, in most instances, the productive interval does not react instantaneously as would be implied by the strict interpretation of Figure 5.7. Actual reservoir response is illustrated by the classical Horner Plot illustrated in Figure 5.8. As illustrated in Figure 5.8, the response by the reservoir to the introduction of a kill fluid is non-linear.

For example, the multiphase frictional pressure loss (represented by Figure 5.8) initially required to control the well is not that which will control the static reservoir pressure. The multiphase frictional pressure loss required to control the well is that which will control the flowing bottom-hole pressure. The flowing bottomhole pressure may be much less than the static bottomhole pressure.

Further, several minutes to several hours may be required for the reservoir to stabilize at the reservoir pressure. Unfortunately, much of the data needed to understand the productive capabilities of the reservoir in a particular wellbore are not available until after the blowout is controlled. However, data from similar offset wells can be considered.

Consider the well control operation at the Williford Energy Company Rainwater No. 2-14 in Pope County near Russellville, Arkansas.[4] The wellbore schematic is presented as Figure 5.9. A high-volume gas zone had been penetrated at 4620 feet. On the trip out of the hole, the well kicked. Mechanical problems prevented the well from being shut in and it was soon flowing in excess of 20 mmscfpd through the rotary table. The drillpipe was stripped to bottom and the well was diverted through the choke manifold. By pitot tube, the well was determined to be flowing at a rate of 34.9 mmscfpd with a manifold pressure of 150 psig. The wellbore

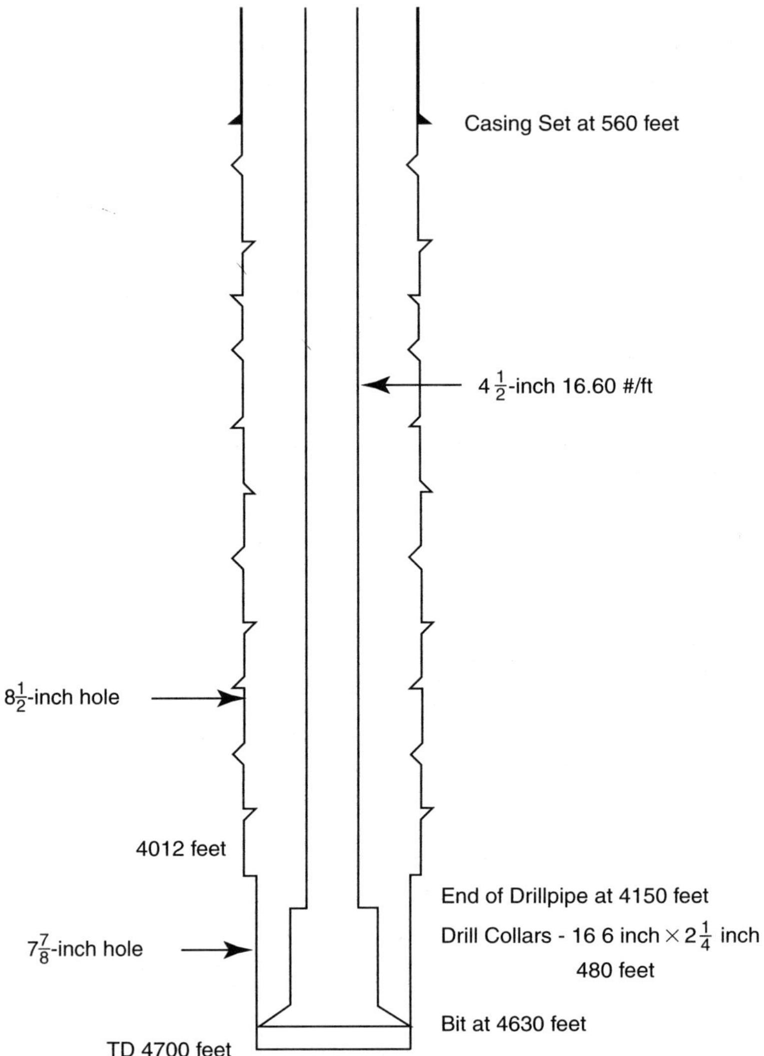

Figure 5.9 *Williford Energy Company, Rainwater No. 1, Pope County, Arkansas.*

schematic, Open Flow Potential Test, and Horner Plot are presented as Figures 5.9, 5.10 and 5.11, respectively.

In this instance, the Orkiszewski method was modified and utilized to predict the multiphase behavior.[5] The well was successfully controlled.

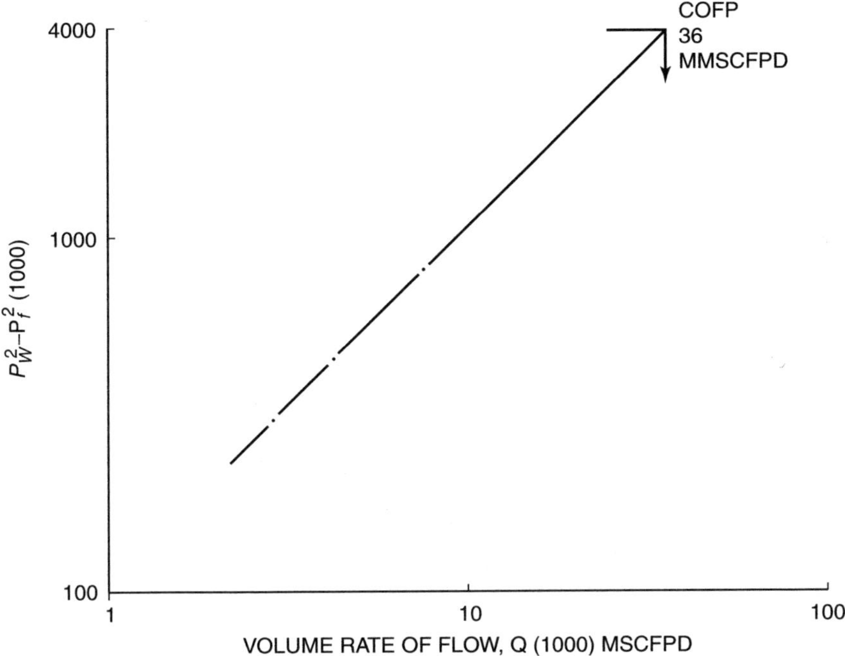

Figure 5.10　　*Calculated Open Flow Potential Test, Williford Energy Company—Rainwater No. 2-14.*

The technique used in the Williford Energy example is very conservative in that it determines the multiphase kill rate required to control the shut-in bottomhole pressure of 1995 psi. Analysis of Figure 5.11 indicates that the static bottomhole pressure will not be reached in the blowout for more than 100 hours. The flowing bottomhole pressure when the kill procedure begins is only 434 psi and is only 1500 psi approximately 20 minutes after the flowing bottomhole pressure has been exceeded. Of course, in this case the kill operation is finished in just over 10 minutes. Including all these variables is more complex but well within the capabilities of modern computing technology. Based on this more complex analysis, the kill rate using 10.7-ppg mud was determined to be 10 bpm.

The model is further complicated by the fact that the Calculated Open Flow Potential curve (COFP) presented as Figure 5.10 is a very optimistic evaluation of sustained productive capacity. An Actual Open Flow Potential curve (AOFP) based on sustained production would be more

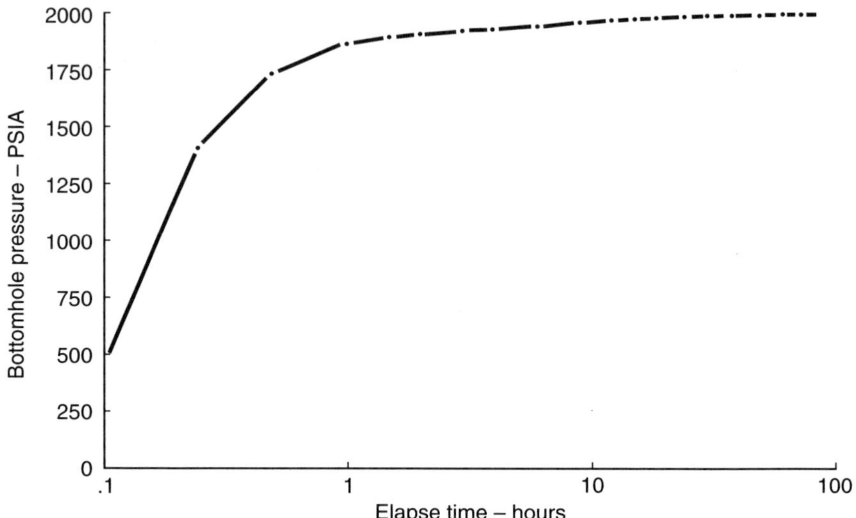

Figure 5.11   Bottomhole Pressure Build Up Test, Williford Energy Company—Rainwater No. 2-14.

appropriate for modeling actual kill requirements. A kill procedure based on the COFP is the more conservative approach. An AOFP curve more accurately reflects the effect of the pressure draw down in the reservoir in the vicinity of the wellbore.

THE MOMENTUM KILL

The Momentum Kill is a procedure where two fluids collide and the one with the greater momentum wins. If the greater momentum belongs to the fluid from the blowout, the blowout continues. If the greater momentum belongs to the kill fluid, the well is controlled. The technology of the Momentum Kill procedure is the newest and least understood of well control procedures. However, the technique itself is not new. In the late fifties and early sixties, the air drillers of eastern Oklahoma thought nothing of pulling into the surface pipe to mud up an air-drilled hole in an effort to avoid the hazards associated with the introduction of mud to the Atoka Shale.

Momentum Kill concepts are best illustrated by Figures 5.12 and 5.13. Figure 5.12 illustrates a situation in which the outcome would never be

Figure 5.12

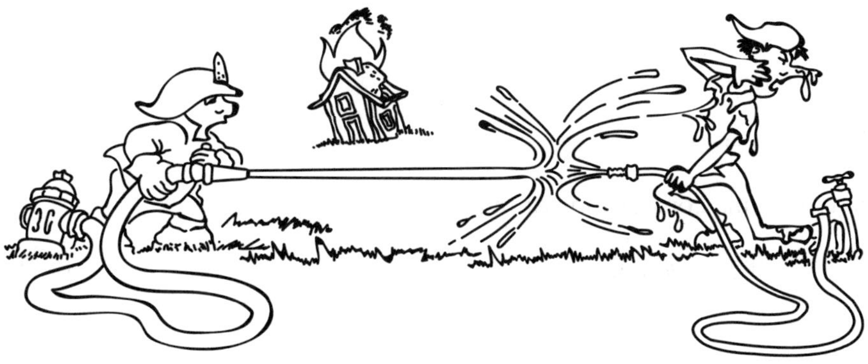

Figure 5.13

in doubt. The most fundamental reasoning would suggest that the occupant of the car is in greater peril than the occupant of the truck. It is most likely that the momentum of the truck will prevail and the direction of the car will be reversed. Conceptually, fluid dynamics are not as easy. However, consider the men in Figure 5.13. They have forgotten the fire and turned their attention to each other. Obviously, the one with the least momentum is destined for a bath.

The dynamics of a blowout are very much the same as those illustrated in Figure 5.13. The fluid flowing from the blowout exhibits a definable quantity of momentum. Therefore, if the kill fluid is introduced at a greater momentum, the flow from the blowout is reversed when the fluids collide. The governing physical principles are not significantly different from those governing the collision of two trains, two cars, or two men. The mass with the greatest momentum will win the encounter.

Newton's Second Law states that the net force acting on a given mass is proportional to the time rate of change of linear momentum of that mass. In other words, the net external force acting on the fluid within a prescribed control volume equals the time rate of change of momentum of the fluid within the control volume plus the net rate of momentum transport out of the surfaces of the control volume.

Consider the following development with all units being basic:

Momentum:

$$M = \frac{mv}{g_c} \tag{5.14}$$

and the mass rate of flow:

$$\omega = \rho vA = \rho q \tag{5.15}$$

and, from the conservation of mass:

$$\rho vA = \rho_i v_i A_i \tag{5.16}$$

Where:

$$m = \text{Mass, } lb_m$$

$$v = \text{Velocity, } \frac{ft}{\sec}$$

$$g_c = \text{Gravitational constant,} \frac{lb_m \cdot ft}{lb_f \cdot \sec^2}$$

$$\rho = \text{Density, } \frac{lb_m}{ft^3}$$

$$q = \text{Volume rate of flow, } \frac{ft^3}{\sec}$$

$$\omega = \text{Mass rate of flow, } \frac{lb_m}{\sec}$$

i = Conditions at any point

A = Cross sectional flow area, ft^2

All variables are at standard conditions unless noted with a subscript, i.

The momentum of the kill fluid is easy to compute because it is essentially an incompressible liquid. The momentum of the kill fluid is given by Equation 5.14:

$$M = \frac{mv}{g_c}$$

Substituting

$$v = \frac{q}{A}$$

$$m = \omega = \rho q$$

results in the momentum of the kill fluid, Equation 5.17:

$$M = \frac{\rho q^2}{g_c A} \tag{5.17}$$

Since the formation fluids are compressible or partially so, the momentum is more difficult to determine. Consider the following development of an expression for the momentum of a compressible fluid.

From the conservation of mass, Equation 5.16:

$$\rho v A = \rho_i v_i A_i$$

And, for a gas, the mass rate of flow from Equation 5.15 is

$$\omega = \rho v A = \rho q$$

Substituting into the momentum equation gives the momentum of the gas as:

$$M = \frac{\rho q v_i}{g_c}$$

Rearranging the equation for the conservation of mass gives an expression for the velocity of the gas, v_i, at any location as follows:

$$v_i = \frac{q_i}{A}$$

$$v_i = \frac{\rho q}{\rho_i A}$$

From the Ideal Gas Law an expression for ρ_i, the density of the gas at any point in the flow stream, is given by Equation 5.18:

$$\rho_i = \frac{S_g M_a P_i}{z_i T_i R} \tag{5.18}$$

Where:

S_g = Specific gravity of the gas
M_a = Molecular weight of air
P_i = Pressure at point i, $\dfrac{lb_f}{ft^2}$
z_i = Compressibility factor at point i
T_i = Temperature at point i, °Rankine
R = Units conversion constant

Substituting results in an expression for v_i, the velocity of the gas at point i as follows:

$$v_i = \frac{\rho q z_i T_i R}{S_g M_a P_i A} \tag{5.19}$$

Making the final substitution gives the final expression for the momentum of the gas:

$$M = \frac{(\rho q)^2 z_i T_i R}{S_g M_a P_i g_c A} \tag{5.20}$$

In this development, all units are BASIC! That means that these equations can be used in any system as long as the variables are entered in their basic units. In the English system, the units would be pounds, feet, and seconds. In the metric system, the units would be grams, centimeters, and seconds. Of course, the units' conversion constants would have to be changed accordingly.

Consider the example at the Pioneer Production Company Martin No. 1-7:

Example 5.6

Given:

Figure 5.14

Bottom pressure, $P_b = 5000$ psi

Volume rate of flow, $q = 10$ mmscfpd

Specific gravity, $S_g = 0.60$

Flowing surface pressure, $P_a = 14.65$ psia

Kill-mud density, $\rho_1 = 15$ ppg

Tubing OD $= 2\frac{3}{8}$ inches

Tubing ID $= 1.995$ inches

Casing ID $= 4.892$ inches

Temperature at 4000 feet, $T_i = 580°$ Rankine

Flowing pressure at 4000 feet, $P_i = 317.6$ psia

Compressibility factor at 4000 feet, $z_i = 1.00$

Required:

Determine the momentum of the gas at 4000 feet and the rate at which the kill mud will have to be pumped in order for the momentum of the kill mud to exceed the momentum of the gas and kill the well.

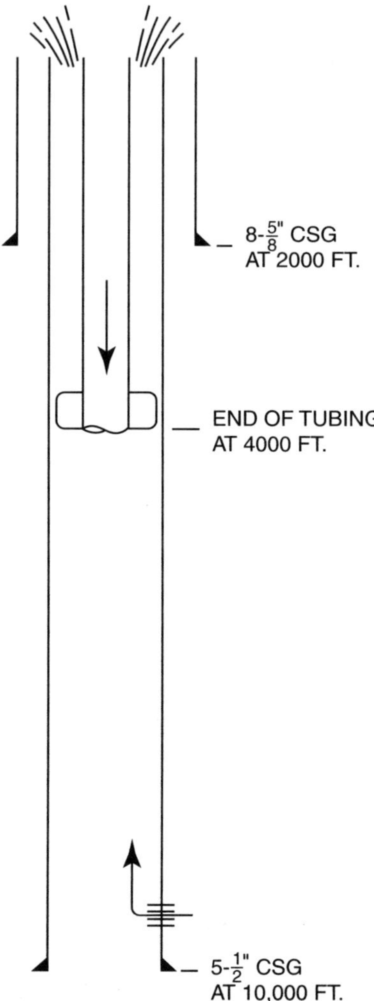

Figure 5.14   *Pioneer Production Company—Martin No. 1-7.*

Solution:

The momentum of the gas at 4000 feet can be calculated from
Equation 5.20 as follows:

$$M = \frac{(\rho q)^2 z_i T_i R}{S_g M_a P_i g_c A}$$

Substituting in the proper units,

$$\rho = 0.0458 \frac{lb_m}{ft^3}$$

$$q = 115.74 \frac{ft^3}{sec}$$

$$z_i = 1.00$$

$$T_i = 580^\circ \text{ Rankine}$$

$$R = 1544 \frac{ft - lb_f}{^\circ R - lb_m}$$

$$S_g = 0.60$$

$$M_a = 28.97$$

$$P_i = 45,734 \frac{lb_f}{ft^2}$$

$$g_c = 32.2 \frac{lb_m - ft}{lb_f - sec^2}$$

$$A = 0.1305 \, ft^2$$

$$M = \frac{[(0.0458)(115.74)]^2 (1.00)(580)(1544)}{(0.60)(28.97)(45,734)(32.2)(0.1305)}$$

$$M = \mathbf{7.53} \, lb_f$$

The rate at which the kill mud must be pumped can be determined be rearranging Equation 5.17, substituting in the proper units, and solving for the volume rate of flow as follows:

$$q = \left[\frac{M g_c A}{\rho} \right]^{\frac{1}{2}}$$

Where:

$$M = 7.53 \, lb_f$$

$$\rho = 112.36 \frac{lb_m}{ft^3}$$

$$A = 0.1305 \, ft^2$$

$$q = \left[\frac{(7.53)(32.2)(0.1305)}{112.36} \right]^{\frac{1}{2}}$$

$$q = \mathbf{0.5307} \, \frac{ft^3}{sec}$$

$$q = \mathbf{5.7 \ bbl/min}$$

The preceding example is only one simple instance of the use of the Momentum Kill technology. The momentum of the wellbore fluids is more difficult to calculate when in multiphase flow. However, the momentum of each component of the flow stream is calculated and the total momentum is the sum of the momentum of each component.

References

1.   Grace, Robert, D., "Fluid Dynamics Kill Wyoming Icicle." *World Oil*, April 1987, page 45.

2.   Blount, E.M. and Soeiinah, E., "Dynamic Kill: Controlling Wild Wells a New Way," *World Oil*, October 1981, page 109.

3.   Kouba, G.E., et al. "Advancements in Dynamic Kill Calculations for Blowout Wells," *SPE Drilling and Completion*, September 1993, page 189.

4.   Courtesy of P.D. Storts and Williford Energy.

5.   Orkiszewski, J., "Predicting Two-Phase Pressure Drops in Vertical Pipe," *Journal of Petroleum Technology*, June 1967, page 829.

6.   Courtesy of Amoco Production Company.

CHAPTER SIX

SPECIAL SERVICES IN WELL CONTROL

SNUBBING

Snubbing is the process of running or pulling tubing, drillpipe, or other tubulars in the presence of sufficient surface pressure to cause the tubular to be forced out of the hole. That is, in snubbing the force due to formation pressure's acting to eject the tubular exceeds the buoyed weight of the tubular. As illustrated in Figure 6.1, the well force, F_w, is greater than the weight of the pipe. The well force, F_w, is a combination of the pressure force, buoyant force, and friction force.

Stripping is similar to snubbing in that the tubular is being run into or pulled out of the hole under pressure; however, in stripping operations the force resulting from the surface pressure is insufficient to overcome the weight of the string and force the tubular out of the hole (Figure 6.2).

Snubbing or stripping operations through rams can be performed at any pressure. Snubbing or stripping operations through a good quality annular preventer are generally limited to pressures less than 2000 psi. Operations conducted through a stripper rubber or rotating head should be limited to pressures less than 250 psi. Although slower, ram-to-ram is the safest procedure for conducting operations under pressure.

Some of the more common snubbing applications are as follows:

- Tripping tubulars under pressure
- Pressure control/well-killing operations
- Fishing, milling, or drilling under pressure
- Completion operations under pressure

There are some significant advantages to snubbing operations. Snubbing may be the only option in critical well control operations. In general, high pressure operations are conducted more safely. For completion operations, the procedures can be performed without kill fluids, thereby eliminating the potential for formation damage.

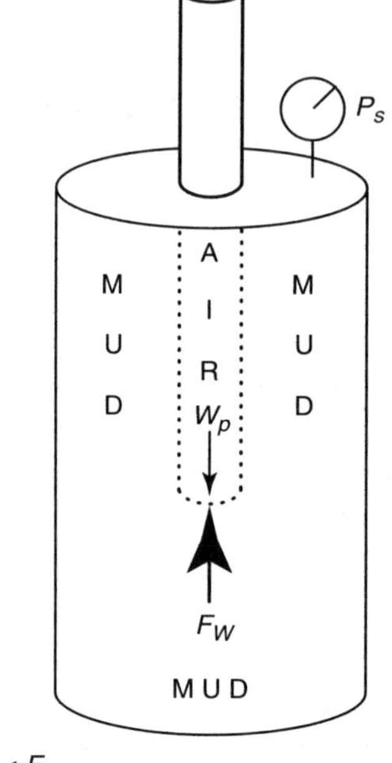

$W_p < F_W$
Where:

$$F_W = F_f - P_s A_{cs} + F_b$$
$$W_p = \text{Nominal pipe weight, lb/ft}$$

Figure 6.1 *Snubbing.*

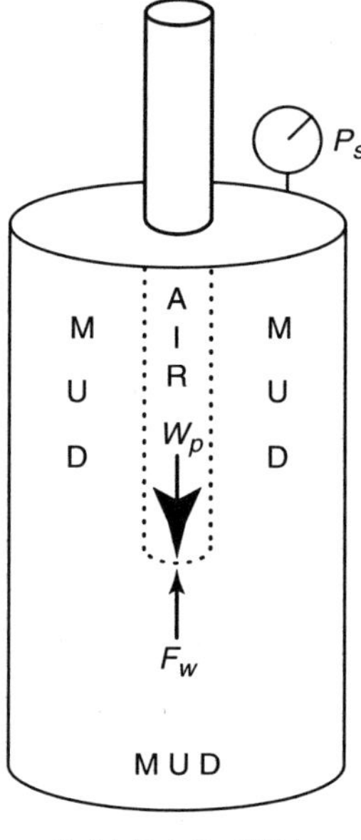

$$P_s A + F_b + F_f < W_p L$$

Figure 6.2 *Stripping.*

There are, however, some disadvantages and risks associated with snubbing. Usually, the procedures and operations are more complex. Snubbing is also slower than stripping or conventional tripping. Finally, during snubbing operations there is always pressure and usually gas at the surface.

EQUIPMENT AND PROCEDURES

The Snubbing Stack

There are many acceptable snubbing stack arrangements. The basic snubbing stack is illustrated in Figure 6.3. As illustrated, the lowermost rams are blind safety rams. Above the blind safety rams are the pipe safety rams. Above the pipe safety rams is the bottom snubbing ram, followed by a spacer spool and the upper snubbing ram. Since a ram preventer should not be operated with a pressure differential across the ram, an equalizing loop is required to equalize the pressure across the snubbing rams during the snubbing operation. The pipe safety rams are used only when the snubbing rams become worn and require changing.

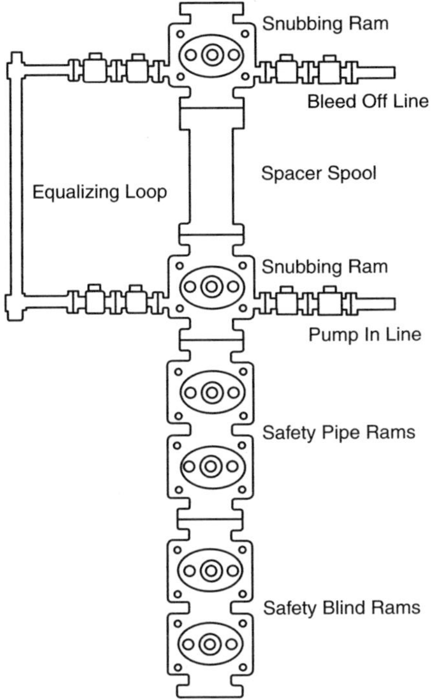

Figure 6.3 *Basic Snubbing Stack.*

When a snubbing ram begins to leak, the upper safety ram is closed and the pressure above the upper safety ram is released through the bleed-off line. The snubbing ram is then repaired. The pump-in line can be used to equalize the pressure across the safety ram and the snubbing operation continued. Since all rams hold pressure from below, an inverted ram must be included below the stack if the snubbing stack is to be tested to pressures greater than well pressure.

The Snubbing Procedure

The snubbing procedure is illustrated beginning with Figure 6.4. As illustrated in Figure 6.4, when snubbing into the hole, the tool joint or connection is above the uppermost snubbing ram, which is closed. Therefore, the well pressure is confined below the upper snubbing ram.

When the tool joint reaches the upper snubbing ram, the lower snubbing ram and equalizing loop are closed, which confines the well pressure below the lower snubbing ram. The pressure above the lower snubbing ram is released through the bleed-off line as shown in Figure 6.5.

After the pressure is released above the lower snubbing ram, the upper snubbing ram is opened, the bleed-off line is closed, and the connection is lowered to a position immediately above the closed lower snubbing ram as illustrated in Figure 6.6. The upper snubbing ram is then closed and the equalizing loop is opened, which equalizes the pressure across the lower snubbing ram (Figure 6.7).

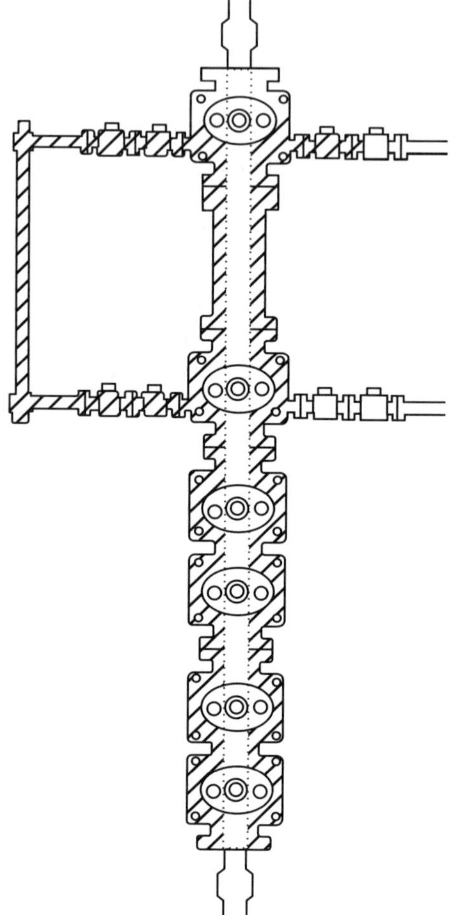

Figure 6.4   Snubbing into the Hole.

The lower snubbing ram is then opened and the pipe is lowered through the closed upper snubbing ram until the next connection is immediately above the upper snubbing ram. With the next connection above the upper snubbing ram, the procedure is repeated.

Snubbing Equipment

If a rig is on the hole, it can be used to snub the pipe into the hole. The rig-assisted snubbing equipment is illustrated in Figure 6.8. With the stationary slips released and the traveling slips engaged, the traveling block is raised and the pipe is forced into the hole. At the bottom of the stroke, the stationary slips are engaged and the traveling slips are released.

The counterbalance weights raise the traveling slips as the traveling block is lowered. At the top of the stroke, the traveling slips are engaged, the stationary slips are released, and the procedure is repeated. The conventional snubbing system moves the pipe. If drilling operations under pressure are required, a power swivel must be included.

In the absence of a rig, a hydraulic snubbing unit can be used. A hydraulic snubbing unit is illustrated in Figure 6.9. With a hydraulic snubbing unit, all work is done from the work basket with the hydraulic system, replacing the rig. The hydraulic system has the capability to circulate and rotate for cleaning out or drilling.

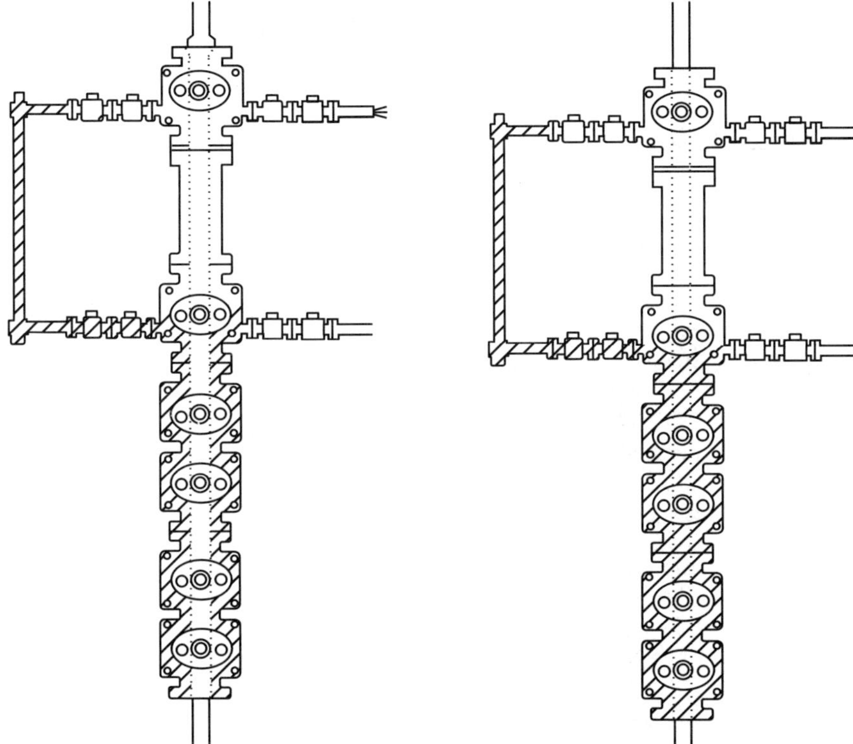

Figure 6.5 *Snubbing into the Hole.*   **Figure 6.6** *Snubbing into the Hole.*

Theoretical Considerations

As shown in Figure 6.1, snubbing is required when the well force, F_w, exceeds the total weight of the tubular. The snubbing force is equal to the net upward force as illustrated in Equation 6.1 and Figure 6.1:

$$F_{sn} = W_p L - (F_f + F_B + F_{wp}) \tag{6.1}$$

Where:

W_p = Nominal weight of the pipe, #/ft
L = Length of pipe, feet
F_f = Friction force, lb_f
F_B = Buoyant force, lb_f
F_{wp} = Well pressure force, lb_f

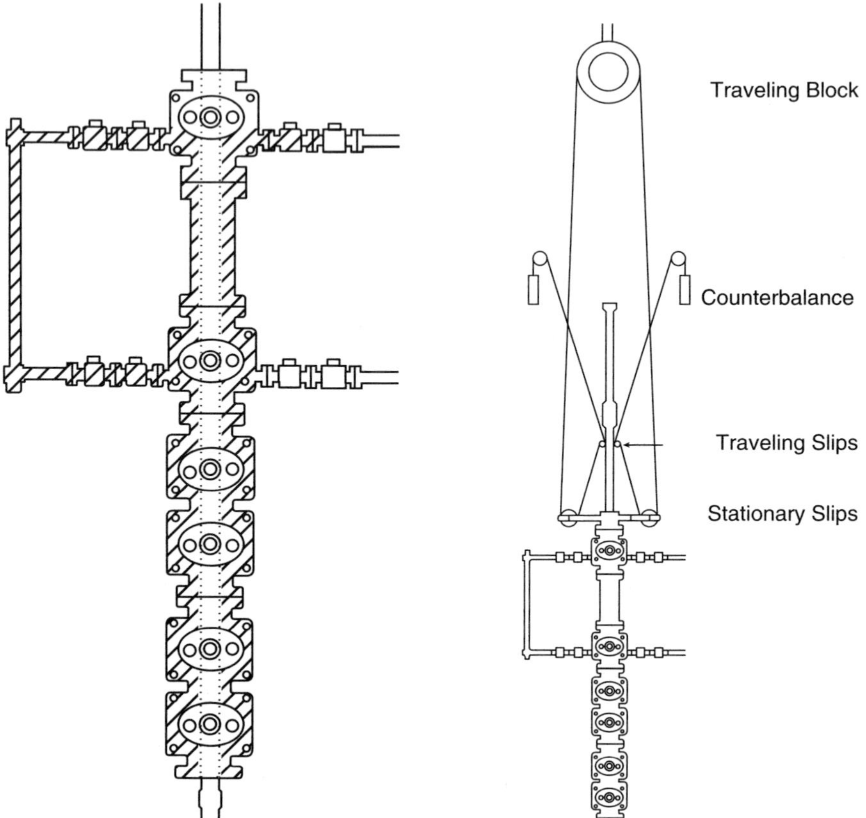

Figure 6.7   *Snubbing into the Hole.*

Figure 6.8   *Conventional or Rig Assisted Snubbing Unit.*

The well pressure force, F_{wp}, is given by Equation 6.2:

$$F_{wp} = 0.7854 D_p^2 P_s \tag{6.2}$$

Where:

$P_s =$ Surface pressure, psi
$D_p =$ Outside diameter of tubular exposed to P_s, inches

As shown in Equation 6.2, the diameter of the pipe within the seal element must be considered. When running pipe through an annular or

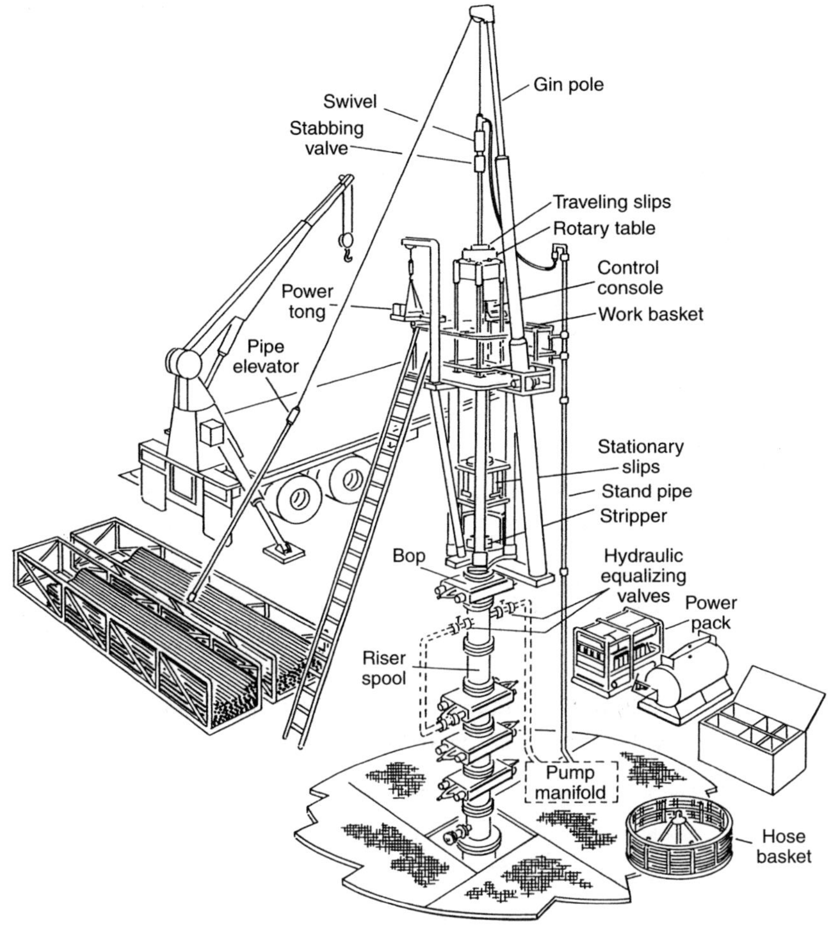

Figure 6.9

stripper, the outside diameter of the connection is the determining variable. When stripping or snubbing pipe from ram to ram, only the pipe body is contained within the seal elements; therefore, the outside diameter of the tube will determine the force required to push the pipe into the well. With drillpipe, there is a significant difference between the diameter of the pipe body and the tool joint.

Example 6.1 illustrates the calculation of the wellhead pressure force:

Example 6.1

Given:

Surface pressure, $P_s = 1500$ psi

Work string $= 4.5$-inch drillpipe

Pipe OD, $D_p = 4.5$ inches

Connection OD, $D_{pc} = 6.5$ inches

Required:

The well pressure force when the annular is closed on

1.   The tube (Figure 6.10)

2.   The connection (Figure 6.11)

Solution:

1.   When the annular is closed on the tube, the force associated with the pressure can be determined using Equation 6.2:

$$F_{wp} = 0.7854 D_p^2 P_s$$

$$F_{wp} = 0.7854(4.5^2)(1500)$$

$$F_{wp} = \mathbf{23{,}857\ lb_f}$$

2.   When the annular is closed on a tool joint, the force is calculated using the diameter of the connection:

$$F_{wp} = 0.7854(6.5^2)(1500)$$

$$F_{wp} = \mathbf{49{,}775\ lb_f}$$

In addition to the pressure area force, the friction force must be considered. Friction is that force which is tangent to the surface of contact between two bodies and resisting movement. Static friction is the force that resists the initiation of movement. Kinetic friction is the force resisting movement when one body is in motion relative to the other. The force

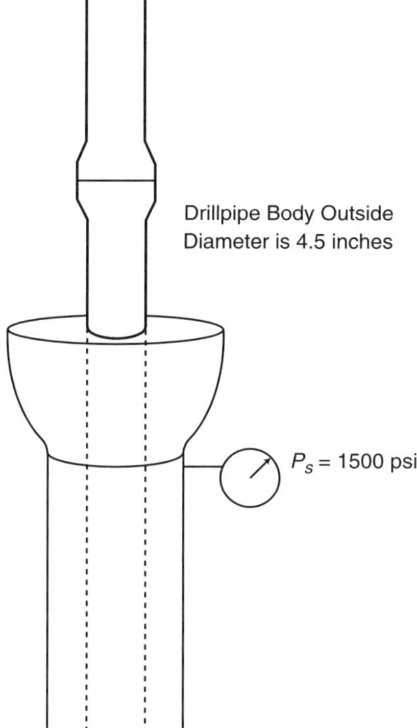

Figure 6.10   *Snubbing Drillpipe through the Annular.*

required to overcome static friction is always greater than that required to maintain movement (kinetic friction).

Since friction is a resistance to motion, it acts in the direction opposite the pipe movement. Friction acts upward when snubbing or stripping into a well and downward when snubbing or stripping out of a well. The magnitude of the force required to overcome friction is a function of the roughness of the surface areas in contact, total surface area, the lubricant being used, and the closing force applied to the BOP.

Additional friction or drag may result between the snubbing string and the wall of the hole. In general, the larger the dogleg severity, inclination, and tension (or compression) in the snubbing string, the greater the friction due to drag.

In addition to the forces associated with pressure and friction, the buoyant force affects the snubbing operation. Buoyancy is the force exerted

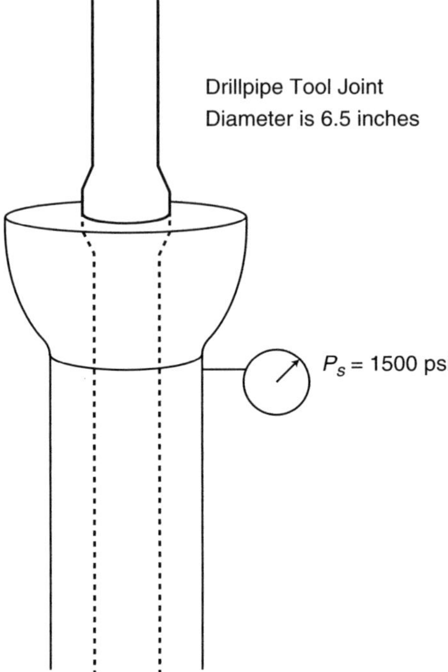

Figure 6.11   *Snubbing the Tool Joint through the Annular.*

by a fluid (either gas or liquid) on a body wholly or partly immersed and is equal to the weight of the fluid displaced by the body.

As illustrated in Figure 6.12, the buoyant force, F_B, is given by Equation 6.3:

$$F_B = 0.7854(\rho_m D_p^2 L - \rho_i D_i^2 L_i) \tag{6.3}$$

Where:

$\rho_m =$ Mud gradient in annulus, psi/ft
$\rho_i =$ Fluid gradient inside pipe, psi/ft
$D_p =$ Outside diameter of pipe, inches
$D_i =$ Inside diameter of pipe, inches
$L =$ Length of pipe below BOP, feet
$L_i =$ Length of column inside pipe, feet

If the pipe is being snubbed into the hole dry, the density of the air is negligible and the $\rho_i D_i^2 L_i$ term is negligible. If the inside of the

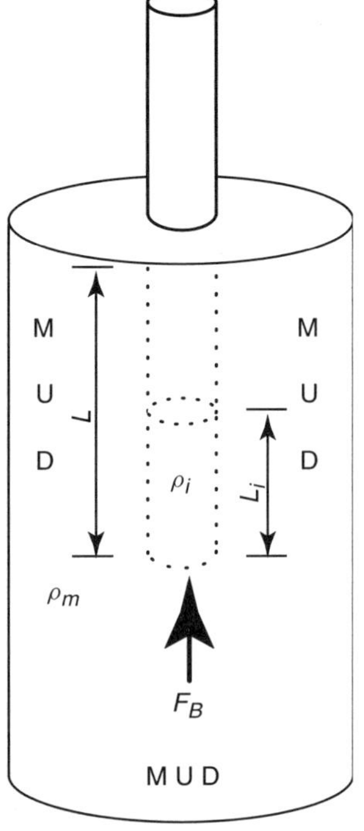

$$F_B = \rho_m L \left(\frac{\rho_i}{4} D_p^2\right) - \rho_i L_i \left(\frac{\rho_i}{4} D_i^2\right)$$

Figure 6.12   *The Buoyant Force.*

pipe is full or partially full, the $\rho_i D_i^2 L_i$ term cannot be ignored. If the annulus is partially filled with gas, the $\rho_m D_p^2 L$ term must be broken into its component parts. If the annulus contains muds of different densities, each must be considered. The determination of the buoyant force is illustrated in Example 6.2, and Equation 6.3 becomes

$$F_B = 0.7854\left[\left(\rho_{m1}L_1 + \rho_{m2}L_2 + \rho_{m3}L_3 + \ldots + \rho_{mx}L_x\right)D_p^2 - \rho_i L_i D_i^2\right]$$

Where:

L_1 = Column length of fluid having a density gradient ρ_{m1}
L_2 = Column length of fluid having a density gradient ρ_{m2}
L_3 = Column length of fluid having a density gradient ρ_{m3}
L_x = Column length of fluid having a density gradient ρ_{mx}

Example 6.2

> **Given:**
>> Schematic $=$ Figure 6.12
>>
>> Mud gradient, $\rho_m = 0.624$ psi/ft
>>
>> Length of pipe, $L = 2000$ feet
>>
>> Tubular $= 4\frac{1}{2}$-inch 16.6 #/ft drillpipe
>>
>> Tubular is dry.
>
> **Required:**
>> The buoyant force.
>
> **Solution:**
>> The buoyant force is given by Equation 6.3:
>>
>> $$F_B = 0.7854(\rho_m D_p^2 L - \rho_i D_i^2 L_i)$$
>>
>> With dry pipe, Equation 6.3 reduces to
>>
>> $$F_B = 0.7854\rho_m D_p^2 L$$
>>
>> $$F_B = 0.7854(0.624)(4.5^2)(2000)$$
>>
>> $$F_B = \mathbf{19{,}849}\ \boldsymbol{lb_f}$$

In this example, the buoyant force is calculated to be 19,849 lb_f. The buoyant force acts across the exposed cross-sectional area which is the end of the drillpipe and reduces the effective weight of the pipe. Without the well pressure force, F_{wp}, and the friction force, F_f, the effective weight of the 2000 feet of drillpipe would be given by Equation 6.4:

$$W_{eff} = W_p L - F_B \tag{6.4}$$

Example 6.3

> **Given:**
>> Example 6.2

Required:

Determine the effective weight of the $4\frac{1}{2}$-inch drillpipe.

Solution:

The effective weight, W_{eff}, is given by Equation 6.4:

$$W_{eff} = W_p L - F_B$$

$$W_{eff} = 16.6(2000) - 19,849$$

$$W_{eff} = \mathbf{13,351\ lbs}$$

As illustrated in this example, the weight of drillpipe is reduced from 33,200 pounds to 13,351 pounds by the buoyant force.

The maximum snubbing or stripping force required occurs when the string is first started, provided the pressure remains constant. At this point, the weight of the string and the buoyant force are minimal and may generally be ignored. Therefore, the maximum snubbing force, F_{snmx}, can be calculated from Equation 6.5:

$$F_{snmx} = F_{wp} + F_f \tag{6.5}$$

Where:

$F_{snmx} = $ Maximum snubbing force, lb_f
$F_{wp} = $ Well pressure force, lb_f
$F_f = $ Frictional pressure force, lb_f

As additional pipe is run in the hole, the downward force attributable to the buoyed weight of the string increases until it is equal to the well pressure force, F_{wp}. This is generally referred to as the balance point and is the point at which the snubbing string will no longer be forced out of the hole by well pressure. That is, as illustrated in Figure 6.13, at the balance point the well force, F_w, is exactly equal to the weight of the tubular being snubbed into the hole. The length of empty pipe at the balance point is given by Equation 6.6:

$$L_{bp} = \frac{F_{snmx}}{W_p - 0.0408\rho D_p} \tag{6.6}$$

Where:

$L_{bp} = $ Length at balance point, feet

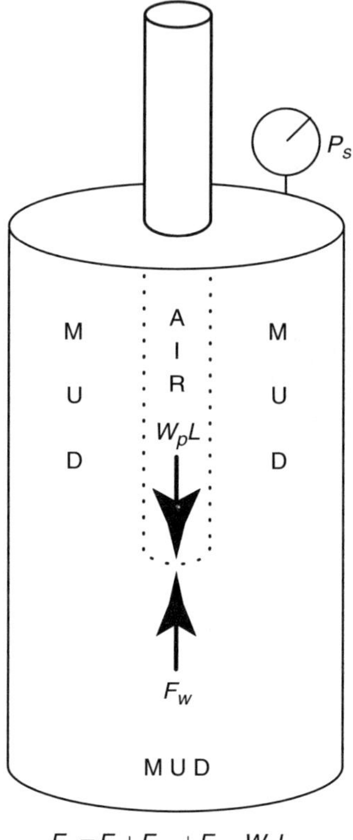

$$F_w = F_f + F_{wp} + F_b - W_pL$$

Figure 6.13   *Balance Point.*

F_{snmx} = Maximum snubbing force, lb_f
W_p = Nominal pipe weight, #/ft
ρ = Mud density, ppg
D_p = Outside diameter of tubular, inches

After the pipe is filled, the net downward force is a positive snubbing force as given by Equation 6.1.

In a normal snubbing situation, the work string is run to a point just above the balance point without filling the work string. While snubbing, the well force must be sufficiently greater than the weight of the pipe to cause the slips to grip the pipe firmly. After the pipe is filled, the weight of the pipe should be sufficient to cause the slips to grip the pipe firmly. This practice increases the string weight and reduces the risk of dropping the work string near the balance point.

The determination of the balance point is illustrated in Example 6.4:

Example 6.4

Given:

$4\frac{1}{2}$-inch 16.6 #/ft drillpipe is to be snubbed ram to ram into a well containing 12-ppg mud with a shut-in wellhead pressure of 2500 psi. The friction contributable to the BOP ram is 3000 lb_f. The internal diameter of the drillpipe is 3.826 inches.

Required:

1. The maximum snubbing force required.

2. Length of empty pipe to reach the balance point.

3. The net downward force after the pipe is filled at the balance point.

Solution:

1. The maximum snubbing force is given by Equation 6.5:

$$F_{snmx} = F_{wp} + F_f$$

   Combining Equations 6.5 and 6.2:

$$F_{snmx} = 0.7854 D_p^2 P_s + F_f$$

$$F_{snmx} = 0.7854(4.5^2)(2500) + 3000$$

$$F_{snmx} = \textbf{42,761}\ \boldsymbol{lb_f}$$

2. The length of empty pipe at the balance point is given by Equation 6.6:

$$L_{bp} = \frac{F_{snmx}}{W_p - 0.0408\rho D_p^2}$$

$$L_{bp} = \frac{42,761}{16.60 - 0.0408(12)(4.5^2)}$$

$$L_{bp} = \textbf{6396 feet}$$

3. The net force after the pipe is filled is given by Equation 6.1:

$$F_{sn} = W_p L - (F_f + F_B + F_{wp})$$

Since $F_{snmx} = F_f + F_{wp}$,

$$F_{sn} + F_{wp} = \textbf{85,958 } \textbf{\textit{lb}}_f$$

The buoyant force, F_B, is given by Equation 6.3:

$$F_B = 0.7854(\rho_m D_p^2 L - \rho_m D_i^2 L_i)$$

$$F_B = 0.7854\big[(0.624)(4.5^2)(6396)$$

$$- (0.624)(3.826^2)(6396)\big]$$

$$F_B = \textbf{17,591 lbs}$$

Therefore,

$$F_{sn} = 6396(16.6) - 42,761 - 17,591$$

$$F_{sn} = \textbf{45,822 lbs}$$

EQUIPMENT SPECIFICATIONS

In hydraulic snubbing operations, the hoisting power required is produced by pressure applied to a multi-cylinder hydraulic jack. The jack cylinder is represented in Figure 6.14. Pressure is applied to different sides of the jack cylinder depending on whether snubbing or stripping is being done. During snubbing, the jack cylinders are pressurized on the piston rod side and on the opposite side for lifting or stripping.

Once the effective area of the jack is known, the force required to lift or snub a work string can be calculated using Equations 6.7 and 6.8:

$$F_{snub} = 0.7854 P_{hyd} N_c (D_{pst}^2 - D_r^2) \tag{6.7}$$

$$F_{lift} = 0.7854 P_{hyd} N_c D_{pst}^2 \tag{6.8}$$

Where:

$F_{snub} =$ Snubbing force, lb_f
$F_{lift} =$ Lifting force, lb_f
$D_{pst} =$ Outside diameter of piston rod in jack cylinder, inches

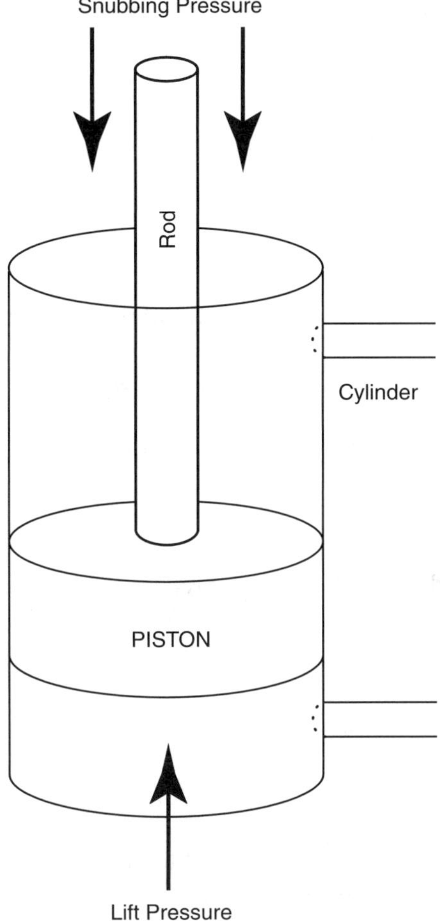

Figure 6.14

N_c = Number of active jack cylinders

P_{hyd} = Hydraulic pressure needed on jacks to snub/lift, psi

The determination of the snubbing and lifting force is illustrated in Example 6.5:

Example 6.5

Given:

A hydraulic snubbing unit Model 225 with four jack cylinders. Each cylinder has a 5-inch diameter bore and a 3.5-inch diameter piston rod. The maximum hydraulic pressure is 2500 psi.

Required:

1.   The snubbing force, F_{snub}, at the maximum pressure.

2.   The lifting force, F_{lift}, at the maximum pressure.

Solution:

1.   The snubbing force at 2500 psi is given by Equation 6.7:

$$F_{snub} = 0.7854\,P_{hyd}\,N_c\,(D_{pst}^2 - D_r^2)$$

$$F_{snub} = 0.7854(2500)(4)(5^2 - 3.5^2)$$

$$F_{snub} = \mathbf{100{,}139}\ \boldsymbol{lb_f}$$

2.   Calculate the lifting force at 2500 psi using Equation 6.8:

$$F_{lift} = 0.7854\,P_{hyd}\,N_c\,D_{pst}^2$$

$$F_{lift} = 0.7854(2500)(4)(5^2)$$

$$F_{lift} = \mathbf{196{,}350}\ \boldsymbol{lb_f}$$

The hydraulic pressure required to snub or lift in the hole can be calculated by rearranging Equation 6.8.

Example 6.6 illustrates the determination of the hydraulic pressure required for a specific lifting or snubbing force.

Example 6.6

Given:

The same hydraulic snubbing unit as given in Example 6.5. The hydraulic jacks have an effective snubbing area of 40.06 in^2 and an effective lifting area of 78.54 in^2.

Required:

1.   The hydraulic jack pressure required to produce a snubbing force of 50,000 lbs.

2.   The hydraulic jack pressure required to produce a lifting force of 50,000 lbs.

Solution:

1.   The hydraulic pressure required for snubbing is determined by rearranging Equation 6.7:

$$P_{shyd} = \frac{F_{snub}}{0.7854(D_{pst}^2 - D_r^2)N_c} \tag{6.9}$$

$$P_{shyd} = \frac{50,000}{0.7854(5^2 - 3.5^2)4}$$

$$P_{shyd} = \frac{50,000}{40.06}$$

$$P_{shyd} = \textbf{1248 psi}$$

2.   The hydraulic pressure required for lifting is determined by rearranging Equation 6.8:

$$P_{lhyd} = \frac{F_{lift}}{0.7854 D_{pst}^2 N_c} \tag{6.10}$$

$$P_{lhyd} = \frac{50,000}{0.7854(5^2)4}$$

$$P_{lhyd} = \frac{50,000}{78.54}$$

$$P_{lhyd} = \textbf{637 psi}$$

Table 6.1 is a listing of the dimensions and capacity of snubbing units normally utilized.

BUCKLING CONSIDERATIONS

After determining the required snubbing force, this force must be compared with the compressive load that the work string can support without buckling. Pipe buckling occurs when the compressive force placed on the work string exceeds the resistance of the pipe to buckling. The smallest force at which a buckled shape is possible is the critical force. Buckling

Table 6.1

Dimensions and Capacities of Snubbing Units

Model	150	225	340	600
Number of Cylinders	4	4	4	4
Cylinder Diameter (in)	4.0	5.0	6.0	8.0
Piston Rod Diameter (in)	3.0	3.5	4.0	6.0
Effective Lift Area (in^2)	50.27	78.54	113.10	201.06
Lifting Capacity at 3000 psi (lbs)	150,796	235,619	339,292	603,186
Effective Snub Area (in^2)	21.99	40.06	62.83	87.96
Snubbing Capacity at 3000 psi (lbs)	65,973	120,166	188,496	263,894
Effective Regenerated Lift Area (in^2)	28.27	38.48	50.27	113.10
Regenerated Lift Capacity at 3000 psi (lbs)	84,810	115,440	150,810	339,300
Block Speed Down (fpm)	361	280	178	137
Block Speed Up (fpm)	281	291	223	112
Bore Through Unit (in)	8	11	11	14
Stroke (in)	116	116	116	168
Rotary Torque (ft-lbs)	1000	2800	2800	4000
Jack Weight (lbs)	5800	8500	9600	34,000
Power Unit Weight (lbs)	7875	8750	8750	11,000

occurs first in the maximum unsupported length of the work string, which is usually in the window area of the snubbing unit if a window guide is not installed.

In snubbing operations, buckling must be avoided at all costs. Once the pipe buckles, catastrophic failure will certainly follow. When the pipe fails, the remainder of the string is usually ejected from the well. The flying steel can seriously injure or kill those in the work area. After that, wells have been known to blow out of control and ignite.

When the work string is subjected to a compressive load, two types of buckling may occur. Elastic or long-column buckling occurs along the major axis of the work string. The pipe bows out from the center line of the wellbore as shown in Figure 6.15a. Inelastic or local-intermediate buckling occurs along the longitudinal axis of the work string as shown in Figure 6.15b.

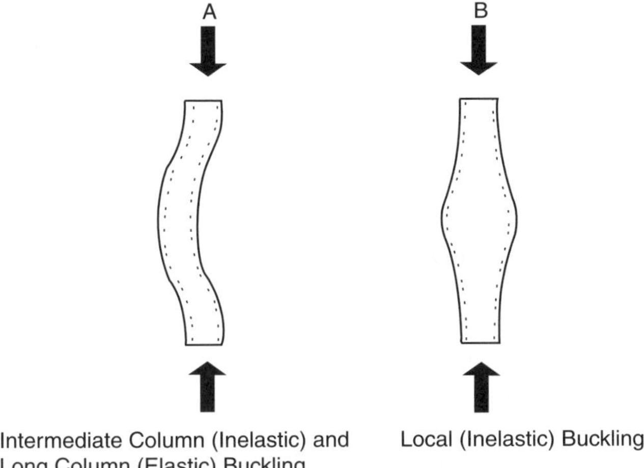

Figure 6.15

Equations describing critical buckling loads were derived by the great mathematician Leonhard Euler in 1757. His original concepts remain valid. However, in oil field work, these concepts have been expanded somewhat.

As illustrated in Figure 6.16, the buckling load is a function of the slenderness ratio. In order to determine the type of buckling which may occur in the work string, the column slenderness ratio, S_{rc}, is compared to the effective slenderness ratio, S_{re}, of the work string. If the effective slenderness ratio, S_{re}, is greater than the column slenderness ratio,

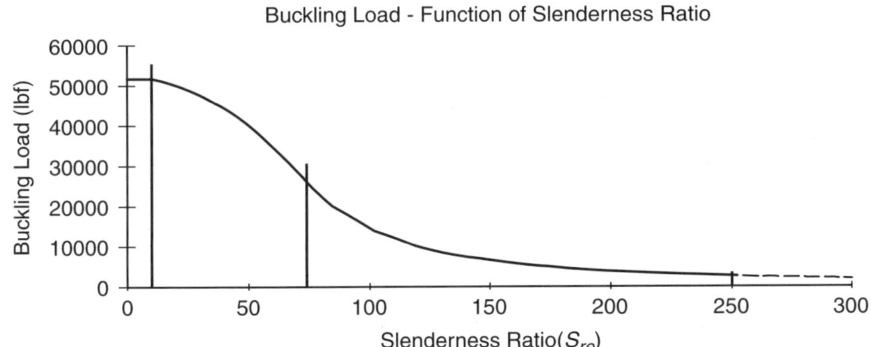

Figure 6.16

S_{rc} ($S_{re} > S_{rc}$), elastic or long-column buckling will occur. If the column slenderness ratio, S_{rc}, is greater than the effective slenderness ratio, S_{rc} ($S_{rc} > S_{re}$), inelastic or local-intermediate buckling will occur. The column slenderness ratio, S_{rc}, divides elastic and inelastic buckling.

The column slenderness ratio, S_{rc}, is given by Equation 6.11:

$$S_{rc} = 4.44 \left(\frac{E}{F_y} \right)^{\frac{1}{2}} \tag{6.11}$$

Where:

$\qquad E$ = Modulus of elasticity, psi
$\qquad F_y$ = Yield strength, psi

The effective slenderness ratio, S_{re}, is given by the larger result of Equations 6.12 and 6.13:

$$S_{re} = \frac{4U_L}{\left(D_p^2 + D_i^2 \right)^{\frac{1}{2}}} \tag{6.12}$$

$$S_{re} = \left(4.8 + \frac{D_i + t}{450t} \right) \left(\frac{D_i + t}{2t} \right)^{\frac{1}{2}} \tag{6.13}$$

Where:

$\qquad U_L$ = Unsupported length, inches
$\qquad t$ = Wall thickness, inches
$\qquad D_p$ = Outside diameter of the tubular, inches
$\qquad D_i$ = Inside diameter of the tubular, inches

Inelastic column buckling can occur if the effective slenderness ratio, S_{re}, is less than the column slenderness ratio, S_{rc}, and is equal to or less than 250 ($S_{re} < S_{rc}$). Inelastic column buckling can be either local or intermediate. Whether inelastic buckling is local or intermediate is determined by a comparison of the effective slenderness ratios determined from Equations 6.12 and 6.13.

If Equation 6.12 results in an effective slenderness ratio less than that obtained from Equation 6.13, local buckling occurs. If Equation 6.13 results in an effective slenderness ratio less than Equation 6.12 (and also less than S_{rc}) ($S_{rc} > S_{re12} > S_{re13}$), intermediate-column buckling occurs.

In either situation, a compressive load, which will cause a work string to buckle, is known as the buckling load, P_{bkl}, and is defined by Equation 6.14:

$$P_{bkl} = F_y\left(D_p^2 - D_i^2\right)\left[\frac{0.7854S_{rc}^2 - 0.3927S_{re}^2}{S_{rc}^2}\right] \tag{6.14}$$

for:

$S_{re} < S_{rc}$—Inelastic buckling
$S_{rel2} < S_{rel3}$—Local buckling
$S_{rc} > S_{rel2} > S_{rel3}$—Intermediate buckling

Where:

F_y = Yield strength, psi
D_i = Inside diameter of the tubular, inches
D_p = Outside diameter of the tubular, inches
S_{re} = Effective slenderness ratio, dimensionless
S_{rc} = Column slenderness ratio, dimensionless

In inelastic buckling, the buckling load, P_{bkl}, can be increased by increasing the yield strength, size, and weight of the work string or decreasing the unsupported section length.

Elastic (long-column) buckling is critical if the effective slenderness ratio, S_{re}, is greater than the column slenderness ratio, S_{rc}, and the effective slenderness ratio is equal to or less than $250(S_{re} \leq 250)$. When these conditions exist, the buckling load, P_{bkl}, is defined by Equation 6.15:

$$P_{bkl} = \frac{225(10^6)\left(D_p^2 - D_i^2\right)}{S_{re}^2} \tag{6.15}$$

for:

$S_{re} > S_{rc}$ and $S_{re} \leq 250$—Long-column buckling

Under this condition, the buckling load, P_{bkl}, can be increased by decreasing the unsupported section length or increasing the size and weight of the work string. Consider the following examples:

Example 6.7

Given:

Work string:

Pipe OD = $2\frac{3}{8}$ inches

$$\text{Nominal Pipe Weight} = 5.95 \text{ lb/ft}$$

$$\text{Pipe Grade} = \text{P-105}$$

$$\text{Unsupported length, } U_L = 23.5 \text{ inches}$$

$$\text{Modulus elasticity, } E = 29 \times 10^6 \text{ psi}$$

$$\text{Yield strength, } F_y = 105,000 \text{ psi}$$

$$\text{Outside diameter, } D_p = 2.375 \text{ inches}$$

$$\text{Inside diameter, } D_i = 1.867 \text{ inches}$$

$$\text{Wall thickness, } t = 0.254 \text{ inch}$$

Required:

The buckling load.

Solution:

The column slenderness ratio is given by Equation 6.11:

$$S_{rc} = 4.44 \left(\frac{E}{F_y} \right)^{\frac{1}{2}}$$

$$S_{rc} = 4.44 \left(\frac{29(10^6)}{105,000} \right)^{\frac{1}{2}}$$

$$S_{rc} = \mathbf{73.79}$$

The effective slenderness ratio, S_{re}, will be the greater value as calculated from Equations 6.12 and 6.13.

Equation 6.12:

$$S_{re} = \frac{4U_L}{\left(D_p^2 + D_i^2 \right)^{\frac{1}{2}}}$$

$$S_{re} = \frac{4(23.5)}{(2.375^2 + 1.867^2)^{\frac{1}{2}}}$$

$$S_{re} = \mathbf{31.12}$$

Equation 6.13:

$$S_{re} = \left(4.8 + \frac{D_i + t}{450t}\right)\left(\frac{D_i + t}{2t}\right)^{\frac{1}{2}}$$

$$S_{re} = \left(4.8 + \frac{1.867 + 0.254}{450(0.254)}\right)\left(\frac{1.867 + 0.254}{2(0.254)}\right)^{\frac{1}{2}}$$

$$S_{re} = \mathbf{9.85}$$

Therefore, the correct effective slenderness ratio is the greater and is given by Equation 6.12 as 31.12.

Since S_{re} (31.12) is $< S_{rc}$ (73.79) and S_{re} is ≤ 250, failure will be in the intermediate (inelastic) mode and the buckling load is given by Equation 6.14:

$$P_{bkl} = F_y \left(D_p^2 - D_i^2\right)\left[\frac{0.7854 S_{rc}^2 - 0.3927 S_{re}^2}{S_{rc}^2}\right]$$

$$P_{bkl} = (105,000)(2.375^2 - 1.867^2)$$

$$\times \left[\frac{0.7854(73.79^2) - 0.3927(31.12^2)}{73.79^2}\right]$$

$$P_{bkl} = \mathbf{161,907\ \mathit{lb_f}}$$

Consider the following example of a buckling load due to long-column mode failure:

Example 6.8

Given:

Work string:

Nominal Pipe OD $= 1$ inch

Nominal Pipe Weight $= 1.80\ \text{lb/ft}$

Pipe Grade $= $ P-105

Unsupported length, $U_L = 36.0$ inches

Modulus of elasticity, $E = 29 \times 10^6$ psi

Yield strength, $F_y = 105{,}000$ psi

Outside diameter, $D_p = 1.315$ inches

Inside diameter, $D_i = 1.049$ inches

Wall thickness, $t = 0.133$ inch

Required:

The buckling load.

Solution:

The column slenderness ratio is calculated using Equation 6.11:

$$S_{rc} = 4.44\left(\frac{E}{F_y}\right)^{\frac{1}{2}}$$

$$S_{rc} = 4.44\left(\frac{29(10^6)}{105{,}000}\right)^{\frac{1}{2}}$$

$$S_{rc} = \mathbf{73.79}$$

The effective slenderness ratio, S_{re}, will be the greater value as calculated from Equations 6.12 and 6.13. Equation 6.12 gives

$$S_{re} = \frac{4U_L}{\left(D_p^2 + D_i^2\right)^{\frac{1}{2}}}$$

$$S_{re} = \frac{4(36)}{\left(1.315^2 + 1.049^2\right)^{\frac{1}{2}}}$$

$$S_{re} = \mathbf{85.60}$$

Equation 6.13 gives

$$S_{re} = \left(4.8 + \frac{D_i + t}{450t}\right)\left(\frac{D_i + t}{2t}\right)^{\frac{1}{2}}$$

$$S_{re} = \left(4.8 + \frac{1.049 + 0.133}{450(0.133)}\right)\left(\frac{1.049 + 0.133}{2(0.133)}\right)^{\frac{1}{2}}$$

$$S_{re} = \mathbf{10.16}$$

The greater effective slenderness ratio is given by Equation 6.12 and is 85.60.

Since S_{rc} (73.79) is $< S_{re}$ (85.60) and S_{re} is ≤ 250, failure will be in the long-column mode and Equation 6.15 will be used to determine the buckling load:

$$P_{bkl} = \frac{225(10^6)\left(D_p^2 - D_i^2\right)}{S_{re}^2}$$

$$P_{bkl} = \frac{225(10^6)\left(1.315^2 - 1.049^2\right)}{85.60^2}$$

$$P_{bkl} = \mathbf{19{,}309\ lbs}$$

Local inelastic buckling is illustrated by Example 6.9.

Example 6.9

Given:

Example 6.8, except that the unsupported length, U_L, is 4 inches.

From Example 6.8:

$$S_{rc} = 73.79$$

$$S_{re13} = 10.16$$

Required:

The buckling load and mode of failure.

Solution:

The slenderness ratio is given by Equation 6.12:

$$S_{re12} = \frac{4U_L}{\left(D_p{}^2 + D_i^2\right)^{\frac{1}{2}}}$$

$$S_{re12} = \frac{4(4)}{(1.315^2 + 1.049^2)^{\frac{1}{2}}}$$

$$S_{re12} = \mathbf{9.51}$$

Since $S_{re12} < S_{re13} < S_{re}$, the buckling mode is local inelastic.

The buckling load is given by Equation 6.14:

$$P_{bkl} = F_y\left(D_p^2 - D_i^2\right)\left(\frac{0.7854S_{cr}^2 - 0.3927S_{re}^2}{S_{cr}^2}\right)$$

$$P_{bkl} = 105{,}000(1.315^2 - 1.049^2)$$

$$\times \left(\frac{0.7854(73.79^2) - 0.3927(10.16^2)}{73.79^2}\right)$$

$$P_{bkl} = \mathbf{51{,}366\ \textit{lb}_f}$$

SPECIAL BUCKLING CONSIDERATIONS: VARIABLE DIAMETERS

In oil field snubbing operations, the most frequently encountered problems involve long column buckling. In these situations, the classical Euler solution is applicable. There are several solutions for the Euler equations depending on the end conditions. The classical approach assumes that the ends are pinned and free to rotate without any restriction due to friction. If the ends are fixed and cannot move, the critical load will be approximately four times that calculated using pinned ends. With one end fixed and the other pinned, the critical load will be approximately twice that determined with pinned ends. If one end is fixed and the other end completely free to move, the critical load will be one half of that calculated assuming pinned ends. For oil field operations, the assumption of pinned ends is reasonable for most operations. However, field personnel should be aware of the assumptions made and remain alert for changes in the end conditions that could significantly reduce the critical load. **It must be remembered that once the critical load for a column is exceeded, failure is imminent and catastrophic.** The classic Euler equation for pinned ends is given as Equation 6.16:

$$P_{cr} = \frac{\pi^2 EI}{L^2} \tag{6.16}$$

Where:

P_{cr} = Critical buckling load, lb_f

E = Young's Modulus, $30(10)^6$

l = Moment of inertia, $\dfrac{\pi}{64}\left(D_o^4 - D_i^4\right)$

D_o = Outside diameter, inches

D_i = Inside diameter, inches

L = Column length, inches

The previous discussion involved only loading of columns of constant dimensions. The problems, which arise when different diameters are involved, have not been addressed in oil field operations. The exact solution of the differential equations is very complicated. Timoshenko[1] described a numerical solution. Only the methodology will be presented. For the theoretical aspects, please refer to the reference.

Consider a symmetrical beam as shown in Figure 6.17. Assume a series of beams are to be snubbed into the hole. To determine the critical buckling load, it is assumed that the deflection of the beam can be described by a sine curve. The critical buckling load is determined pursuant to the methodology presented as Table 6.2.

Consider Example 6.10:

Example 6.10

A series of 4-inch OD blast joints are to be included in a string of $2\frac{7}{8}$-inch tubing and snubbed into the hole. The well head pressure is 7500 psi. Using the results developed in Table 6.2 and the following, determine a safe snubbing procedure.

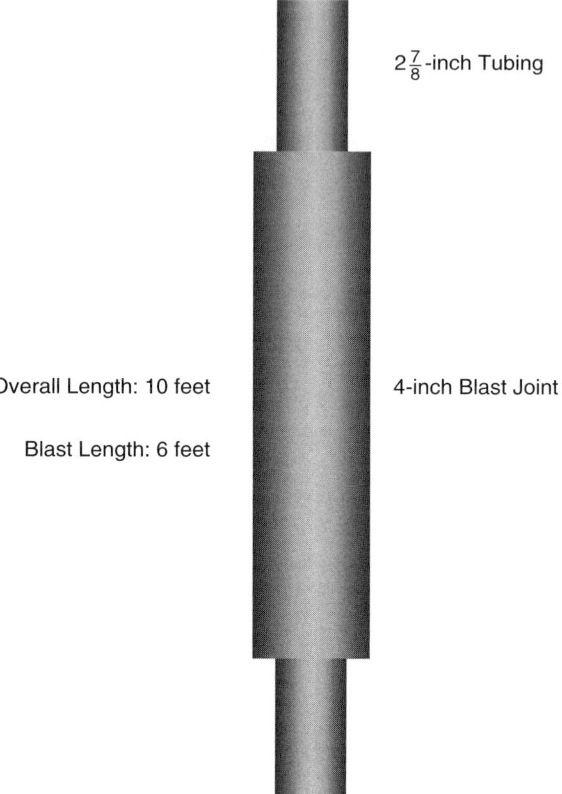

Figure 6.17

Given:
   Tubing dimensions:

   $$OD = 2.875 \text{ inches}$$

   $$ID = 2.323 \text{ inches}$$

   $$\text{Moment of inertia} = 1.924 \text{ inches}^4$$

   $$\text{Cross-sectional area} = 6.492 \text{ inches}^2$$

   Blast joint dimensions:

   $$OD = 4.000 \text{ inches}$$

Table 6.2

Determining Critical Load for Beams with Varying Cross Sections

Column Number	0	1	2	3	4	5	6	7	8	9	10		
Station Number	0	1	2	3	4	5	6	7	8	9	10		
Y_1	0	31	59	81	95	100	95	81	59	31	0	632	Delta 1/100
M_1/EI	0.00	77.50	147.50	81.00	95.00	100.00	95.00	81.00	59.00	77.50	0.00		PDelta 1/100E12
			59.00						147.50				
R		7.69	9.59	8.03	9.43	9.92	9.43	8.03	9.59	7.69	0.00		
Average Slope	39.69	32.01	22.42	14.38	4.96	0.00	4.96	14.38	22.42	32.01	39.69		
Y_2	0.00	3.97	7.17	9.41	10.85	11.35	10.85	9.41	7.17	3.97	0.00	74.15	
Y_1/Y_2		7.81	8.23	8.61	8.76	8.81	8.76	8.61	8.23	7.81			
$Y_{1\text{avg}}/Y_{2\text{avg}}$	8.52												

Therefore, $P_{cr} = 8.52 \times EI_2/I^2$.

Assume, $I_1 = 0.4I_2$.

$Y_1 = 100 \times \text{sine} (180 \times \text{station number/number of stations})$

$f = M_1/EI = Y_1/.4$ (for members with I_1, geometry) $= Y_1$ (for members with I_2 geometry)

$R_n = (1/\text{number of stations}) \times (f_{n-1} + 10f_n + f_{n+1})/12 -$ for constant geometry

$R_n = (1/\text{number of stations}) \times (7f_n + 6f_{n-1} - f_{n-2})/24 + (1/\text{number of stations})$
$\times (7f_n + 6f_{n+1} - f_{n+2})/24$ for changing geometry

For Column 3; $R_3 = 0.1 \times (7 \times 147.5 + 6 \times 77.6 - 0)/24 + 0.1$
$\times (7 \times 59 + 6 \times 81 - 95)/24$

Continued

Table 6.2 (continued)

Column Number	0	1	2	3	4	5	6	7	8	9	10

Average Slope, $A_n = (R_0 + R_1 + R_2 + R_3 + R_4 + R_5/2) - R_n - R_{n-1} - R_{n-2} \ldots$

For Column 3; $A_3 = 39.69 - 8.03 - 9.59 - 7.69 = 14.38$

$Y_{2n} = (1/\text{number of stations}) \times A_n$

Pipe 1

OD, inches	2.875
ID, inches	2.323
Moment of inertia, inches^4	1.92
Cross sectional area, in^2	6.49

Pipe 2

OD, inches	4
ID, inches	3.548
Moment of inertia, inches^4	4.79
Cross sectional area, in^2	12.57
I_1/I_2	0.40

Young's Modulus	30,000,000
Length, feet	10
Well pressure, psi	7500
Ram friction, lb_f	10,000
Snub Force, Pipe 1, lb_f	**58,689**
Snub Force, Pipe 2, lb_f	**104,248**
Critical Load, lb_f	**85,017**
Euler Critical Load—$2\frac{7}{8}$	**39,565.49**
Euler Critical Load—4 inch	**98,443.52**

$$ID = 3.548 \text{ inches}$$

$$\text{Moment of inertia} = 4.788 \text{ inches}^4$$

$$\text{Cross-sectional area} = 12.566 \text{ inches}^2$$

$$\text{Young's Modulus} = 30{,}000{,}000$$

$$\text{Unsupported stroke length} = 10{,}000 \text{ feet}$$

$$\text{Ram friction} = 10{,}000 \; lb_f$$

Solution:

$$\textit{Snubbing force on } 2\tfrac{7}{8}, \; F_{snub} = \frac{P}{A} = \frac{7500 \;\text{psi}}{6.49 \;\text{in}^2} = 58{,}689 \; lb_f$$

$$\textit{Snubbing force on 4-inch}, \; F_{snub} = \frac{P}{A} = \frac{7500 \;\text{psi}}{12.57 \;\text{in}^2}$$

$$= 104{,}248 \; lb_f$$

$$\textit{Critical load}, \; P_{cr} = 8.52\frac{EI_2}{L^2} = 8.52 \times \frac{(30{,}000{,}000 \times 4.79)}{(10 \times 12)^2}$$

$$= 85{,}017 \; lb_f$$

As a check on the critical load determined by numerical analysis in Table 6.2, the value determined should be between those obtained using the classical Euler equation (6.16) to calculate the critical load for each member.

$$\textit{Critical load for } 2\tfrac{7}{8}$$

$$= \frac{\pi^2 EI}{L^2} = \frac{3.14^2 \times 30{,}000{,}000 \times 1.92}{(10 \times 12)^2} = 39{,}565 \; lb_f$$

and

$$\textit{Critical load for 4-inch}$$

$$= \frac{\pi^2 EI}{L^2} = \frac{3.14^2 \times 30{,}000{,}000 \times 4.79}{(10 \times 12)^2} = 97{,}444 \; lb_f$$

Therefore, the numerical solution is between the Euler solutions and is reasonable. These calculations indicate the assembly can be safely snubbed into the hole **only if the snubbing rams are closed on the $2\frac{7}{8}$ and not on the 4-inch. If the snubbing rams must be closed on the 4-inch due to spacing problems, the well head pressure must be lowered or the stroke length reduced. If the stroke length cannot be reduced due to spacing problems, the only solution is to reduce the well head pressure.**

FIRE FIGHTING AND CAPPING

Oil well fire fighting is as much an art as a science. Fire fighters from the United States are heavily influenced by the tradition and practices developed by Myron Kinley, the father of oil well fire fighters. Those from outside the United States follow the same general procedure, which is to remove the remnants of the rig or other equipment until the fire is burning through one orifice straight into the air.

FIRE FIGHTING

The equipment used to remove the rig or other equipment from the fire may differ slightly. The fire fighters trained in the tradition of Myron Kinley rely heavily on the Athey Wagon such as illustrated in Figures 6.18 and 6.19. The Athey Wagon was originally developed for the pipeline industry. As shown, it is a boom on a track. The Athey Wagon tongue is attached to a dozer with a winch (Figure 6.20). The Athey Wagon is maneuvered into position for a particular operation utilizing the dozer. The Athey Wagon boom is about 60 feet long and is raised and lowered by the winch on the dozer, and the end of the Athey Wagon (Figure 6.19) may be changed to adapt to different requirements. For example, the hook shown on the end of the Athey Wagon in Figure 6.19 is routinely used to drag pieces of a melted drilling rig from the fire around the well.

Protecting men and equipment from the heat of an oil well fire is difficult. As in fighting any fire, water is used to cool the fire and provide protection from the heat. Oil well fire fighters from the United States use skid-mounted centrifugal pumps such as the one illustrated in Figure 6.21. These pumps are capable of pumping as much as of 4800 gallons (more than 100 barrels) per minute. At this rate, water supply becomes a critical factor. In order to support a full day of operations, pits are constructed with a

Figure 6.18

Figure 6.19

Figure 6.20

Figure 6.21

Figure 6.22

typical capacity of approximately 25,000 barrels. The pumps and monitors (see Figure 6.23) are often connected by hard lines or a combination of hard lines and fire hoses. Rig up can be time-consuming.

Safety Boss, the Canadian oil well fire fighting company, has perfected pumping equipment based on specially designed and modified fire trucks. Their fire trucks (Figure 6.22) are equipped with pumps capable of delivering water at a maximum rate of 2100 gallons (50 barrels) per minute. In addition, these trucks are capable of delivering a variety of fire retardant chemicals in addition to or in along with the water. Utilizing this equipment, water requirements can be reduced to an available capacity of approximately 3000–4000 barrels. All connections between the fire trucks, the tanks, and the monitors are made with fire hoses, which reduces rig-up time.

Due to their mobility, fire truck response time is significantly reduced in a localized environment. In Kuwait, for example, this mobile equipment was to the fire fighting effort while the cavalry was to the army. That is, in most cases utilizing mobile equipment, the fire would be out and the well capped before the skid-mounted equipment could be moved and

Figure 6.23

rigged up. The mobility factor contributed significantly to the fact that the Canadian team fixed approximately 50 percent more wells than the nearest other team. In addition, in Kuwait, the mobile fire fighting equipment did not require as much support as the skidded equipment.

All equipment such as dozers, trac hoes, front-end loaders, etc. and their operators are required to work near the fire and must be protected from the heat. The hydraulic systems for the equipment are protected by a covering with reflective shielding and insulating material. The personnel are protected with heat shields constructed from reflective metal. Reflective protection for a water monitor is shown in Figure 6.23. In addition, heat shields and staging houses constructed with reflective metal offer personnel relief from the heat in the proximity of the fire.

Most organizations require that all personnel wear long-sleeved coveralls made of fire-retardant materials. Around an oil well fire, ordinary cotton coveralls are a hazard. Some utilize more conventional fire fighting protective clothing such as the bunker suit commonly used by local fire departments. A fewer number use the perimeter suits, which can be worn into a fire.

In a typical fire fighting operation, the crew will approach the fire from the same direction as the prevailing wind. The pumps will be spotted approximately 300 feet from the fire. Water will provide protection as dozers on front-end loaders are used to move the monitor houses toward the fire. Once the monitor houses are within approximately 50 feet of the burning well, other equipment, such as the Athey Wagon, is brought into the proximity of the well to remove remnants of the rig and potential re-ignition sources. Work continues until the fire is burning straight in the air. Once the fire is burning straight into the air, the fire can be extinguished and the well capped. If conditions require, the well can be capped with the fire burning.

EXTINGUISHING THE FIRE

In most instances, the fire is extinguished prior to the capping operation. However, in some cases conditions dictate that the fire continue to burn until after the capping operation. For example, environmental concerns may dictate that the well be permitted to burn until the wellbore fluids can be contained or the flow stopped. Further, in some areas the regulatory agency requires that sour gas wells be ignited and that control operations be conducted with the well burning.

There are several alternatives commonly utilized to extinguish an oil well fire. Explosives are the most famous and glamorous technique used. Myron Kinley's father, Karl, was the first to extinguish an oil well fire with explosives. In 1913, Mr. Kinley walked up to a well fire near Taft, California, dropped a dynamite bomb onto the well head and ran.[2] The subsequent explosion extinguished a fire that had been burning for several months.

Today, generally between 100 and 1000 pounds of dynamite, with the lower being more common, are packed into a 55-gallon drum. Fire-retarding powders are included in the drum, which is subsequently wrapped with insulating material. The loaded drum is attached to the end of an Athey Wagon. The water monitors are concentrated on the drum as the Athey Wagon is backed into the fire. With the drum positioned at the base of the fire, the driver and the shooter take cover in the blade of the dozer and the charge is detonated. The explosion momentarily deprives the fire of oxygen and, as a result, the fire is extinguished. The water monitors are then concentrated on the well head in an effort to prevent re-ignition.

In Kuwait, the fires were usually extinguished with water. Several monitors were concentrated on the base of the fire. Usually in a matter of minutes, the fire was cooled below the ignition point.

Safety Boss, the company of Canadian oil well fire fighters, has relied upon and perfected the use of fire-retardant chemicals and powders. Custom-designed and -constructed fire-extinguishing equipment is used to spray these chemicals and powders directly on the fire. These techniques have proven to be very reliable. In Canada, sour gas fires often have to be extinguished and re-ignited several times during the course of a day—a process which would not be possible using explosives or water monitors.

Countries associated with the former Soviet Union utilized mounted jet engines to literally blow the fire out. Most often, the fire extinguishing technique included a MIG engine mounted on a flat bed trailer. Water would be sprayed on the fire and the engine engaged. Using only one jet engine, the time to extinguish the fire would be extended and often exceeded an hour. In the opinion of this writer, one jet engine would not extinguish a large oil well fire. The most impressive wind machine was designed and utilized in Kuwait by the Hungarian fire fighters (see Figure 11.19). The Hungarian "Big Wind" used two MIG engines on a tank trac. The crew had the capability to inject water and fire-retardant chemicals into the flow stream. The "Big Wind" quickly extinguished the fire in every instance in Kuwait. However, it was never used in Kuwait on one of the bigger fires.

CAPPING THE WELL

Once the fire is out, the capping operation begins. The well is capped on an available flange or on bare pipe, utilizing a capping stack. The capping stack is composed of one or more blind rams on top followed by a flow cross with diverter lines. The configuration of the bottom of the capping stack depends upon the configuration of the remaining well components.

If a flange is available, the bottom of the capping stack below the flow cross will be an adapter flange. A flanged capping stack is illustrated in Figure 6.24. If bare pipe is exposed, the bottom of the capping stack below the flow cross will be composed of an inverted pipe ram followed by a slip ram. A capping stack with an inverted pipe ram and a slip ram is

Figure 6.24

depicted in Figure 6.25. The capping stacks are placed on the well with a crane or an Athey Wagon.

In the case of exposed pipe, an alternative to the inverted pipe ram and slip ram is to install a casing flange. As illustrated in Figure 6.26, an ordinary casing flange is slipped over the exposed tubular. A crane or hydraulic jacks, supported by a wooden foundation composed of short lengths 4 × 4's, are used to set the slips on the casing head. Concrete is then poured around the jacks and foundation to the bottom of the casing head. Once the casing head is set, the excess casing is cut off using a pneumatic cutter. A capping stack can then be nippled up on the casing flange.

Another common technique used on bare pipe is to install a weld-on flange on the bottom of the capping stack. The stack is then lowered over and onto the bare pipe. If the fire has been extinguished, the stack is lowered with a crane. If the fire has not been extinguished, the stack is installed with an Athey Wagon. With the stack in place, the slip-on flange is welded to the bare pipe. The flow from the well is far enough above the welding operation to prevent re-ignition.

Figure 6.25

Figure 6.26

Figure 6.27

In all instances, it is important that the capping stack be larger than the wellhead. The larger stack will produce a chimney effect at the capping point. A smaller stack would result in back pressure and flow at the capping point.

In the worst cases, guides are required to bring the capping stack over the flow. If the flow is strong, the stack has to be snubbed onto the well (Figure 6.27). There is almost always a period of time during capping when visual contact is impossible (Figure 6.28).

Once the stack is landed, the vent lines are connected and the blind ram is closed, causing the flow to be vented to a pit which should be at least 300 feet from the wellhead. With the well vented, the capping operation is complete and the control and killing operation commences.

FREEZING

Freezing is a very useful tool in well control. Invariably, the top ball valve in the drill string will be too small to permit the running of a plug.

Figure 6.28

In order to remove the valve with pressure on the drillpipe, the drillpipe would have to be frozen. A wooden box is constructed around the area to be frozen. Then, a very viscous mixture of bentonite and water is pumped into the drillpipe and spotted across the area to be frozen.

Next, the freeze box is filled with dry ice (solid carbon dioxide). Nitrogen should never be used to freeze because it is too cold. The steel becomes very brittle and may shatter upon impact. Several hours may be required to obtain a good plug. The rule of thumb is one hour for each inch in diameter to be frozen. Finally, the pressure is bled from above the faulty valve; it is removed and replaced and the plug is permitted to thaw. Almost everything imaginable has been frozen, including valves and blowout preventers.

HOT TAPPING

Hot tapping is another useful tool in well control. Hot tapping consists of simply flanging or saddling to the object to be tapped and drilling into the pressure. Almost anything can be hot tapped. For example, an inoperable valve can be hot tapped or a plugged joint of drillpipe can be hot tapped and the pressure safely bled to the atmosphere. In other instances, a joint of drillpipe has been hot tapped and kill fluid injected through the tap.

JET CUTTING

Abrasive jet cutting technology has been used in well control as well as other industries for many years. However, since the extensive use of abrasive jet cutting during the Al-Awda Project (Kuwait), service providers for the well control industry have designed and utilized much more sophisticated equipment.

An abrasive jet cutter used in well control is shown in Figure 6.29. These cutters attach to the end of the Athey Wagon. The jet nozzle is positioned adjacent to the object to be severed using the Athey Wagon attached to a dozer. Frac sand at a concentration of about 2 ppg is transported

Figure 6.29   *Jet Cutting. (Courtesy of BJ Services.)*

by either water or a mixture of bentonite and water. The mixture is pumped through nozzles at a rate of approximately 2 bpm and a pressure between 5000 psi and 8000 psi. The nozzle size normally used is approximately $\frac{3}{16}$ inch.

Just as often, a jet nozzle is attached to the end of a joint of pipe and the assembly is moved into position by a crane or trac hoe. Cutting can be just as effective. The big advantage is mobility and availability. It is not necessary to import a large piece of equipment. The nozzles can be transported in a briefcase and the remaining required equipment is readily available.

This technology has been used to cut drillpipe, drill collars, casing strings, and well heads. The time required to make the cut depends on the object to be cut and the operating conditions. A single piece of pipe can be cut in a few minutes. Abrasive jet cutting is preferable to other methods in most instances.

References

1.   Timoshenko, S., *Theory of Elastic Stability*, McGraw-Hill, 1936.

2.   Kinley, J.D. and Whitworth, E.A., *Call Kinley: Adventures of an Oil Well Fire Fighter*, Cock a Hoop Publishers, 1996, p. 18.

3.   Ibid.

CHAPTER SEVEN

RELIEF WELL DESIGN AND OPERATIONS

The industry has long considered the relief well option a last resort in well control. The problems are obvious. Even with the best surveying techniques, the bottom of the hole was unknown to any degree of certainty. The ability to communicate with the bottom of the hole was very limited and most often governed by the principle of trial and error. However, relief well technology has advanced in the past 10 years to the point that a relief well is now a viable alternative. Modern technology has made intercepting the blowout a certainty and controlling the blowout from the relief well a predictable engineering event.

HISTORY

ULSEL AND MAGNETIC INTERPRETATION INTRODUCED

On March 25, 1970, a blowout occurred at the Shell Oil Corporation Cox No. 1 at Piney Woods, Rankin County, Mississippi.[1] The well had been drilled into the Smackover at a total depth of 21,122 feet and cased to 20,607 feet. The well flowed at rates estimated between 30 and 80 million standard cubic feet of gas per day plus 14,000 to 20,000 barrels of water per day.

The hydrogen sulfide concentration in the gas stream made the gas deadly toxic to humans and, combined with the saline, produced water deadly corrosive to steels. Shortly after the well kicked on the morning of the disaster, the blowout preventer stack rose and fell over, releasing a stream of gas and invert oil-emulsion mud. Within minutes the well ignited and the derrick fell. The well had cratered.

This combination of events and circumstances made surface control at the Cox No. 1 impossible. Therefore, an all-out effort was made to control the blowout from a relief well. A conventional relief well, Cox No. 2, was spudded on May 3, 1970, at a surface location 3500 feet west of the blowout. This well was designed to be drilled straight to 9000 feet and from there directionally to 21,000 feet to be bottomed close to the Cox No. 1.

There was considerable skepticism about the effectiveness of this approach. Was directional drilling possible at these depths? Could the bottom of the blowout be determined with any reasonable accuracy? Could solids-laden fluid be directed to the blowout through the Smackover with its relatively low porosity and permeability? Because of these uncertainties, a special task force was formed to explore new techniques. On May 16, 1970, the Cox No. 4 was spudded 1050 feet east of the blowout. Its mission was to intercept the blowout in the interval between 9000 and 13,000 feet and effect a kill. The resulting work was to pioneer modern relief well technology.

Two methods of evaluation were introduced and developed and formed the basis for modern technology. One technique involved the use of resistivity measurements to determine the distance between the relief well and the blowout. The use of electrical logs for locating various drillpipe and casing fishes in wellbores was well known and many examples could be cited. However, little effort had been expanded to utilize resistivity devices as a direct means of determining distance between wells. Also, there were very few examples of intersecting wellbores. Due to the nature and depth of the reservoir rocks, a wellbore intercept at or about 10,000 feet was necessary to effect a kill at the Cox No. 1.

In response to this disaster, Shell and Schlumberger developed and reported the use of ultra-long, spaced electrical logs, commercially known as ULSEL, to determine the distance between wells. The ULSEL technology was developed primarily for the profiling of salt structures or other resistive anomalies. The ULSEL tool is merely the old short-normal technique with electrode spacings of 150 feet for AM and 600 feet for the AN electrodes.

An example of induction-electric log response in an intersecting well is shown in Figure 7.1. Rather inaccurate estimates of distances between wellbores are made by curve fitting techniques designed to model the short-circuiting effect of the casing in the blowout. At the Amoco R. L. Bergeron No. 1 in the Moore-Sams Field near Baton Rouge, Louisiana, the ULSEL technology predicted a blowout in early 1980 to be

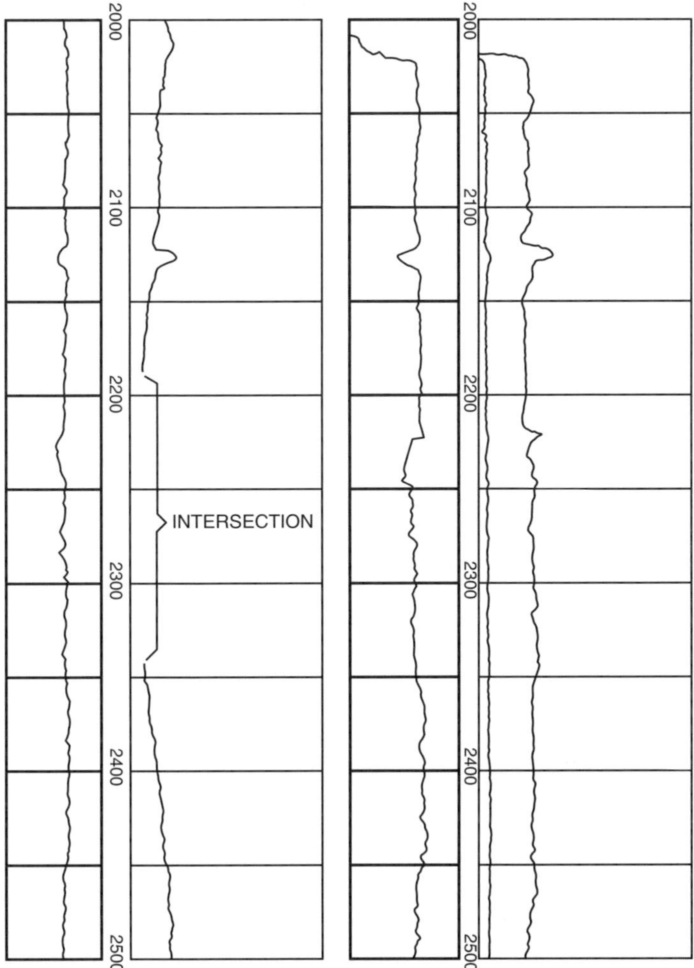

Figure 7.1   *Short Circuit Effect of Casing Ulsel Response.*

12 to 18 feet from the relief well at 9050 feet measured depth. In reality, the blowout was approximately 30 feet from the relief well.[2]

A most interesting technique was developed by Shell Development Company and reported by J. D. Robinson and J. P. Vogiatzis. This technique involved the use of sensitive magnetometers to measure the earth's magnetic field and the distortion of that magnetic field resulting from the presence of casing or other tubulars possessing remnant magnetization.

Measurements were made from the relief well and sidetracks, and models were assumed in an effort to match the relative position of the two wells. These efforts were rewarded when the relief well intercepted the Cox No. 1 at approximately 10,000 feet at a position estimated to be between 3 and 9 inches from the position of the axis predicted by the calculations.[2] As a result, United States Patent #3,725,777 was awarded in recognition of this new advance in technology.

SCHAD'S CONTRIBUTION

A very significant and heretofore relatively unnoticed contribution was made by Charles A. Schad and is documented in United States Patent #3,731,752 filed June 25, 1971 and issued May 8, 1973.[3] Schad developed a highly sensitive magnetometer consisting of at least one pair of generally rectangular core elements having square hypothesis loops for use in a guidance system for off-vertical drilling. Magnetometers were well known prior to this work. However, their usefulness was limited to relatively strong magnetic fields.

In wellbore detection, a sensitivity of 0.05 gammas would be required and would need to be distinguished from the horizontal component of the earth's magnetic field, which is between 14,000 and 28,000 gammas. Obviously, the development of a magnetometer with sensitivity such that the earth's magnetic field could be nulled in order to study a small magnetic field caused by a ferromagnetic source within the earth was a tremendous contribution to relief well technology.

It is this magnetometer concept that is used in modern wellbore proximity logging. In addition, Schad envisioned placing a magnetic field in one wellbore and using his magnetometer to guide a second wellbore to an intercept with the first. This technology later proved to be extremely useful in relief well operations.

MAGRANGE DEVELOPED

Unfortunately, no commercial service resulted from the work of Schad or Robinson. Schlumberger continued to offer the ULSEL technology; however, the distances were subject to interpretation and no concept of direction was available. Therefore, when Houston Oil and Minerals experienced a blowout in Galveston Bay in 1975, reliable wellbore proximity logging services were not commercially available. As a result, Houston Oil and Minerals Corporation commissioned Tensor, Inc. to develop a system for making such measurements.

In response to that need and commission, Tensor developed the MAGRANGE service. The MAGRANGE service is based on United States Patent #4,072,200 which was filed May 12, 1976 and issued to Fred J. Morris, et al. on February 7, 1978.[4] The Morris technology was similar to that of Robinson and Schad in that highly sensitive magnetometers were to be used to detect distortions in the earth's magnetic field caused by the presence of remnant magnetism in a ferromagnetic body.[5]

However, it differed in that Morris envisioned measuring the change in magnetic gradient along a wellbore. He reasoned that the magnetic gradient of the earth's magnetic field is small and uniform and could be differentiated from the gradient caused by a ferrous target in the blowout wellbore.

The MAGRANGE service then made a continuum of measurements along the wellbore of the relief well and analyzed the change in gradient to determine the distance and direction to the blowout. This technology was state of the art for several years following the blowout at Galveston Bay and was used in many relief well operations. However, interpretation of the data proved less reliable than needed for accurate determination of the distance and direction to a blowout. In addition, detection was limited to approximately 35 feet.

WELLSPOT DEVELOPED

In early 1980, the R. L. Bergeron No. 1 was being drilled by Amoco Production Co. as a Tuscaloosa development well in the Moore-Sams Field near Baton Rouge, Louisiana, when it blew out at a depth of 18,562 feet.[9] Systematic survey errors and the limited depth of reliable investigation of the available commercial borehole proximity logs prompted the operator to seek alternate techniques. To that end, the operator contacted Dr. Arthur F. Kuckes, professor of physics at Cornell University in Ithaca, New York. In response to that challenge, new technology was developed that provided reliability never before available in relief well operations.

This technology is currently marketed by Vector Magnetics, Inc. under the trade name WELLSPOT. The theoretical aspects are fully

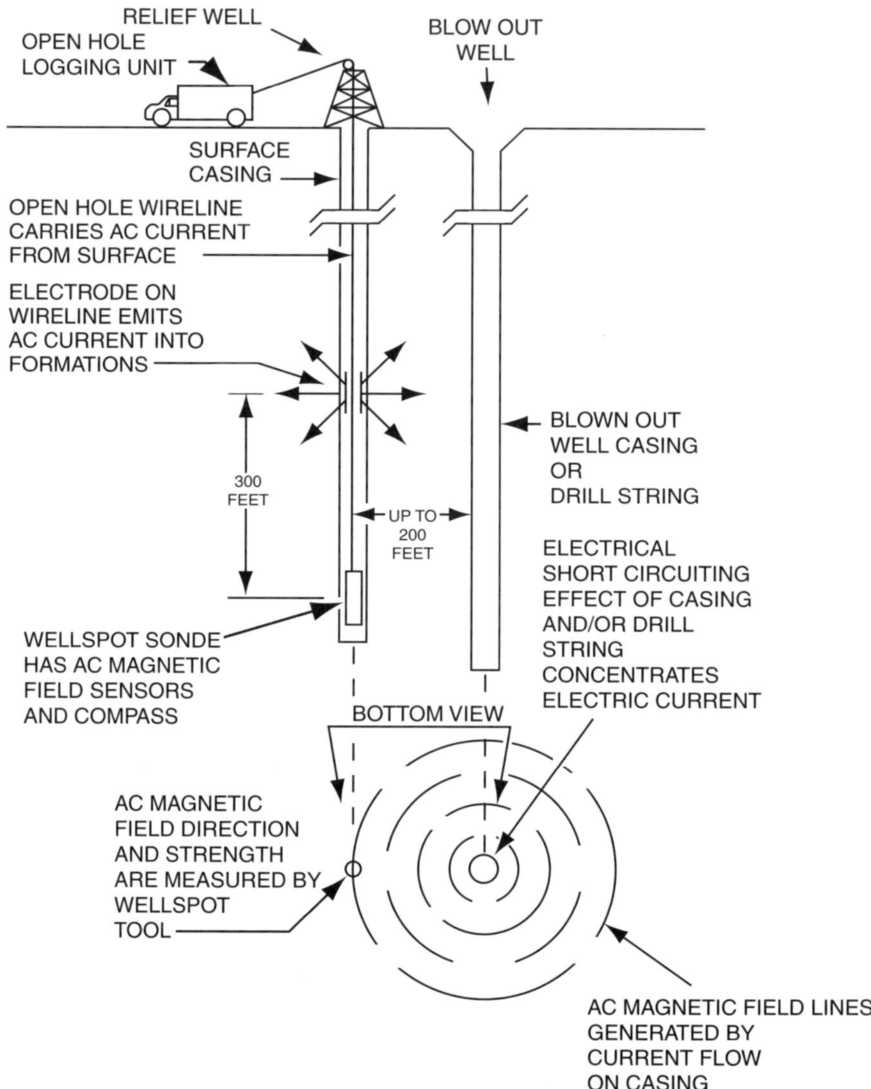

Figure 7.2   *Wellspot Operations.*

described in the referenced material.[6] The approach is quite simple and straightforward. As illustrated in Figure 7.2, an electrode is run on a conventional electric line 300 feet above a tool consisting of four magnetometers. Two AC magnetometers respond to the two components of an

AC magnetic field perpendicular to the axis of the tool, and two fluxgates measure the two components of the earth's magnetic field perpendicular to the tool axis. The fluxgates act as a magnetic compass so that the tool's orientation can be determined.

Between the tool and the wireline from the logging truck, there is an insulating bridle approximately 400 feet long. On this bridle cable 300 feet above the tool, an electrode emits AC electric current into the formations. In the absence of a nearby ferrous material, such as the casing in the blowout, the current flows symmetrically into the ground and dissipates. In the presence of a ferromagnetic body, such as the casing or drill string in the blowout, the flow of electric current is short-circuited, creating a magnetic field. The intensity and direction of the magnetic field are measured at the tool. The various parameters are analyzed using generally routine and theoretically straightforward mathematical analysis.

Depending on the conductivity of the formations in the wellbore relative to the conductivity of the ferromagnetic body, the method can detect the blowout from the relief well at distances greater than 200 feet. As will be discussed, the method is more accurate as the relief well approaches the blowout. Typical WELLSPOT data are illustrated as Figures 7.3 and 7.4.

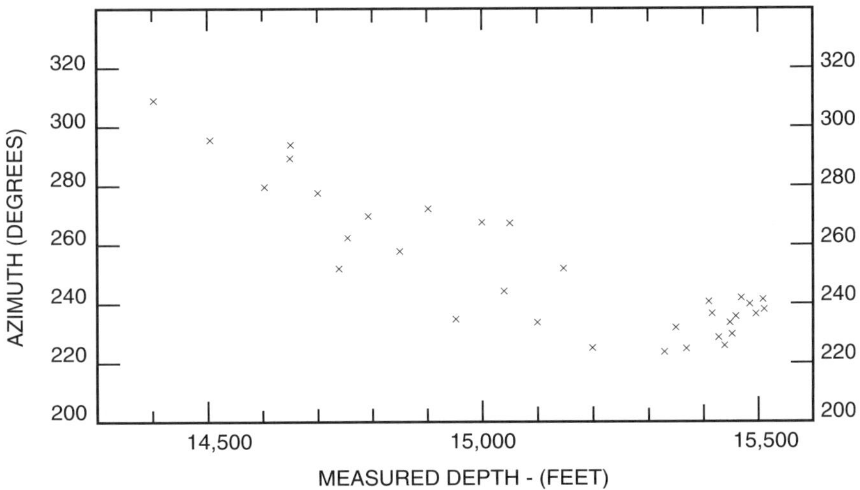

Figure 7.3   *Wellspot Blowout Azimuth Direction from Relief Well.*

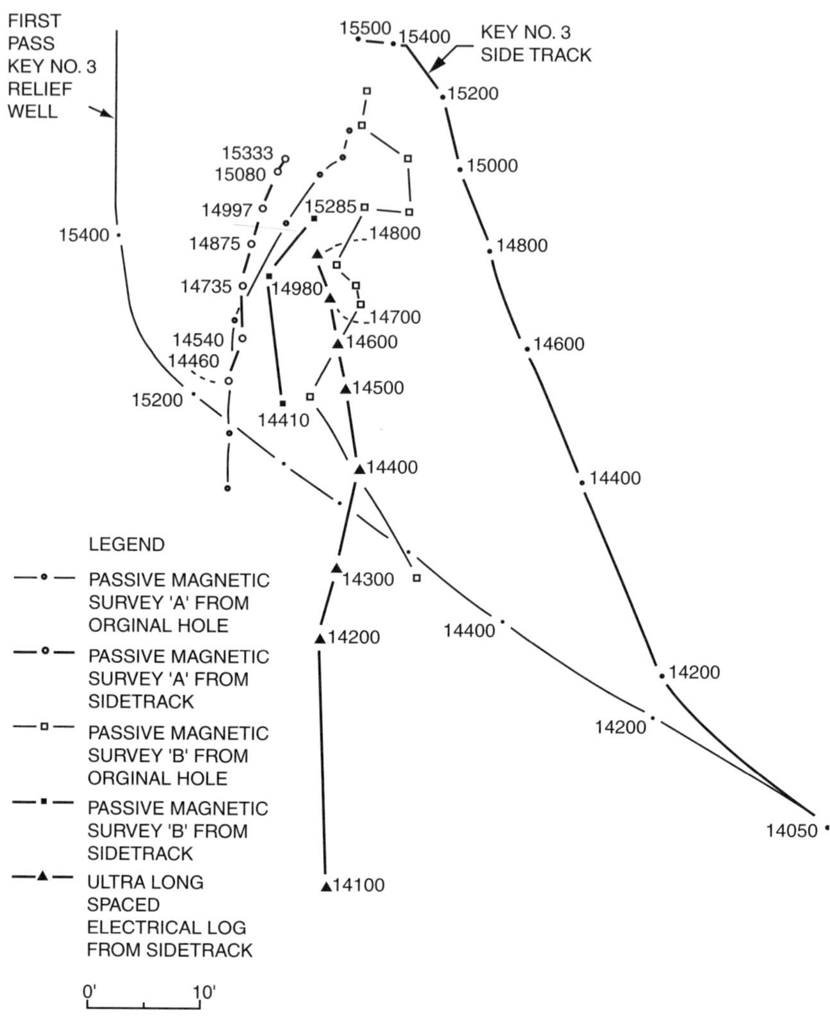

Figure 7.4   *Comparison of Various Proximity Logging Techniques.*

MAGRANGE AND WELLSPOT COMPARED

In June 1981, Apache Corporation completed the Key 1-11 in the Upper Morrow at 16,000 feet in eastern Wheeler County of the Texas Panhandle.[7] The Key 1-11 was one of the best wells ever drilled in the Anadarko basin, having 90 feet of porosity in excess of 20 percent. The original open flow potential was in excess of 90 mmscfpd.

On Sunday afternoon, October 4, 1981, after being shut in for 78 days waiting on pipeline connection, the well inexplicably erupted. The blowout that was known as the biggest in the history of the state of Texas was controlled on February 8, 1983 when the Key 3 relief well intercepted the 5-inch liner in the blowout at 15,941 feet true vertical depth (TVD)—the deepest intercept in the history of the industry. The intercept was made possible by the alternating current–active magnetic technology.

The passive magnetic technology of MAGRANGE was utilized first at the Key to direct the relief well effort. The early MAGRANGE interpretations conflicted with available data concerning the bottomhole location of the blowout. However, operations continued pursuant to the MAGRANGE interpretations and, at 15,570 feet measured depth, the relief well was discovered to be so far off course that the relief well had to be plugged back to 14,050 feet measured depth and re-drilled.

At a measured depth of 15,540 feet in the sidetrack, the MAGRANGE interpretation again conflicted with other available data. Based on the MAGRANGE interpretation, the recommendation was made to turn the relief well south to intercept the blowout (Figure 7.4). Realizing that a mistake would result in yet another sidetrack, the crew employed alternate techniques.

One alternative was WELLSPOT. The WELLSPOT interpretation was that the blowout was not 12 feet south as interpreted by MAGRANGE but 1.5 feet south. Further, it was interpreted from WELLSPOT data that the blowout would pass to the north of the relief well within the next 60 feet. Obviously, both interpretations could not be correct.

If the WELLSPOT interpretation was correct and the relief well was turned to the south as suggested by the MAGRANGE interpretation, a plug back would be required. If the MAGRANGE interpretation was correct and the relief well was not turned to the south, a side track would result. In order to resolve the conflict, 60 feet of additional hole were made and both surveys were re-run.

Both interpretations agreed that the blowout wellbore had passed from south to north of the relief well wellbore, confirming the WELLSPOT interpretation. Both techniques were used throughout the remainder of the relief well operation at the Apache Key. Neither was given benefit of the other's interpretation prior to offering its own interpretation.

The MAGRANGE and WELLSPOT interpretations conflicted in every aspect except direction at each subsequent logging point. Without exception, the WELLSPOT interpretation was proven to be correct.

RELIABILITY OF PROXIMITY LOGGING

How reliable is wellbore proximity logging? This question was investigated at the TXO Marshall relief well operation.[8] In this case, the first indications of the blowout wellbore were received at the relief well wellbore while the blowout was 200 feet away. The relief well intercepted the blowout at 13,355 feet. The plan view from interpretation of a gyro previously run in the blowout was utilized to evaluate the accuracy of the WELLSPOT interpretations. Table 7.1 and Figure 7.5 summarize the proximity logging runs. For the early runs, the blowout was about 30 percent further away than the interpretations predicted, although the direction given was always correct.

The greatest discrepancy occurred on the second run at 12,068 feet TVD. Pursuant to the plan view, the relief well was 122 feet away from the

Table 7.1

Proximity Logging Summary

Run No.	Date	Calculated distance, ft A-5 to A-1	Calculated direction, ° A-5 to A-1	Depth, ft MD	Depth, ft TVD	Actual distance, ft A-5 to A-1
1	Aug. 18	90 ± 15	318 ± 6	12,000	11,957	138
2	Aug. 22	50 ± 10	319 ± 4	12,112	12,068	122
3	Aug. 28	50 ± 10	317 ± 3	12,212	12,168	109
4	Aug. 31	65 ± 20	318 ± 4	12,341	12,295	94
5	Sept. 4	68 ± 10	319 ± 4	12,488	12,442	78
6	Sept. 8	48 ± 7	320 ± 5	12,625	12,578	65
7	Sept. 11	32 ± 7	319 ± 4	12,771	12,723	48
8	Sept. 15	26 ± 7	320 ± 4	12,884	12,836	37
9	Sept. 18	24 ± 5	316 ± 4	13,012	12,963	27
10	Sept. 23	11 ± 2	309 ± 4	13,142	13,093	17
11	Sept. 29	11 ± 2	285 ± 4	13,248	13,199	14
12	Oct. 2	6.5 ± 1.5	292 ± 4	13,300	13,250	8.5
13	Oct. 5	2 ± 0.75	295 ± 5	13,360	13,310	2.5

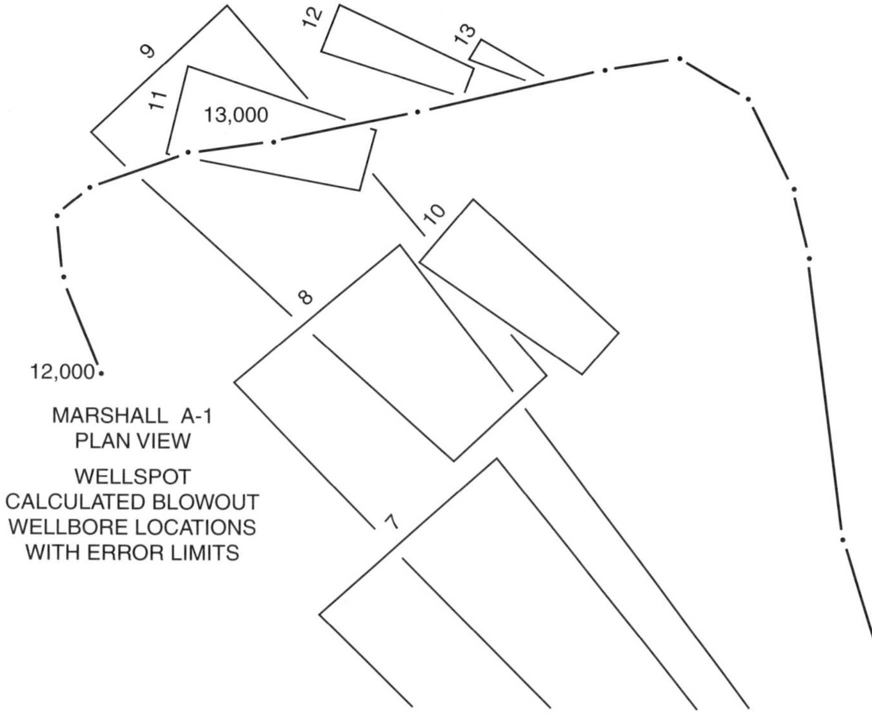

Figure 7.5

blowout as opposed to the 50 feet predicted from the proximity log interpretation. It was not until the ninth run and the wells had closed to within 30 feet that the two wellbores were actually within the prescribed error limits of the interpretation (see Figure 7.6). Further, two of the remaining four interpretations were in error in excess of the prescribed error limits. Interpretation is continuing to improve. However, the current state of the art dictates that the wellbore proximity log should be run frequently to insure the intercept, particularly as the wellbores converge.

At the Marshall, for example, 13 surveys were run in 1336 feet of drilled hole for an average of approximately 100 feet per survey. Given the present state of the art, survey intervals should not exceed 200 feet in order to accomplish a wellbore intercept.

As of this writing, the active magnetic technology has been improved by the inclusion of a gradient tool. This tool is particularly accurate within a few feet from the blowout wellbore.

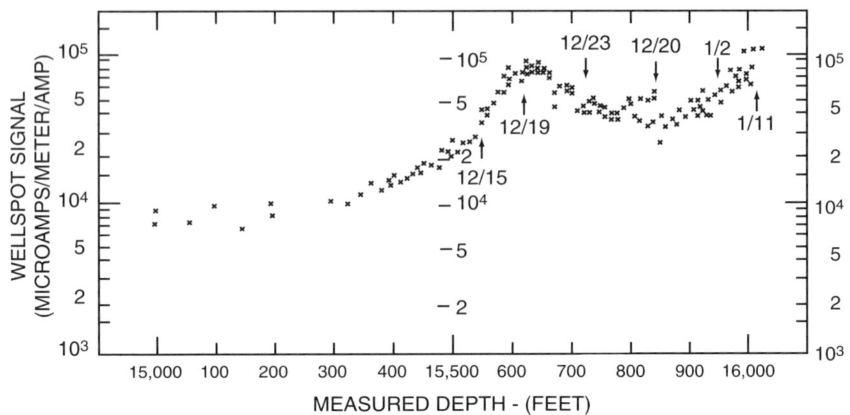

Figure 7.6   *Bottomhole Wellspot Signal Strength.*

Currently, all proximity logging requires that the drill string be removed from the hole, and the logs run on a conventional electric line. Currently, the technology to perform the proximity logging operation while drilling is evolving. The drawback is that the passive magnetic technology is used. As previously discussed, passive magnetic technology is much less reliable than active magnetics.

RELIABILITY OF COMMERCIAL WELLBORE SURVEY INSTRUMENTS

The accuracy and reliability of survey instruments are of great interest in relief well operations. For various reasons, the relief well sometimes has to be plugged back and re-drilled after contact has been made with the blowout. With what degree of accuracy, reliability, and repeatability can the relief well be plugged back and drilled to the position of the blowout as described by available survey equipment?

Several directional surveys were available at the TXO Marshall. Unfortunately, there were serious discrepancies among them. As illustrated in Figure 7.7, two north-seeking gyro surveys (3-21 and 3-26) and a magnetic single-shot were tied together at 11,800 feet and plotted. The two gyro surveys, which represent the state of the art as of this writing, were very discordant with one another even though they were run with the SAME TOOL, by the SAME OPERATOR and using the SAME WIRELINE.

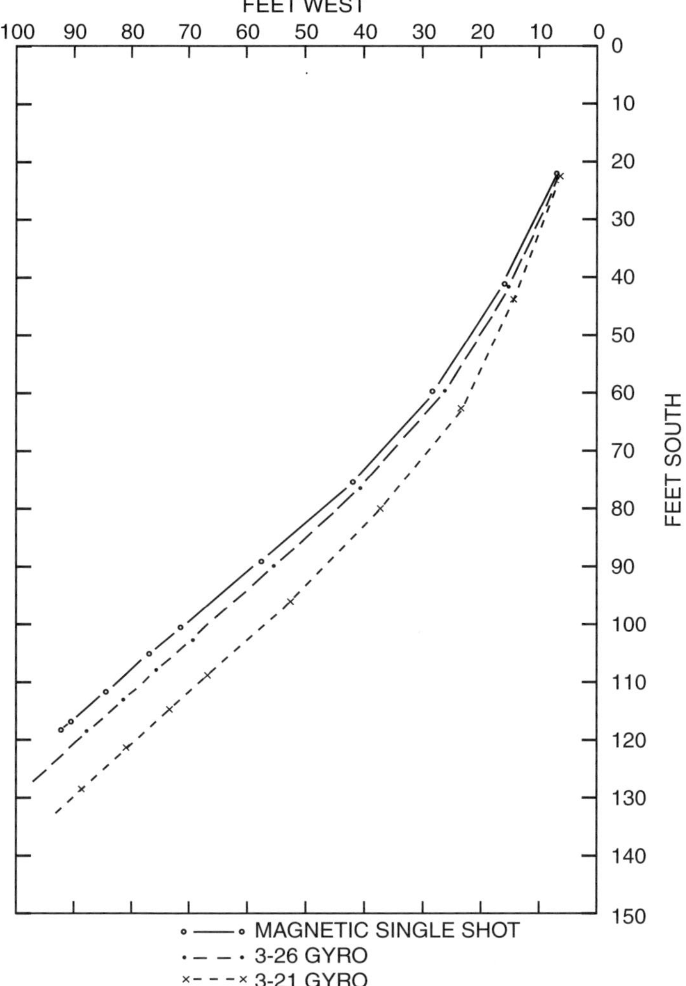

Figure 7.7　*Marshall A-1X Plan View At 12,600 Feet Relative to 11,800 Feet.*

After only 650 feet MD of surveying from 11,800 feet to 12,450 feet, the north-seeking gyros disagreed by 7 feet for an overall reliability of only 11 feet per 1000 feet. Further, the 3-21 gyro survey disagreed with the single-shot data by 10 feet in the 650 feet for an overall reliability of 15.7 feet per 1000 feet. The deviation in this portion of the hole ranged from 10 to 14 degrees. These values for reliability and repeatability are far worse than those normally quoted within the industry.

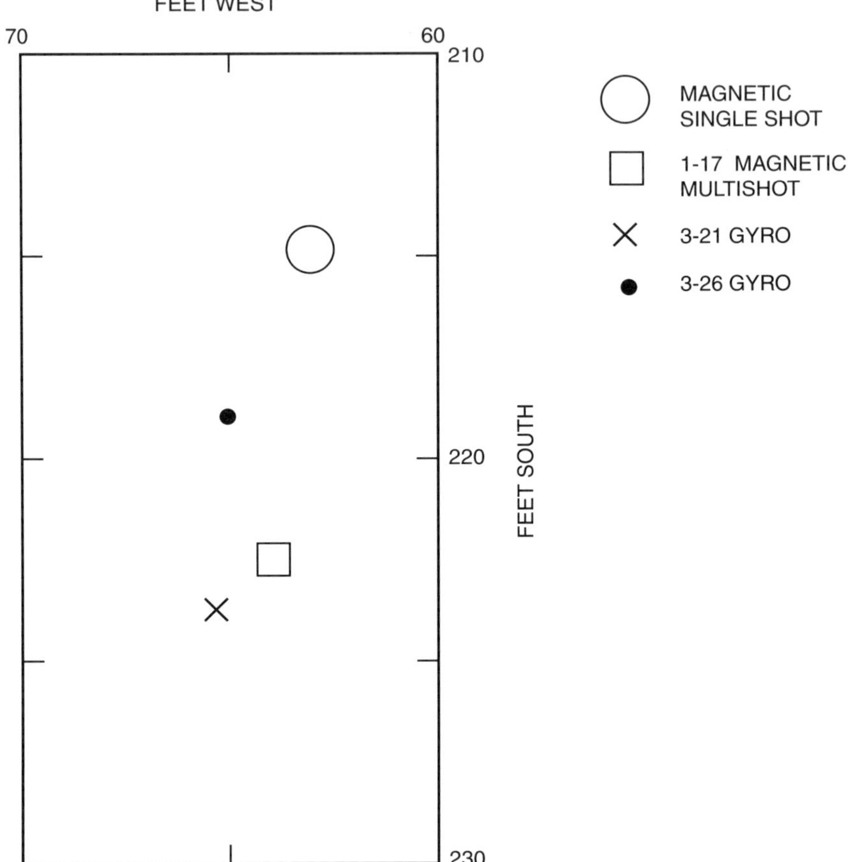

Figure 7.8   *Apparent Marshall A-1X Well Location at 11,500 Feet Relative to 10,650 Feet.*

Survey data were also available from other portions of the Marshall for comparison. In the upper part of the hole between 10,650 feet and 11,500 feet, there were four sets of survey data: a magnetic single-shot, a magnetic multi-shot (1-17), and the two north-seeking gyros (3-21 and 3-26). These data were plotted together at 10,650 feet and their deviations are illustrated as Figure 7.8.

Analysis of Figure 7.8 indicates similar systematic relationships between the two gyro runs and the magnetic single-shot data. The magnetic multi-shot results seem to support the 3-21 gyro data. The maximum

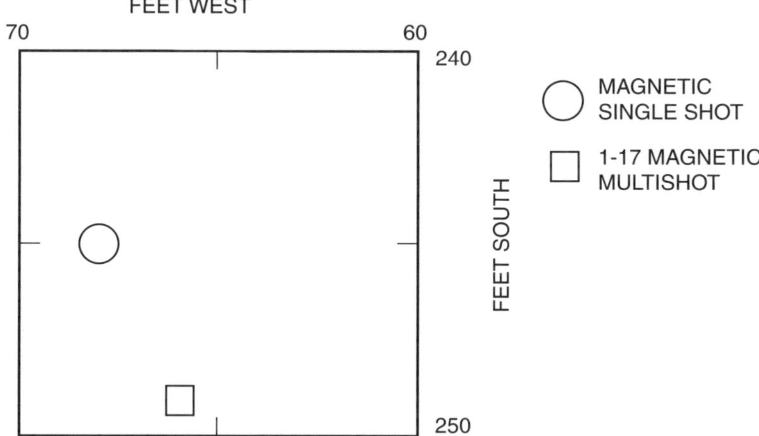

Figure 7.9　*Apparent Marshall A-1X Well Location at 12,300 Feet (Old Hole) Relative to 11,400 Feet.*

discrepancy is between the magnetic single-shot and the 3-26 gyro. As illustrated, these two surveys disagree by approximately 9 feet over the 850-foot interval for an overall reliability of 10.6 feet per 1000 feet.

In some instances, magnetic data compared better than the more expensive north-seeking gyro data. For example, the 1-17 magnetic multi-shot and the magnetic single-shot data were compared in the interval between 11,400 feet and 12,300 feet (see Figure 7.9). A 5-foot uncertainty is indicated over the 863-foot interval for a reliability of 5.8 feet per 1000 feet.

Several other comparisons were made and are presented in reference 9. Careful study of the survey data obtained in this instance makes it difficult to ascribe a survey precision of better than 10 feet of lateral movement per 1000 feet of survey. Further, since the survey errors were primarily systematic there was no reason to prefer the more expensive north-seeking gyroscopic survey over the more conventional magnetic survey. Typical north-seeking gyros are the Seeker, Finder, and Gyro Data gyros. Typical MWD magnetic instruments are Teleco and Anadril.

Each relief well operation is different and must be evaluated on its merits. However, in most cases, magnetic survey instruments are more appropriate for relief well operations than the more expensive north-seeking gyros. The MWD systems are more economical and expeditious during the directional operations. Further, as the well gets deeper, the MWD system offers several advantages. Surveys can record every joint, if necessary, to monitor deviation, direction, and bottomhole assembly behavior. Another significant advantage is that it is not necessary to wait idly for long periods without moving the drill string while a survey is run on a slick or braided wireline.

SUBSURFACE DISTANCE BETWEEN RELIEF WELL AND BLOWOUT

A critical question in relief well operations is how close to the blowout the relief well has to be. Prior to the previously discussed technology, the position of the relief well relative to the blowout was only a poorly educated guess. Consequently, successful relief well operations could not be assured. The standard procedure was to drill into the zone that was being produced in the blowout, manifold all the pumps in the world together, and pump like hell. Sometimes it worked and sometimes it didn't. A typical relief well pump pad where "close was good enough" is shown in Figure 7.10.

A fire broke out at Shell Oil Co.'s Platform "B" at Bay Marchand, offshore Louisiana, on December 1, 1970.[9] Of the 22 completed wells, 11 caught fire. Subsequent operations to control the blowouts were reported by Miller.[10] Production was from approximately 12,000 feet with an initial reservoir pressure of 6000 psia. The porosity in the reservoir was 29 percent with a permeability of 400 millidarcies. Conventional directional surveys were available on all the wells.

The distances between wells were determined by analyzing the conventional directional surveys and by running ULSEL logs in the relief wells. A simplified reservoir model was used to predict performance of the relief well control operations. Eleven relief wells were drilled. The distance between the relief well and the blowout ranged from 12 feet to 150 feet. In each instance the blowout zone was penetrated and water was pumped through the producing interval. Of the 11, four compared favorably with predicted performance. The relief wells were 12 to 18 feet from the blowouts. Water volumes to establish communication ranged from 450 barrels to 1340 barrels. One well was killed with water while the other three required as much as 1300 barrels of mud.

Figure 7.10

A second four compared less favorably but experienced no major problems in killing. The distances varied from 13 feet to 82 feet. Water volumes varied between 1250 barrels and 13,000 barrels. All four wells required mud with the volumes varying between 1187 barrels and 3326 barrels.

The remaining three wellbores had unfavorable comparisons with predicted performance. The water volumes exceeded 100,000 barrels. In one instance, pumping into the relief well resulted in no noticeable effect on the blowout and control had to be regained from the surface.

The Bay Marchand operation was one of the more successful efforts. Often, when an operation was successful, the success was due to luck as much as skill. The reasons are fairly obvious. Only the most prolific reservoir rocks demonstrate sufficient permeability to permit the flow of kill fluids containing mud solids of every imaginable size, including large quantities of barite guaranteed to be approximately 44 microns in size.

If the zone fractured, there was no good reason for the fracture to extend to the blowout. The blowout had to be killed with water before kill mud could be pumped. If the kill mud was unsuccessful in controlling the blowout, communication with the blowout was usually lost because of kill-mud gel strengths and barite settling. Under these conditions, communication might not be regained.

As wells were drilled deeper, the reservoir parameters, such as permeability and porosity, were even less favorable in terms of their ability to permit the flow of solids-laden kill fluids. In wells below 15,000 feet, it is not reasonable even to think about pumping kill fluids through the formations.

With the current technology, intercepting the wellbore is the preferred approach. With an interception, killing the blowout becomes predictable. Further, the predictability permits more precise design of the relief well. Kill rates and pressures can be determined accurately and tubulars can be designed to accomplish specific objectives. With an intercept, it is no longer necessary to manifold all the pumps in captivity, as many blowouts can be controlled with the rig pumps. At the Apache Key, for example, approximately 100 barrels of mud per hour were being lost to the blowout from the relief well, and control was instantaneous. With a planned intercept, it is no longer necessary to drill the largest diameter hole imaginable in order to pump large volumes of kill fluid at high rates.

Perforating between wellbores has drawn renewed interest with the introduction of tubing-conveyed perforating guns. There are very few charges available for conventional wireline perforating of sufficient size to perforate meaningful distances. In addition, orientation can be shown mathematically to be difficult depending on the size of the casings involved and the distance between casing.

However, using tubing-conveyed perforating guns, large charges can be run and oriented. Such a system was successfully used at Corpoven's Tejero blowout,[1] in northeast Venezuela. In that instance, it was reported that three attempts were required to communicate the relief well with the blowout.

In the first attempt, two 6-inch guns with 14,300-gram charges in each gun were run. The shots were aligned along the gun in three rows with a displacement between each row of 5 degrees. The two guns were connected with the center rows displaced 10 degrees. On the second attempt, two $4\frac{5}{8}$-inch oriented TCP guns were run with 72 27-gram charges. The center row was displaced 33 degrees. The relief well was believed to be within 2 feet of the blowout.

With a planned wellbore intercept, it is no longer mandatory that the relief well be drilled into the blowout zone. Other intervals in the blowout wellbore may offer more attractive targets. At both the Apache Key and the TXO Marshall, the intercepts were accomplished above the producing formations. At the Shell Cox, the intercept was accomplished approximately midway between the surface and total depth. At a recent North Sea operation, the intercept was in open hole approximately 100 feet below the end of the bit. At the Amerada Hess - Mil-Vid #3, the drill string was intercepted several hundred feet above the blowout zone and the relief well proceeded in direct contact with the drill string in the blowout until the final bridge was drilled and the blowout was killed. In that instance, less than 20 barrels of mud controlled the blowout. A planned intercept offers significant advantages.

SURFACE DISTANCE BETWEEN RELIEF WELL AND BLOWOUT

The distance at the surface between the relief well and the blowout is a function of overall project management; however, the closer the better. The cost of relief well operations is exponential with displacement. However, some projects are managed in such a fashion that the relief well project manager has no choice but to put the relief well a mile away in a bad direction. Ideally, overall project management will permit the relief well to be drilled within 1000 feet of the blowout with deviation angles below 15 degrees. The best direction is that which takes advantage of the regional drift and fracture orientation tendencies.

Some recommend that the relief well trajectory include a pass by the blowout prior to dropping angle for an intercept. The objective in this approach is to gain a better picture of the blowout trajectory and facilitate the intercept. As the blowout wellbore is passed by the relief well, its position can be triangulated from the relief well, thereby better establishing its position at that particular depth. Typically, ranging tools have more accurately determined direction than distance.

In my opinion, this is not the best strategy. In recent years a magnetic gradient tool has been developed which more accurately predicts distance when the separation is less than 5 feet. Therefore, the distance between the two well bores can be sufficiently determined without triangulation.

If a pass is to be made, the relief well design must include a higher angle. At the higher angle, the distance between the two wells decreases more rapidly. Since distance determination is less precise, there is less time to react. With less time to react, the relief well may get too close to the blowout too fast and with too much relative angle. In that event, drastic measures may be required or the relief well could pass too far beyond the blowout and not be able to achieve the planned intercept.

In addition, relief wells have been known to communicate with the blowout when the ranging tool predicted that the separation was several feet. This is not to say that the ranging tool was inaccurate since it could have been that the hole was enlarged at the point of intercept. In instances such as this, the outcome of the worst case scenario was that communication was established sooner than expected. In some instances, the premature communication made the blowout difficult to control. In one instance, it is doubtful that the blowout was controlled. (Because the flow was underground rather than at the surface, it was difficult to know for sure.)

SUMMARY

There are many other considerations in relief well operations that are related to overall project management. Large sums of money can be saved by the operator if overall management is considered. The relief well operations need to be coordinated with the total control effort. This will affect relief well location, for example, which can make or break an operation. Also, a relief well operation is not just a directional operation. It is a drilling operation, a well control operation, and a logging operation, as well as a directional operation. Needless to say, these considerations can impact the total cost significantly.

In summary, relief well technology has advanced to the extent that relief well operations are now a viable, reliable alternative in well control operations and should be considered in the overall planning and management of a blowout. A recent blowout at a deep, high-pressure well in the North Sea is a good example. More than a year of expensive surface work failed to provide a solution to the problem. After many expensive months, the blowout was finally controlled from the relief well. There is a good chance that the relief well would have been just as successful in the first 60 days of the operation.

RELIEF WELL PLAN OVERVIEW

1. Prepare a wellbore schematic pursuant to Figure 7.11.
2. Determine the approximate bottomhole location and preferential direction of drift as illustrated in Figure 7.12.
3. Select and build a location for the relief well approximately 820 feet northeast of the blowout.
4. Drill $17\frac{1}{2}$-inch hole to 3933 feet true vertical depth and reduce hole size to $12\frac{1}{4}$ inches for directional work.
5. Pick up motor, $1\frac{1}{2}$ degree bent sub, and MWD and build angle to 15 degrees at approximately 1.5 degrees per 100 feet.
6. Lay down motor, bent sub, and MWD and pick up holding assembly.
7. Drill to intermediate casing point at 4888 feet true vertical depth, holding angle at 15 degrees.
8. Open $12\frac{1}{4}$-inch hole to $17\frac{1}{2}$ inches.
9. Set and cement $13\frac{3}{8}$ inch casing. If kill considerations permit , reduce hole size to $12\frac{1}{4}$ inches and casing size to $9\frac{5}{8}$ inches.
10. Hold 15 degrees in the $12\frac{1}{4}$-inch hole to 6561 feet measured depth.
11. Run wellbore proximity log to detect blowout wellbore.
12. Make course corrections as required by the wellbore proximity log using motor, bent sub, and MWD.
13. Drill to 7090 feet measured depth and run wellbore proximity log.
14. Make course corrections as required by the wellbore proximity log using motor, bent sub, and MWD.

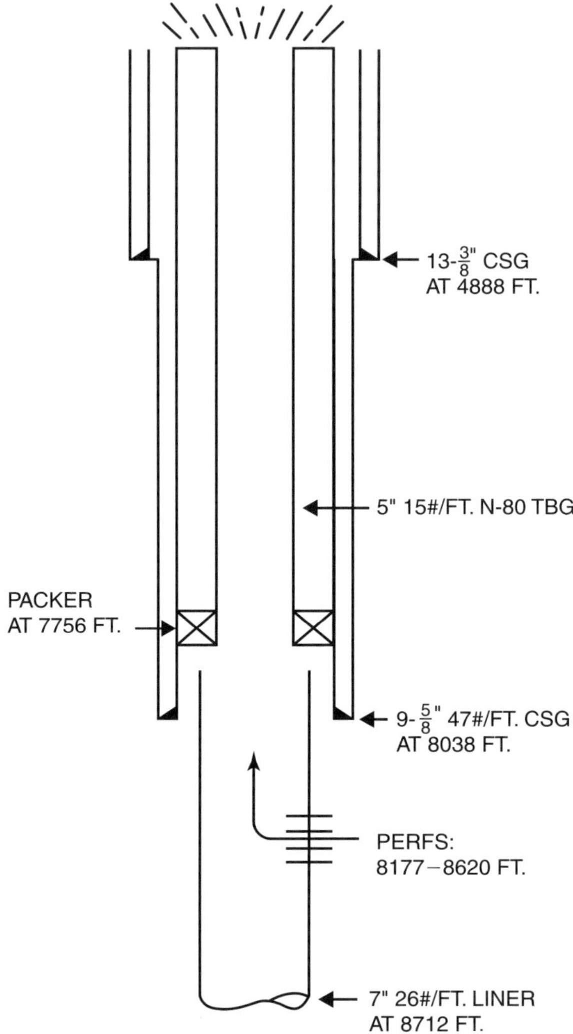

Figure 7.11 *Blowout Wellbore Schematic.*

15.   Using pendulum bottomhole assembly, drop angle to vertical. Run wellbore proximity log at 200-foot intervals until blowout wellbore is confirmed.

16.   Intercept the casing in the blowout well at 8000 feet true vertical depth.

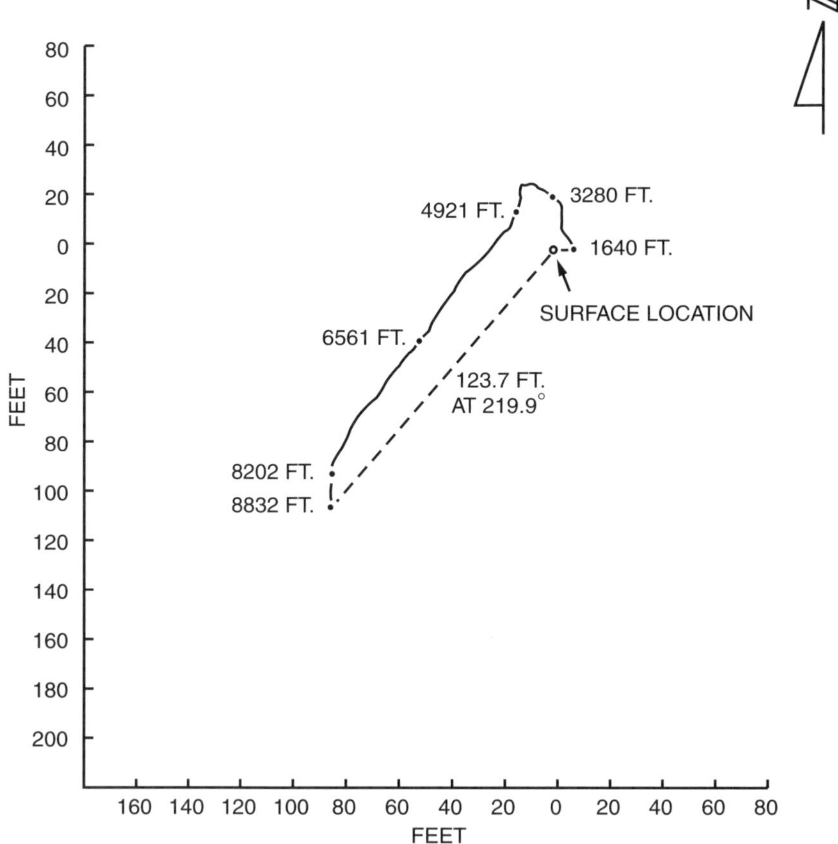

Figure 7.12   *Blowout Plan View.*

17.   Drill to casing point at 8038 feet true vertical depth and run the wellbore proximity log.

18.   Run $9\frac{5}{8}$-inch casing to 8038 feet true vertical depth. If kill considerations permit, reduce hole size to $8\frac{1}{2}$ inches and casing size to 7 inches.

19.   Drill out with $8\frac{1}{2}$-inch (or 6-inch) bit and run wellbore proximity log as required to intercept the blowout wellbore.

20.   Drill into the blowout wellbore (see Figure 7.13).

21.   Kill the blowout pursuant to plan and procedure.

820 FT.
KOP
AT 3933 FT.
13-3/8" CSG.
AT 4888 FT.
13-3/8" CSG.
AT 4888 FT.
MAXIMUM
ANGLE 15°
5" 15#/FT.
N-80 TBG
9-5/8" CSG.
AT 8038 FT.
TVD
PACKER
AT 7756 FT.
9-5/8" 47#/FT.
CSG AT 8038 FT.
PERFS:
8177-8620 FT.
7" 26#/FT. LINER
AT 8172 FT.

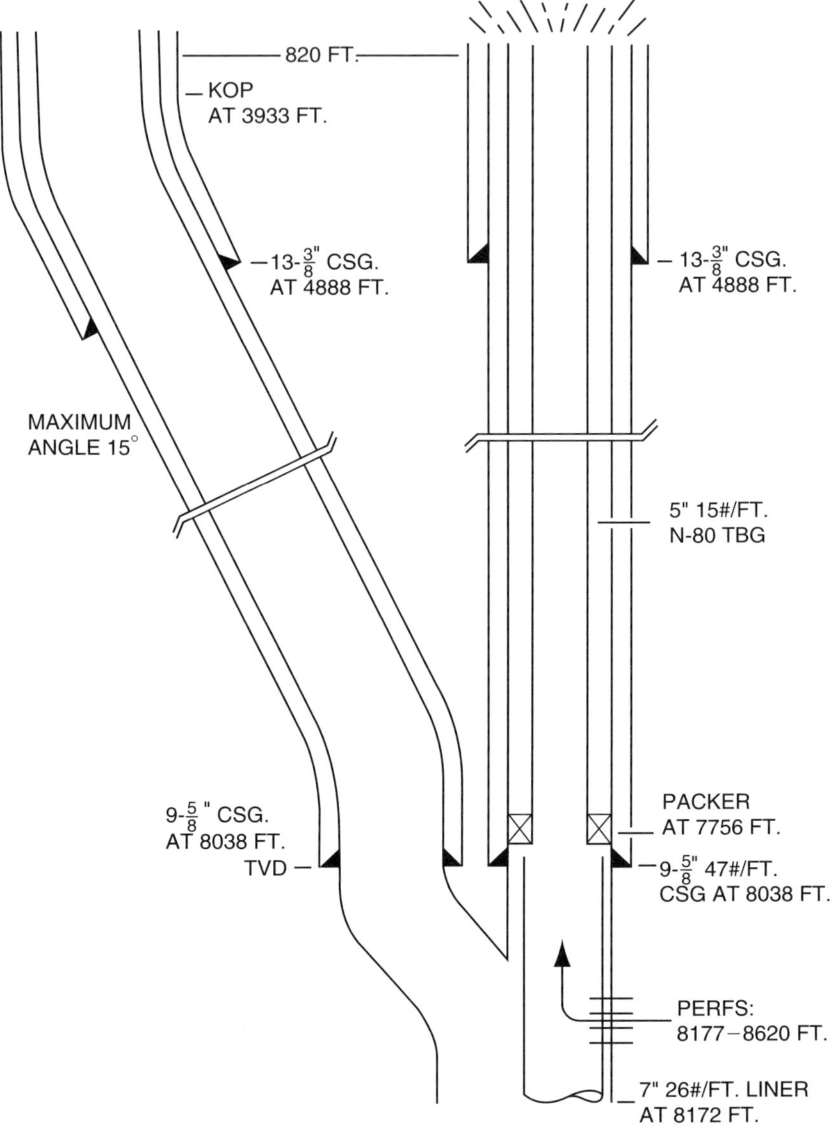

Figure 7.13

References

1.   Robinson, J.D. and Vogiatzis, J.P., "Magnetostatic Methods for Estimating Distance and Direction from a Relief Well to a Cased Wellbore," *Journal of Petroleum Technology*, June 1972, pages 741–749.

2.   Warren, Tommy M., "Directional Survey and Proximity Log Analysis of a Downhole Well Intersection," Society of Petroleum Engineers Number 10055, October 1981.

3.   Schad, Charles A., United States Patent #3,731,752 issued May 8, 1973, "Magnetic Detection and Magnetometer System Therefore."

4.   Morris, F.J., et al. United States Patent #4,072,200 issued February 7, 1978, "Surveying of Subterranean Magnetic Bodies from an Adjacent Off-Vertical Borehole."

5.   Morris, F.J., et al., "A New Method of Determining Range and Direction from a Relief Well to a Blowout," SPE 6781, 1977.

6.   Kuckes, Arthur F., United States Patent #4,372,398 issued February 8, 1983, "Method of Determining the Location of a Deep-Well Casing by Magnetic Field Sensing."

7.   Grace, Robert D., et al. "Case History of Texas' Largest Blowout Shows Successful Techniques on Deepest Relief Well," *Oil and Gas Journal*, May 20, 1985, page 68.

8.   Grace, Robert D., et al. "Operations at a Deep Relief Well: The TXO Marshall," SPE 18059, October 1988.

9.   Lewis, J.B., "New Uses of Existing Technology for Controlling Blowouts: Chronology of a Blowout Offshore Louisiana," *Journal of Petroleum Technology*, October 1978, page 1473.

10. Miller, Robert T. and Clements, Ronald L., "Reservoir Engineering Techniques Used to Predict Blowout Control During the Bay Marchand Fire," *Journal of Petroleum Technology*, March 1972, page 234.

CHAPTER EIGHT

THE UNDERGROUND BLOWOUT

An underground blowout is defined as the flow of formation fluids from one zone to another. Most commonly, the underground blowout is characterized by a lack of pressure response on the annulus while pumping on the drillpipe or by a general lack of pressure response while pumping. The underground blowout can be most difficult, dangerous, and destructive. It can be difficult because the conditions are hidden and can evade analysis. Often, the pressures associated with an underground blowout are nominal, resulting in a false sense of security.

It can be dangerous because some associate danger with sight. In many instances there is no physical manifestation of the underground blowout. If a well is on fire or blowing out at the surface, it commands respect. However, if the same well is blowing out underground, it is more easily ignored. Since the underground blowout is not seen, it is often not properly respected.

If the flow is to a shallow formation (less than 3000 feet), there is a very real possibility the flow will fracture to the surface. This possibility is greater in young sediments such as those common to coastal areas and offshore.

At a location in Bolivia, the flow broached to the surface and created a crater more than 100 meters in diameter (Figure 8.1). At least one wellhead, a workover rig, mud tanks, and several pump trucks disappeared into the crater and were never recovered.

Shallow underground blowouts in offshore operations can be even more dangerous. There is no place to run. If the flow fractures to the surface, the only safe getaway from the well is by helicopter. If the sea is gasified, boats lose their buoyancy, as do life preservers.

Another reason that shallow underground blowouts are more dangerous offshore is that the sediments are very young and the flow is more likely to surface immediately beneath the rig. Jackups and platforms are

Figure 8.1

the most vulnerable. If the crater is beneath the rig, it can destabilize the structure, causing it to topple.

Floating operations have a unique problem. A floating rig loses buoyancy in a gasified sea. Ship-shaped floating rigs have been lost under the described conditions. Semi-submersible rigs are more stable because the floatation devices are several feet beneath the surface of the sea.

I was on the deck of the rig shown in Figure 8.2 when the flow fractured to the surface. We were in only about 300 feet of water. The flow rate was in the hundreds of millions of cubic feet of gas per day. It is a miracle the rig didn't tip over as it listed several degrees—very near the critical list. The boil was as high as the deck and at times dwarfed the rig. I do not mind admitting that I was scared and thought my time had come.

The forces were so violent that the riser and drillpipe were sheared at the sea floor. After the well bridged, the crater was determined to be 400 meters in diameter and more than 100 meters in depth. There was never any sign of the subsea blowout preventer stack even using magnetometers.

Water depth is an ally in a situation such as this. As the depth increases, so does the likelihood that the plume will be displaced away from the rig by the current. In very deep water, the probability that the plume will be under or near the rig is remote. In that respect, deep water drilling is safer than shallow water drilling or drilling from a platform or jackup.

Figure 8.2

Underground blowouts are generally more challenging than surface blowouts. The volume of influx is not known nor is the composition. Further, the condition of the wellbore and tubulars which are involved are not reliably descriptive. The well control specialist is confronted with the necessity of analyzing and modeling the blowout and preparing a kill procedure. The tools of analysis and modeling are limited.

Additionally, the tools and techniques should be limited to only those absolutely necessary since any wireline operation is potentially critical. With the underground blowout, the condition of the wellbore can never be known with certainty and the risk of sticking or losing wire and tools is significantly increased. Stuck or lost wire and wireline tools can be fatal or at least limit future operational alternatives.

Since the consequences of an underground blowout can be severe, critical questions must be answered:

1. Should the well be shut in?
2. Should the well be vented to the surface?
3. Is the flow fracturing to the surface?
4. Can the losses be confined to a zone underground?

5. Is the casing capable of containing the maximum antici-
    pated pressures?

6. Should the casing annulus be displaced with mud or
    water?

7. If the casing annulus is to be displaced, what should be
    the density of the mud?

8. Is the flow endangering the operation and the personnel?

To vent or not to vent—that is the question! Personally, I love
to vent a well. Yet, apparently most are afraid to let a well blow at the
surface. It almost seems that flowing underground is somehow different
than flowing to the surface. Certainly a well has to be properly equipped
and rigged if it is to be vented at the surface, and most rigs are not. As a
result, re-rigging the choke manifold and flow lines is often the first thing
that has to be done.

At the TXO Marshall, the wellhead pressure was being held such
that part of the flow from the well was to the surface and part beneath
the surface. The quantity being lost was unknown. However, when relief
well operations were commenced, it was impossible to drill below approximately 1500 feet without blowing out. The Marshall was rigged up and
the surface pressure was drawn down below a few hundred pounds. The
measured production increased by about 15 mmscfpd and the underground
charge ceased. Relief well operations were conducted without incident.

Except when diverters are used, venting a well offshore is generally not possible. On platforms and jackups, it is very difficult to rig up
a satisfactory surface system. In floating drilling, any venting operation
would be hampered by the surface system and the small choke lines. In
addition, the sediments are incompetent and would most likely collapse if
the well were flowed very hard.

At the Trintomar Pelican Platform, a drilling well collided with a
producing well at a relatively shallow depth. Loss of the platform was a
legitimate concern. Fortunately, it was possible to vent the well through
the production system and prevent cratering beneath the platform.

The key to venting the well is this: The well must be flowed just
enough to eliminate the cross-flow downhole. How can we determine when
cross-flow is eliminated? It's simple. Just continue to open the well up
at the surface until the surface pressure begins to decline. At that point,

the preponderance, if not all, of the gas being produced will be produced at the surface.

Unlike classical pressure control, there are no solutions that apply to all situations. The underground blowout can normally be analyzed utilizing the surface pressures and temperature surveys. The noise log can be confusing. In all instances, the safety of the personnel working at the surface should be the first concern and the potential for fracturing to the surface must also be considered carefully.

The temperature survey is generally the best tool for analyzing the underground blowout. However, for several reasons analyzing temperature surveys is not always easy.

Typically, in my experience, the geothermal gradient is not well known in a given area. I always begin by establishing the geothermal gradient in the area of interest. This must be done prior to running the temperature survey. If not, everyone will try to fit the temperature survey to the presumed geothermal gradient or vice versa. Therefore, it is preferable to agree on the geothermal gradient prior to running a temperature survey.

In one instance, an engineer attempted to dismiss a temperature anomaly by saying that the geothermal gradient in the area was consistently experiencing a sudden ten-degree shift. Of course, the geothermal gradient does not suddenly shift several degrees naturally. He tried to offer MWD data in support of his premise. The only good source of geothermal gradient data is from a shut-in gas well. If a reliable gradient is not available, shut in a nearby well for 72 hours and run a temperature survey to establish the gradient.

Another common problem is thermal instability. In my experience, the biggest mistake in running temperature surveys is running the survey too soon after pumping. Temperature surveys must not be run until the conditions in the well have stabilized. If the well has not reached thermal equilibrium, any attempt to analyze temperature information is pure speculation. I've seen good engineers argue for hours over a survey run in unstable conditions. If the well is not in thermal equilibrium, anyone's guess is as good as anyone else's guess—but all are just guesses.

Thermal equilibrium can usually be obtained in a matter of a few hours. If in doubt, wait a couple of hours and re-run the survey. The temperature tool should be run first in a suite of logs and the values recorded while logging down. Data obtained by logging up or logging again immediately after the first run are worthless.

Another common and often fatal mistake is to interpret the data from the typical log presentation. In so doing, huge anomalies have been overlooked and, in some instances, small, meaningless fluctuations have been misinterpreted as severe problems. I much prefer to obtain the data digitally—even if I have to read it off the log—and present it on a spread-sheet and in graphic form such as presented in this text. These presentations take a little time and work but are worth the effort.

Usually, there is considerable argument about the appearance of a temperature anomaly. Several variables will affect the appearance of the anomaly. Its appearance is a function of the flow path, the density of the flow stream, the volume rate of flow, and the distance between the flowing zone and the zone of accumulation.

Normally, the flow is outside the drill string. In that case, the temperature profile can be anticipated to look like Figure 8.3. As shown in Figure 8.3, the temperature profile is affected by the temperature of the hot fluid leaving bottom and the temperature of the surrounding environment. The measured temperature will not be as high as the temperature of the flowing fluid and will not be as low as the surrounding environment. Rather, the measured temperature will be a compromise between all the thermal bodies detected by the measuring device. Further, the surrounding environment will never warm to the temperature of the flowing fluid, and the flowing fluid will never completely cool to the temperature of the surrounding environment—as long as the well continues to flow.

The flowing fluid stream will lose heat as it moves up the well according to the established laws of thermodynamics. A small flow will not retain as much heat as a large flow and, as a result, will not deviate as far from the established geothermal gradient. The density of the flow will affect the size of the anomaly. Gas does not retain heat as well as oil or water.

When the flowing fluid reaches the zone of accumulation, the volume of flowing fluid becomes larger, relative to the surrounding environment, and the temperature device will record an increase in temperature. Of course, the flow stream is not warmer at the zone of accumulation. There is simply relatively more fluid volume.

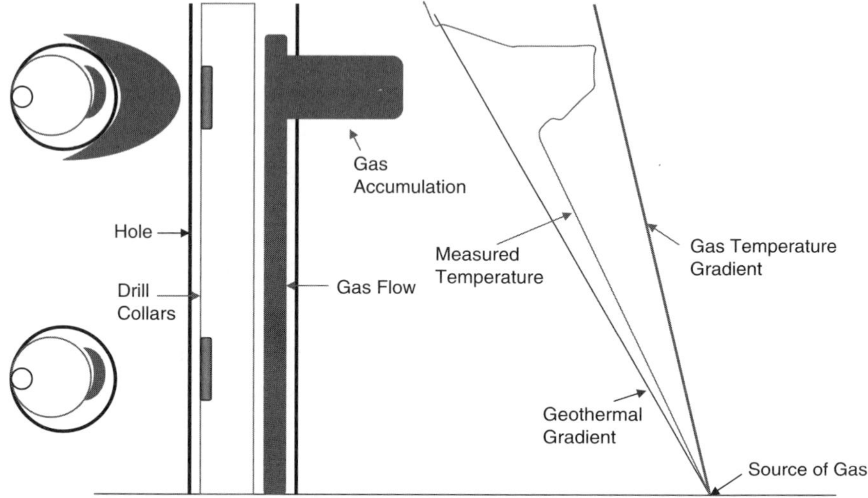

1 - Temperature Tool records a compromise between gas gradient and geothermal gradient. Flow and heat transfer reaches steady state. Recorded temperature is not as high as the temperature of the gas or as low as the temperature of the formation

2 - Temperature Tool records a compromise between gas gradient and geothermal gradient. In this case, the large accumulation of gas causes the temperature tool to record a higher temperature. The recorded temperature is not as high as the temperature of the gas and not as low as the temperature of the formation

3 - Above the zone of loss, the temperature tool is only influenced by the formations surrounding the tool

Figure 8.3   *Theoretical Underground Blowout Temperature Profile.*

The size of the temperature anomaly in the zone of accumulation is also a function of the same variables. The higher the flow rate, the greater the density of the flow stream, the farther the distance between the flowing zone and the zone of accumulation, the bigger the anomaly.

A typical temperature profile for an underground blowout flowing outside the drill string is shown and explained in Figure 8.4. In this instance the anomaly deviated from the established geothermal gradient by more than 50 degrees Fahrenheit. More than 5000 feet separated the zone of accumulation from the blowout zone, and the well was in an area famous for flowing large volumes of gas. Many temperature anomalies are not this dramatic.

The temperature profile for a strong water flow in North Africa is shown in Figure 8.5. Since water retains heat, there is a one-degree centigrade drop in temperature between the source at 700 meters and the surface.

Figure 8.27 (see page 370) illustrates the difference between a temperature profile of a flow outside the drill string compared to that of

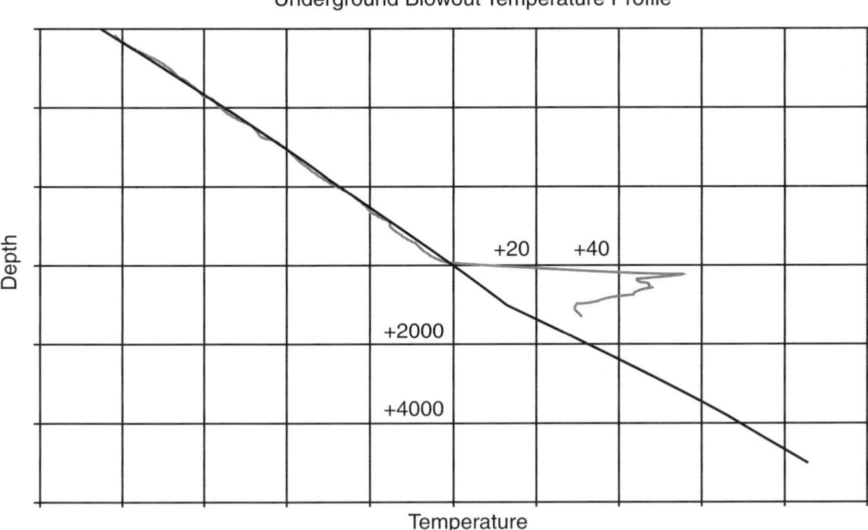

Figure 8.4   *Actual Underground Blowout Temperature Signature.*

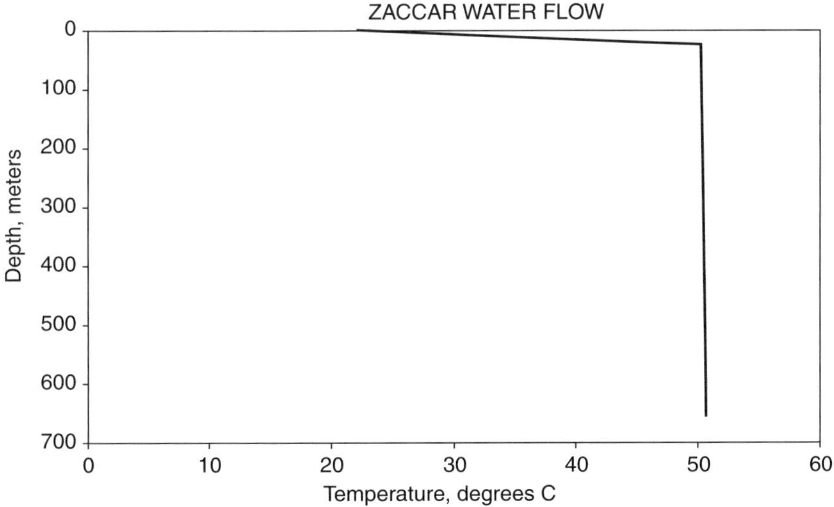

Figure 8.5   *Temperature Profile, Zaccar Water Flow, Algeria.*

a flow inside the drill string. The profile noted as "1st Run" was obtained when the flow was outside the drill string. The well was subsequently killed but the inside of the drill string was not adequately plugged.

Consequently, when the inside of the drill string was cleaned out, the well kicked off up the drill string. As shown in the depiction of the second blowout in Figure 8.27, the temperature anomaly was much greater compared to the first run and had different characteristics. The primary difference is that the local environment does not as greatly influence the measured temperature. Therefore, the measured temperature will simply increase linearly with respect to the maximum, and there will not be a large anomaly in the zone of accumulation.

It is obviously important to assess the hazards associated with the conditions at the blowout. In temperature survey analysis, it should be noted that the temperature of the flowing fluid will be essentially the same as the temperature of the reservoir from which it came. Therefore, if the flow is from a deep formation into a shallow formation, there should be an abnormally high temperature in the zone of loss.

Surface pressures are a reflection of the conditions downhole. If the surface pressure is high, the zone of loss is deep. Conversely, if the surface pressure is low, the zone of loss is shallow. If the density of the annular fluids is known, the depth to the zone of loss can be calculated.

The noise log is helpful in some instances. The flow of fluids can generally be detected with sensitive listening devices. However, in some instances the blowout goes undetected by the noise log while in other instances the interpretation of the noise log indicates the presence of an underground flow when there is none. The application of these principles is best understood by consideration of specific field examples.

CASING LESS THAN 4000 FEET

With the casing set at less than 4000 feet, the primary concern is that the underground blowout will fracture to the surface and create a crater. If the blowout is offshore, it is most probable that the crater will occur immediately under the drilling rig. If the productivity is high, then the crater will be large, and the operation in great peril.

At one operation in the Gulf Coast, several workers were burned to death when the flow fractured to the surface under the rig. At another operation in the Far East, a nine-well platform was lost when the flow fractured to the surface under the platform. At still another operation, a jackup was lost when the crater occurred under one leg. Drill ships have been lost to cratered blowouts.

The blowout at the Pelican Platform offshore Trinidad is a good example. Pelican A-4X was completed at a total depth of 14,235 feet measured depth (13,354 feet true vertical depth). The well was contributing 14 mmscfpd plus 2200 barrels condensate per day at 2800 psi flowing tubing pressure. Bottomhole pressure was reported to be 5960 psi. The wellbore schematic for the A-4X is shown in Figure 8.6.

At the Pelican A-7, $18\frac{5}{8}$-inch surface casing was set at 1013 feet and cemented to the surface, and drilling operations continued with a $12\frac{1}{4}$-inch hole. The wellbore schematic for the Pelican A-7 is presented as Figure 8.7.

The directional data indicated that the A-7 and A-4X were approximately 10 feet apart at 4500 feet. However, in the early morning hours, the A-7 inadvertently intercepted the A-4X at 4583 feet. The bit penetrated the $13\frac{3}{8}$-inch casing, $9\frac{5}{8}$-inch casing and $4\frac{1}{2}$-inch tubing. Pressure was lost at the A-4X wellhead on the production deck and the A-7 began to flow. The A-7 was diverted and bridged almost immediately. The A-4X continued to blowout underground. With only 1013 feet of surface pipe set in the A-7, the entire platform was in danger of being lost if the blowout fractured to the sea floor under the platform.

After the intercept, the shut-in surface pressure on the tubing annulus at the A-4X stabilized at 2200 psi. The zone into which the production is being lost can be approximated by analyzing this shut-in pressure. The pressure may be analyzed as illustrated in Example 8.1:

Example 8.1

Given:

Fracture gradient, $F_g = 0.68$ psi/ft

Compressibility factor, $z = 0.833$

Required:

The depth to the interval being charged.

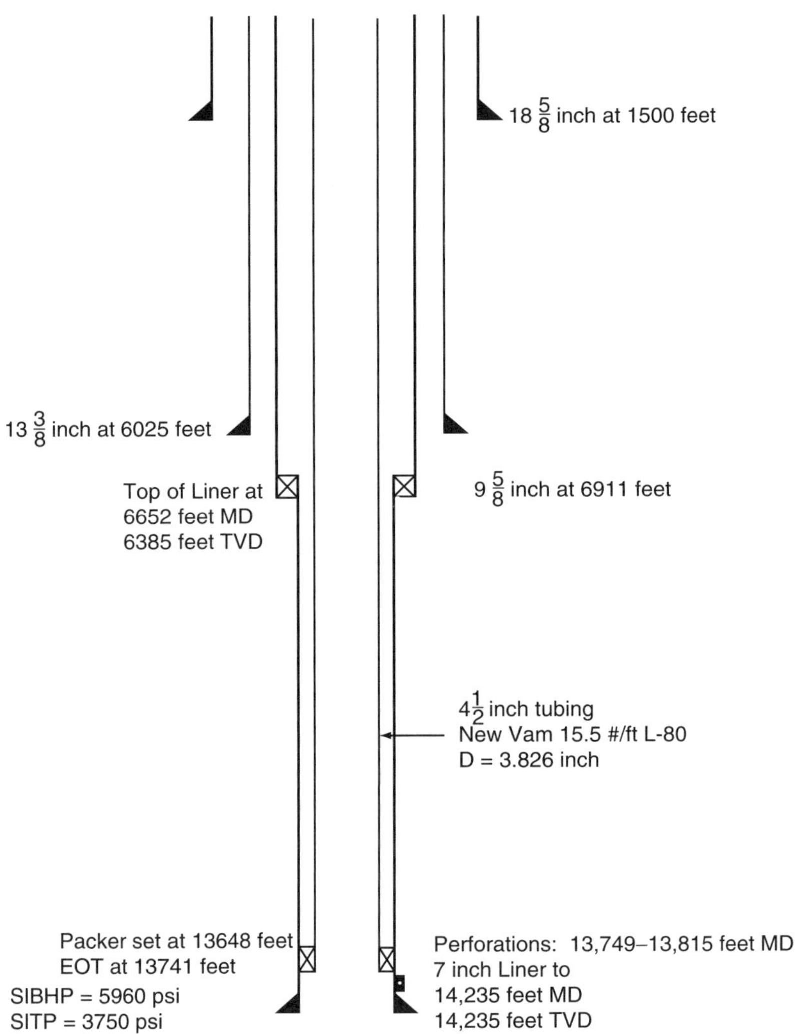

Figure 8.6   Pelican A-4X.

Solution:

The gas gradient, ρ_f, is given by Equation 3.5:

$$\rho_f = \frac{S_g P}{53.3zT}$$

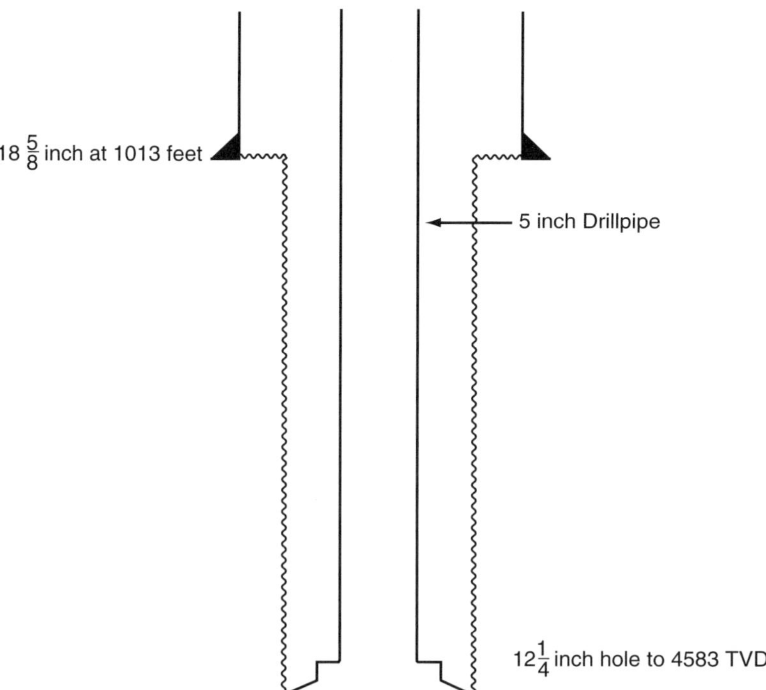

Figure 8.7   *Pelican A-7.*

Where:

$$S_g = \text{Specific gravity of the gas}$$

$$P = \text{Pressure, psia}$$

$$z = \text{Compressibility factor}$$

$$T = \text{Temperature, }^\circ\text{Rankine}$$

$$\rho_f = \frac{0.6(2215)}{53.3(0.833)(580)}$$

$$\rho_f = \textbf{0.052 psi/ft}$$

The depth to the interval being charged is given by Equation 8.1:

$$F_g D = P + \rho_f D \tag{8.1}$$

or, rearranging

$$D = \frac{P}{F_g - \rho_f}$$

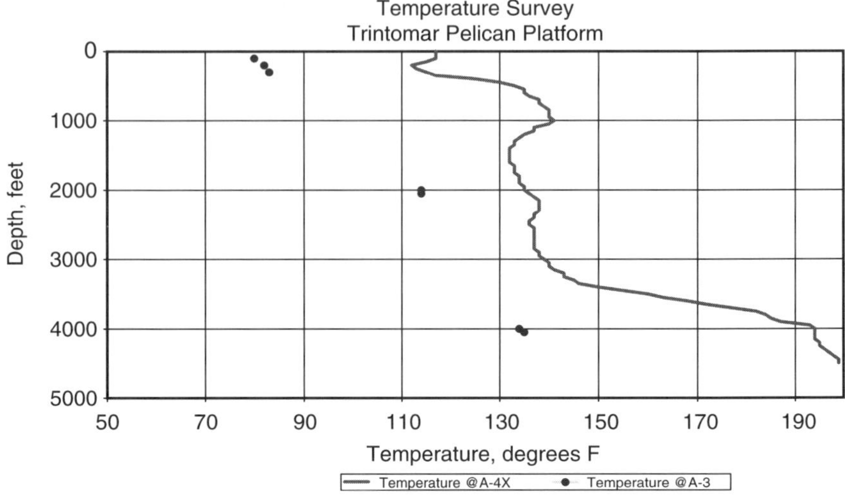

Figure 8.8

$$D = \frac{2215}{0.65 - 0.052}$$

$$D = \textbf{3704 feet}$$

This analysis of the shut-in surface pressure data indicated that the flow was being lost to a zone at approximately 3700 feet.

As confirmation, a temperature survey was run in the A-7 and is presented as Figure 8.8. Also included in Figure 8.8 are static measurements from the A-3 which were utilized to establish the geothermal gradient. The interpretation of the temperature data was complicated by the fact that the flow path of the hydrocarbons being lost was from the A-4X into the A-7 wellbore and ultimately into a zone in the A-7 wellbore.

The high temperatures at 3600 feet shown in Figure 8.8 were as expected and consistent with the pressure data analyses indicating the zone of charge to be at approximately 3700 feet. Pursuant to the analysis of the offset data, the normal temperature at 3600 feet would be anticipated to be approximately 130 degrees Fahrenheit. However, due to the flow of the gas into the interval at 3600 feet, the temperature at that zone had increased 45 degrees to 175 degrees.

By similar analysis, the heating anomaly from 500 to 1000 feet could be interpreted as charging of sands between 1000 feet and the sea

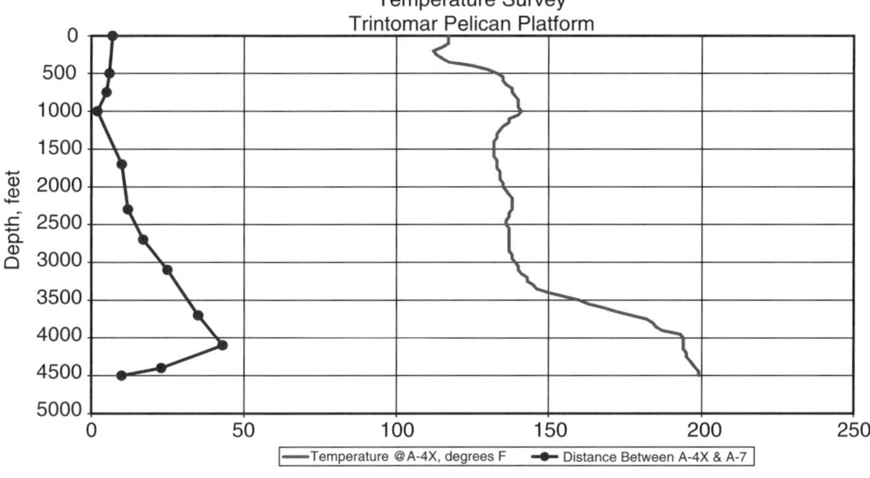

Figure 8.9

floor. Pursuant to that interpretation, cratering and loss of the platform could result.

Further analysis was warranted. When evaluated in conjunction with the relative position of the two wells, the condition became apparent. Figure 8.9 illustrates the temperature profile in the A-7 overlain by the directional survey analysis of the relative distance between the wellbores.

This figure further confirms the previous pressure and temperature analyses. The two wells are the greatest distance apart, which is 45 feet, at 4100 feet. That depth corresponds with the most pronounced anomaly, confirming the conclusion that the thermal primary zone of loss is below 3600 feet.

As illustrated, the wellbores are interpreted to be 2 feet apart at 1000 feet and 5 feet apart at the sea bed. Therefore, the temperature anomaly above 1000 feet was interpreted to be the result of the proximity of the two wellbores and not caused by the flow of gas and condensate to zones near the sea bed.

Based on the analyses of the surface pressure and the temperature data, it was concluded that working on the platform was not hazardous and that the platform did not have to be abandoned. As further confirmation, no gas or condensate was observed in the sea around the platform at any time during or following the kill operation.

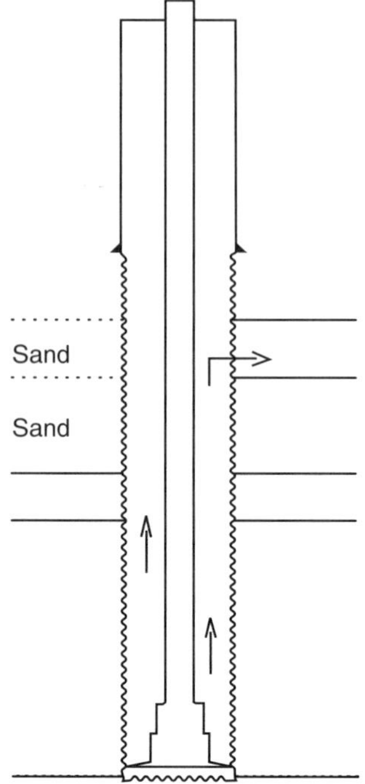

Figure 8.10   *Offshore Underground Blowout.*

At the Pelican Platform the surface pressures remained constant. When the surface pressures remain constant, the condition of the wellbore is also constant. However, when the surface pressures fail to remain constant, the conditions in the wellbore are, in all probability, changing and causing the changes in the surface pressures.

Consider an example of an underground blowout at an offshore drilling operation. With only surface casing set, a kick was taken and an underground blowout ensued. The pressures on the drillpipe and annulus stabilized, and analysis pursuant to the previous example confirmed that the loss was into sands safely below the surface casing shoe (Figure 8.10).

The pressure history is presented as Figure 8.11. As illustrated, after remaining essentially constant for approximately 30 hours, both pressures began to change rapidly and dramatically, which indicated that the conditions in the wellbore were also changing rapidly and dramatically.

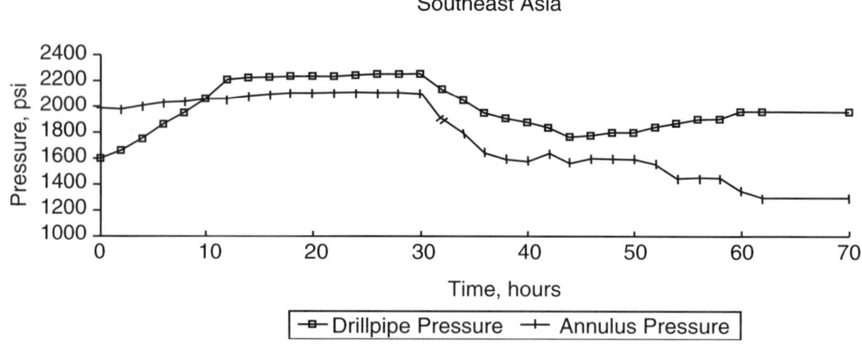

Figure 8.11

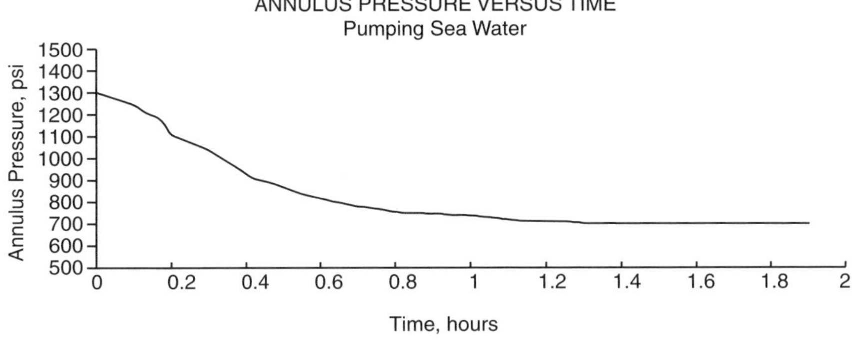

Figure 8.12

The pressure changes were confusing and not readily adaptable to analysis, and several interpretations were possible.

The declining pressures could have indicated that the wellbore was bridging or that the flow was depleting. Further, a change in the composition of the flow could have contributed to the change in the pressures. Finally, a decline in annulus pressure could have been the result of the flow fracturing toward the surface.

In an effort to define the conditions in the wellbore, a more definitive technique was used to determine the precise depth of the loss from the underground blowout. With the well shut in, sea water was pumped down the annulus at rates sufficient to displace the gas. As illustrated in Figure 8.12, while pumping, the annulus pressure declined and stabilized. Once the pumps were stopped, the annulus pressure began to increase. With this data, the depth to the loss zone could be determined using Equation 8.2:

$$D = \frac{\Delta P}{\rho_{sw} - \rho_f} \tag{8.2}$$

Consider Example 8.2:

Example 8.2

Given:

Sea water gradient, $\rho_{sw} = 0.44$ psi/ft

Gas gradient, $\rho_f = 0.04$ psi/ft

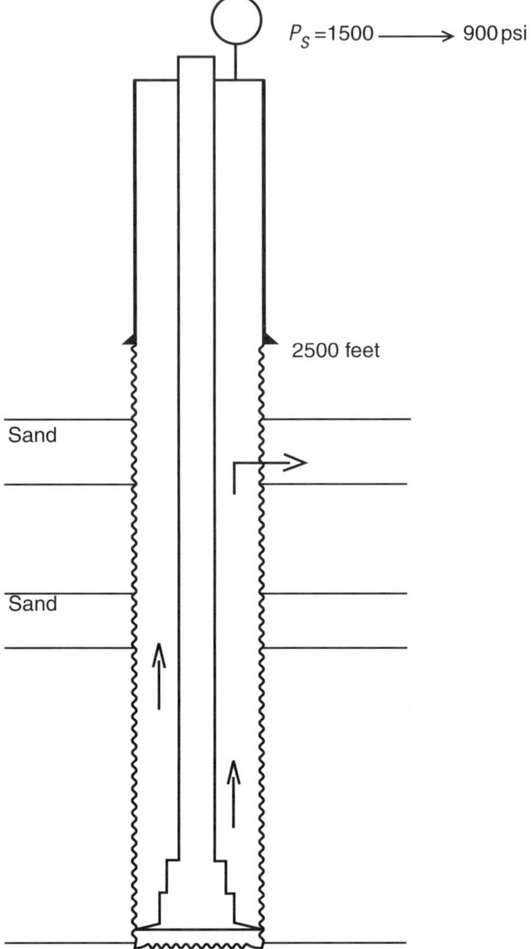

Figure 8.13   *Offshore Underground Blowout.*

Sea water is pumped into the well shown in Figure 8.13 and the surface pressure declines from 1500 to 900 psi.

Required:

The depth to the zone of loss.

Solution:

The depth to the zone of loss is given by Equation 8.2:

$$D = \frac{\Delta P}{\rho_{sw} - \rho_f}$$

$$D = \frac{600}{0.44 - 0.04}$$

$$D = \textbf{1500 feet}$$

As illustrated in Example 8.2, replacing the hydrostatic column of wellbore fluids from the zone of loss to the surface with a hydrostatic column of sea water only reduced the surface pressure by 600 psi. Therefore, the length of the column of sea water between the surface and the zone of loss could be only 1500 feet, which would be 1000 feet higher than the surface casing shoe. The obvious conclusion would be that the flow was fracturing to the surface. Continuing to work on the location would not be safe. In the actual situation, the flow fractured to the sea floor beneath the rig the following day.

In hard rocks the flow may fracture to the surface at any point. At the Apache Key, shown in Figure 8.14, the well cratered at the wellhead. In the Sahara Desert near the community of Rhourde Nouss, Algeria, the flow cratered a water well 127 meters away from the well (Figure 8.15). It is not uncommon in desert environments for the flow to surface in numerous random locations. It is equally common for the gas to percolate through

Figure 8.14

Figure 8.15

the sand. At Rhourde Nouss, the hot gas would auto-ignite when it reached the desert floor, producing an eerie blue glow and small fires in the sand (Figure 8.16).

The wellbore configuration at Rhourde Nouss is illustrated in Figure 8.17. It was critical to know the location of the holes in the tubulars. Since there was flow to the surface, holes had to be present in the 5-inch tubing, $9\frac{5}{8}$-inch casing, and $13\frac{3}{8}$-inch casing. A temperature survey run in the $2\frac{3}{8}$-inch tubing, which had been run into the well in a kill attempt, is illustrated in Figure 8.18.

As can be seen, the temperature survey is not definitive. Often a change in temperature or delta temperature can be definitive when the temperature survey alone is not. The delta temperature survey is usually plotted as the change in temperature over a 100-foot interval. The delta survey is then compared with the normal geothermal gradient, which is usually 1.0 to 1.5 degrees per 100 feet. A greater change than normal denotes a problem area.

At Rhourde Nouss, when the change in temperature as presented in Figure 8.19 was analyzed, the tubular failures became apparent. The hole in the $5\frac{1}{2}$-inch tubing is the dominant anomaly at 560 meters. The holes in

Figure 8.16

the $9\frac{5}{8}$-inch casing and the $13\frac{3}{8}$-inch casing are defined by the anomaly at 200 meters and 60 meters, respectively.

PIPE BELOW 4000 FEET

With pipe set below 4000 feet, there is no reported instance of fracturing to the surface from the casing shoe. There are instances of fracturing to the surface after the casing strings have ruptured. Therefore, maximum permissible casing pressures must be established immediately and honored. The maximum annulus pressure at the surface will be the fracture pressure at the shoe less the hydrostatic column of gas.

The wellbore schematic for the Amerada Hess Mil Vid #3 is presented as Figure 8.20. An underground blowout followed a kick at 13,126 feet. The 5-inch drillpipe parted at the $9\frac{5}{8}$-inch casing shoe at 8730 feet. During the fishing operations that followed, a temperature survey was run inside the fishing string.

The temperature survey is presented as Figure 8.21. The 85-degree temperature anomaly at 8700 feet confirmed the underground blowout. It is interesting to note that the temperature decreased below the top of

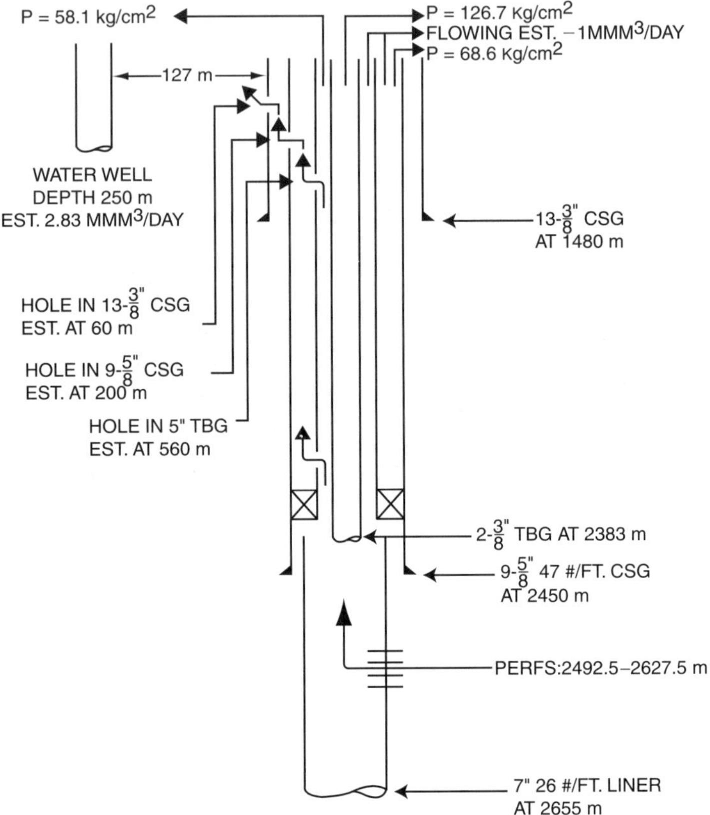

Figure 8.17   *RN-36 Blowout, 24 June 1989.*

the drillpipe fish at 8730 feet. This anomaly established that the flow was through the drillpipe and that the annulus had bridged.

It is equally interesting that numerous noise logs run in the same time period failed to detect the underground flow. The noise log run on the Mil Vid #3 is presented as Figure 8.22.

The fracture gradient was measured during drilling to be 0.9 psi/ft. At offset wells the gas gradient was measured to be 0.190 psi/ft. Utilizing this data, the maximum anticipated surface pressure was determined using Equation 8.3:

$$P_{\max} = (F_g - \rho_f)D \qquad (8.3)$$

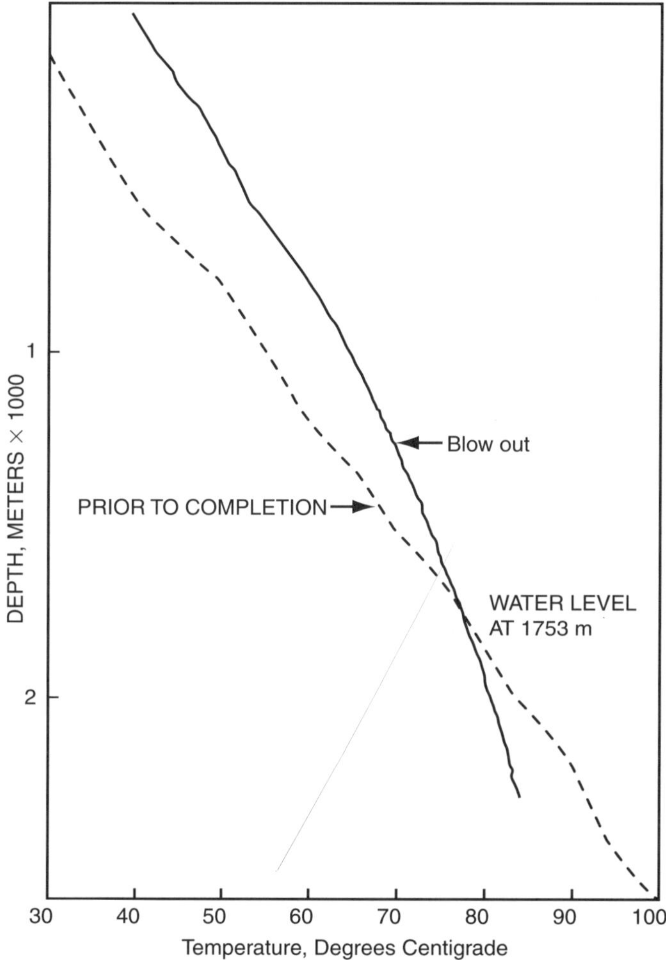

Figure 8.18   *Temperature Survey Comparison.*

$$P_{\max} = (0.90 - 0.19)8730$$

$$P_{\max} = \mathbf{6198\ psi}$$

Although the calculated maximum anticipated surface pressure was only 6200 psi, during subsequent operations the surface pressure was as high as 8000 psi, indicating that the zone of loss was being charged and pressured. Therefore, the actual surface pressure could be much more than the calculated maximum, depending on the volume of flow and the character of the zone of loss.

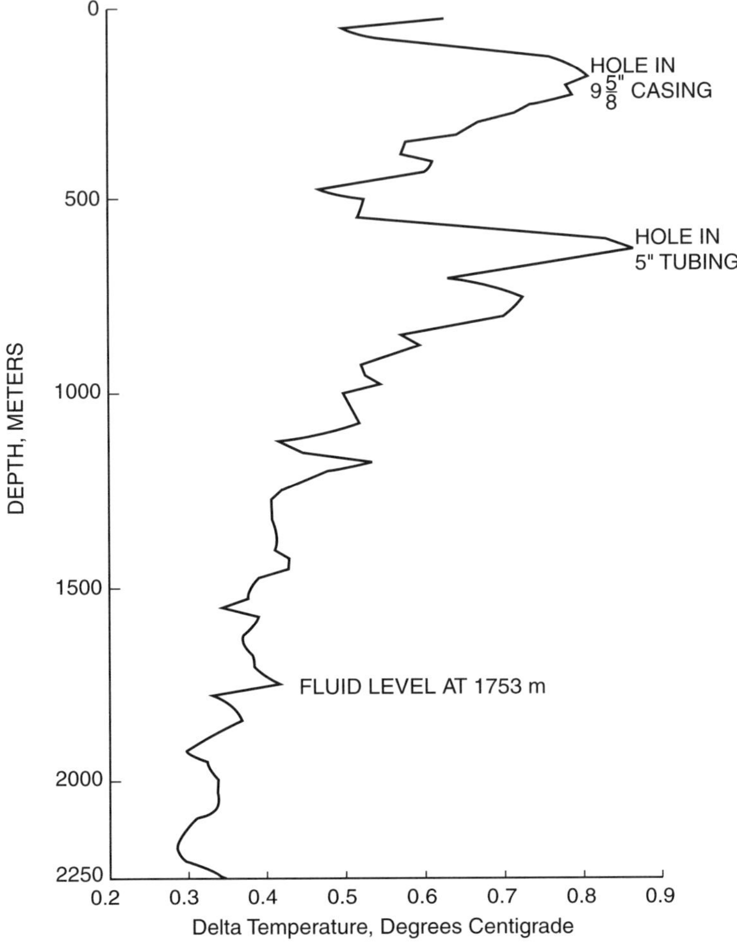

Figure 8.19   *Delta Temperature versus Depth.*

Once the maximum anticipated surface pressure has been determined, there are three alternatives. The well can remain shut in, provided there is no concern for the integrity of the tubulars. If the pressure is higher than can be tolerated, the well can be vented at the surface, provided that the surface facilities have been properly constructed. Finally, mud or water can be pumped down the annulus to maintain the pressure at acceptable values.

Since the anticipated surface pressures were unacceptable and the Mil-Vid #3 was located within the town of Vidor, Texas, the only alternative was to pump mud continuously into the drillpipe annulus. Accordingly,

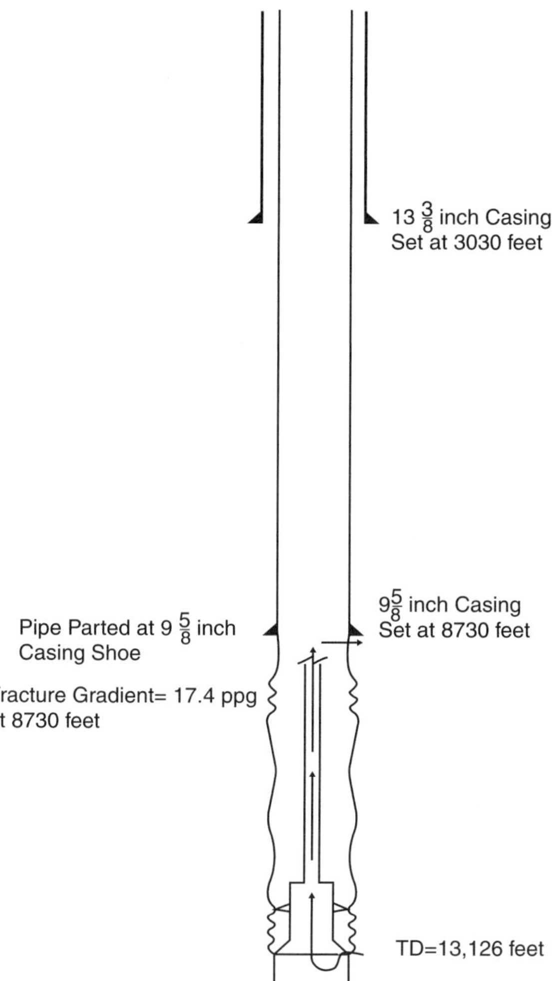

Figure 8.20   *Amerada Hess Mil-Vid #3.*

16-ppg to 20-ppg mud was pumped continuously down the drillpipe annulus for more than 30 days—an expensive but necessary operation. However, the surface pressures were maintained below an acceptable 1000 psi.

For another comparison between the noise log interpretation and the temperature survey, consider the well control problem at the Thermal Exploration Sagebrush No. 42-26 located in Sweetwater County, Wyoming. The wellbore schematic is presented as Figure 8.23. During

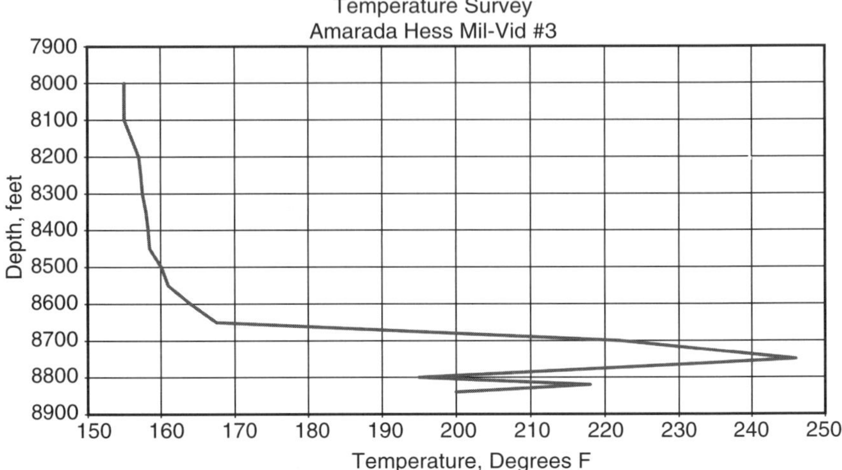

Figure 8.21

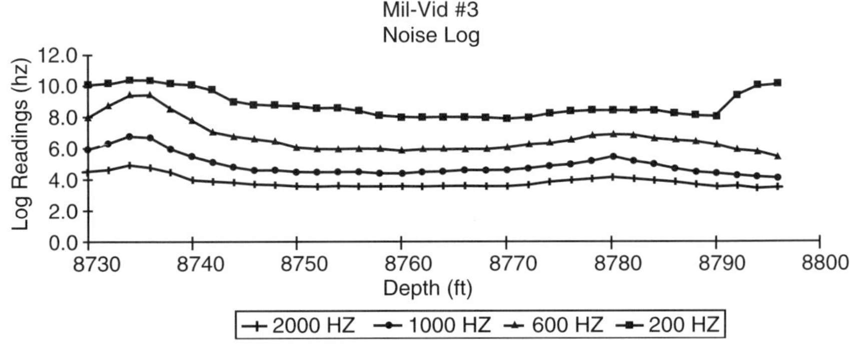

Figure 8.22

drilling at approximately 12,230 feet, a kick was taken, the well was shut in, and an underground blowout ensued. Water flows from intervals above 4000 feet further complicated analysis.

In an effort to understand the problem, a temperature survey was run and is presented as Figure 8.24. As illustrated in Figure 8.24, the temperature gradient between the top of the drill collars at 11,700 feet and 5570 feet was normal at 1.25 degrees per 100 feet. With a normal gradient, it is conclusive that there can be no flow from the interval at 12,230 feet to any interval in the hole. The significant drop in temperature at 5570 feet indicated that the well was flowing from this depth or that a lost circulation

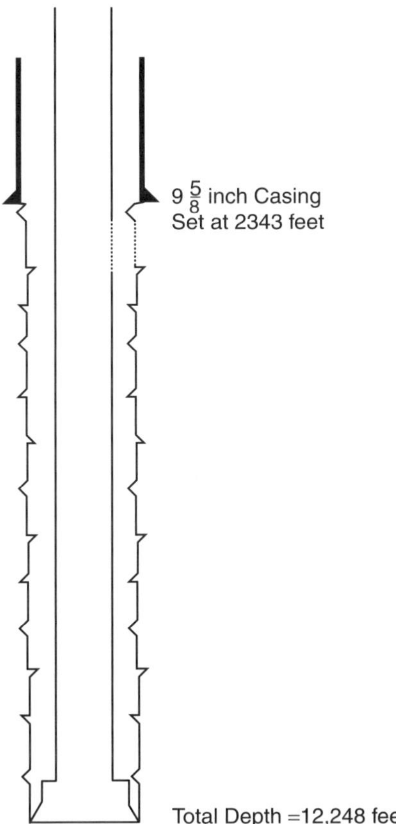

Figure 8.23 *Thermal Exploration–Sagebrush 42-26, Sweetwater County, Wyoming.*

zone was at this depth. The temperature survey was conclusive that the well was flowing above 4000 feet since the gradient above this point is essentially zero.

The noise log is presented as Figure 8.25. As illustrated in Figure 8.25, noise anomalies are indicated at 4000 feet and 6000 feet, which correspond to the temperature interpretation that fluid was moving at these depths. However, in addition, the noise log was interpreted to indicate flow from 12,000 feet to 7500 feet, which was in conflict with the interpretation of the temperature survey. Subsequent operations proved conclusively that there was no flow from the bottom of the hole at the time that these logs were run.

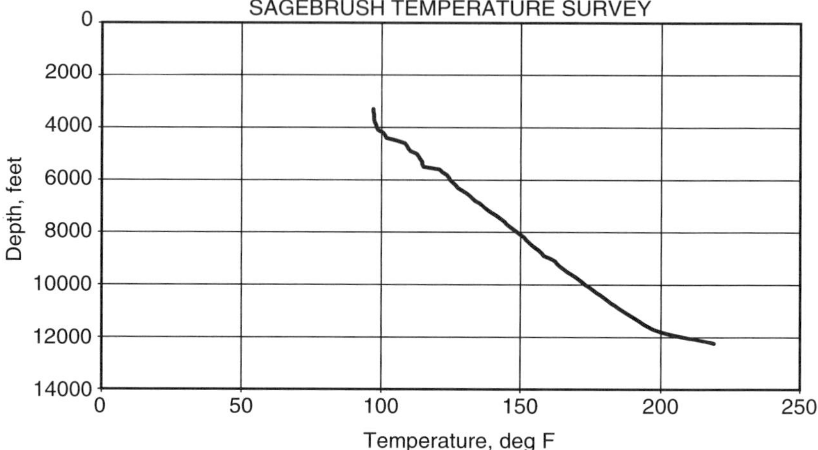

Figure 8.24

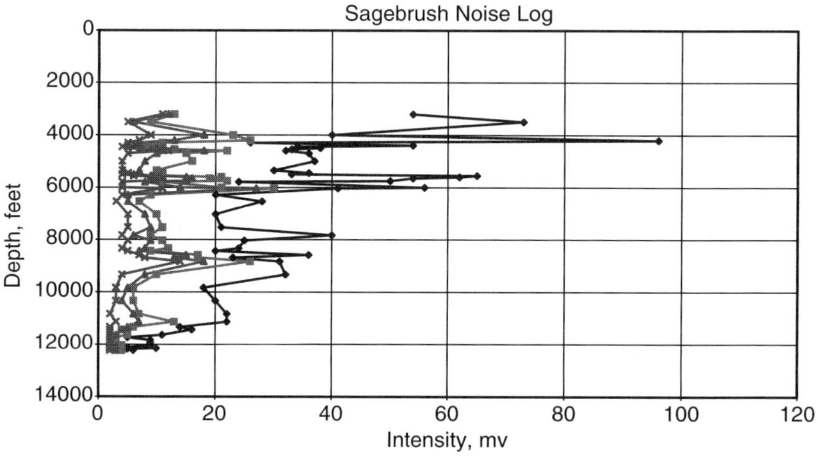

Figure 8.25

A deep well in Leon County, Texas, is an interesting and instructive case history. The wellbore schematic is presented as Figure 8.26. An unanticipated sand at 14,975 feet caused a kick, resulting in an underground blowout. After the initial kill attempt with the rig pumps, a noise log was run. The service provider's interpretation of the noise log was that there was no longer an underground flow.

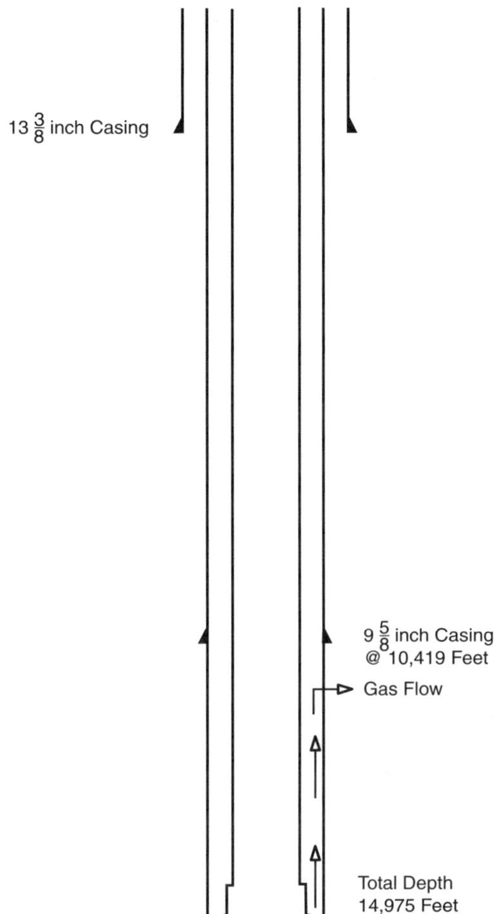

Figure 8.26 *First Underground Blowout Wellbore Schematic, Leon County, Texas.*

In an effort to further define the situation, a temperature survey was run and is presented as Figure 8.27 (1st Run). The temperature log was conclusive. The underground blowout was continuing from bottom into a zone at approximately 11,000 feet. A second temperature survey, also presented in Figure 8.27 (After Kill) indicated the second kill attempt was successful and the drill string was cemented to approximately 13,000 feet. In this instance, it is conclusive that the temperature survey was definitive and the noise log was in error.

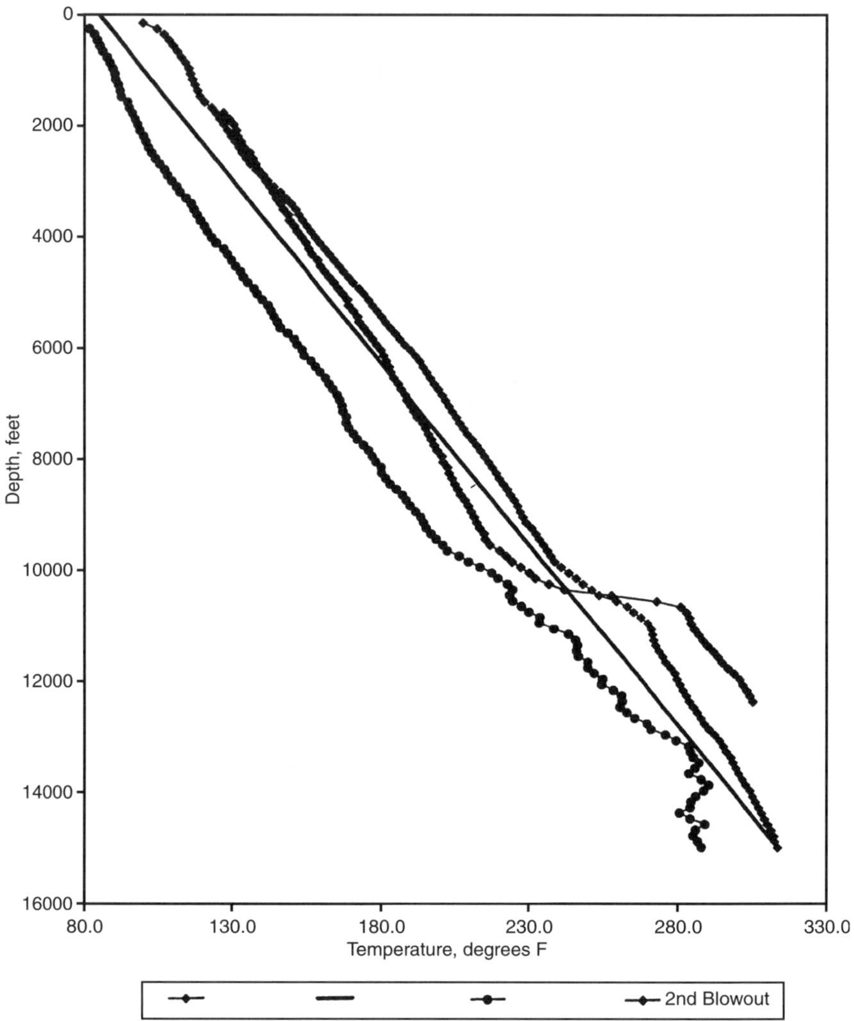

Figure 8.27   *Temperature Profiles, Leon County, Texas, February, 1998.*

Subsequently, a plug was set at 14,708 feet in the drill collars and the drillpipe was perforated at 10,600 feet. The well was circulated without incident. A severing charge was then fired at 12,412 feet. Immediately after the severing charge was fired, the well began to flow and the underground blowout resumed.

At that time, the primary question was whether the flow was in the annulus or through the drill string. An accurate model of the flow path was vital if alternatives were to be determined and evaluated. Since the severing tool failed to sever the drillpipe, further wireline work could be conducted.

Accordingly, another temperature survey was run and is presented in Figure 8.27 (2nd Blowout). As illustrated, the temperature at the charged zone was significantly greater than previously recorded. It was logical that since the temperatures were different than previously recorded, the condition of the well must be different. It was further reasoned that since the temperature was significantly hotter in and below the charge zone than anything previously recorded, the measuring device had to be directly in the flow, which meant that the flow path had to be up the drill string as opposed to up the annulus. Based on these interpretations, it was concluded that the flow was as depicted in Figure 8.28. The well was controlled by setting additional plugs inside the heavy-weight drillpipe.

CHARGED INTERVALS—CLOSE ORDER SEISMIC—VENT WELLS

In underground blowouts, the charging of the zone of loss is an important consideration for relief well operations. A relief well located within the charged interval will encounter the charged interval and experience well control problems. The mud weight required to control the charged interval can approach 1 psi/ft, which usually exceeds the fracture gradient of the intervals immediately above and below the zone of loss. Therefore, the relief well can be lost or another casing string can become a necessity.

In addition, the charging of the zone of loss is an important consideration in analyzing the potential for the influx to fracture to the surface. Nowhere is the question of shallow charging more important than offshore.

In late September, 1984, Mobil experienced a major blowout at its N-91 West Venture gas field, offshore Nova Scotia, Canada. A relief well, the B-92, was spudded approximately 3000 feet from the blowout. During drilling at 2350 feet with conductor set at 635 feet, a gas kick was taken. The gas zone encountered was the result of the charging caused by the N-91 blowout. A shallow seismic survey was conducted to assist in defining the extent of the underground charging. Booth reported that, when

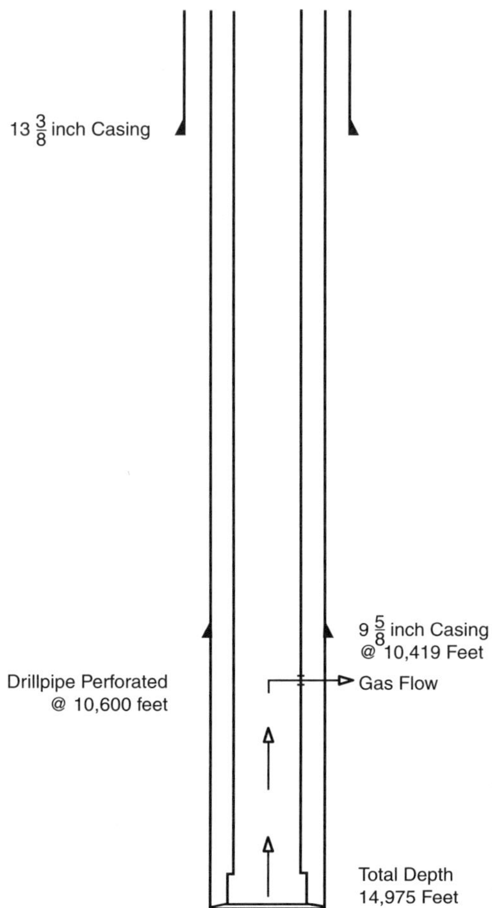

Figure 8.28   *Second Underground Blowout Wellbore Schematic, Leon County,*
Texas.

the seismic data was compared with the original work, two new seismic events had been identified.[1]

The deeper event occurred at about 2200 to 2300 feet, which corresponded to the charged zone in the relief well. However, there was also a second event at approximately 1370 to 1480 feet. The upper interval was interpreted to be approximately 3300 feet in diameter emanating from the N-91. This event was of great concern since only unconsolidated sandstones, gravels, and clays were present between the charged interval and the ocean floor 1100 feet away.

Fortunately, the charged interval never fractured to the surface. Eight additional surveys were conducted between November 5, 1984 and May 9, 1985. Those surveys revealed that the gas in the shallow zone had not grown significantly since the first survey and had migrated only slightly up dip. In addition, the surveys were vital for the selection of safe areas for relief well operations. Finally, the surveys were vital in analyzing the safety and potential hazard of continuing operations onboard the Zapata Scotain with the rig on the blowout.

In the past, it has been customary to drill vent wells into the charged zones in an effort to reduce the charging. Generally, such efforts have not proven successful. The zones of loss are normally not good quality reservoir. Therefore, the amount of gas being lost greatly exceeds that recovered from the vent wells. The result is that the charging is relatively unaffected by the vent wells.

At the TXO Marshall, for example, three vent wells were completed. The blowout was discovered to be losing approximately 15 mmscfpd underground. The three vent wells were producing a total of less than 2 mmscfpd. Experiences such as this are commonly reported.

If charging is a problem, the better alternative may be to vent the blowout at the surface. If charging is to be affected, the volume of gas vented would have to be sufficient to cause the flowing surface pressure to be less than the shut-in surface pressure plus the frictional losses between the zone of loss and the surface. Once the charging is stopped, the operations at the surface can be conducted safely and the relief well, if necessary, can be in the most expeditious position.

SHEAR RAMS

If shear rams have been used, the situation is very similar to that at the Mil-Vid #3 in that the drillpipe has been severed immediately below the shear rams and the flow is usually through the drillpipe, down the drillpipe annulus, and into the formations below the shoe.

A typical example is illustrated in Figure 8.29. As illustrated, when the shear rams are used, the result is often very similar to setting pipe and completing for production. If flow is only through the annulus, the well will often bridge, especially offshore in younger rocks which are more unstable.

With pipe set and open to the shoe, the flow can continue indefinitely. Shear rams should be used only as a last resort.

CEMENT AND BARITE PLUGS

Generally, when an underground blowout occurs, the impulse is to start pumping cement or setting barite pills. Cement can cause terminal damage to the well. At least, such indiscriminate actions can result in deterioration of the condition of the well. It is usually preferable to bring the blowout under control prior to pumping cement. If the problem is not solved, when the cement sets, access to the blowout interval can be lost. Cement should not be considered under most circumstances until the well is under control or in those instances when it is certain that the cement will control the well.

Barite pills can be fatal or cause the condition of the wellbore to deteriorate. A barite pill is simply barite, water, and thinner mixed to approximately 18 ppg. Each mixture can demonstrate different properties and should be pilot-tested to insure proper settling. Intuitively, the heavier, the better for a barite pill. However, consider Figure 8.30. As illustrated, for this particular mixture, barite pills ranging in weight from 14 ppg to 18 ppg would permit the barite to settle.

Mixtures above 20 ppg failed, though, to settle barite. The failure to settle is caused by the interaction of the barite particles. All mixtures can attain a density which will not permit the settling of barite. When setting a barite pill, the total hydrostatic does not have to exceed the reservoir pressure for the barite pill to be successful. However, the greatest success is experienced when the total hydrostatic does exceed the reservoir pressure.

When barite or cement is chosen, the drill string should be sacrificed. Attempting to pull the drill string out of a cement plug or barite pill only retards the setting of the cement or the settling of the barite.

Too often, cement plugs and barite pills only complicate well control problems. For example, in one instance in East Texas, the improper use of cement resulted in the loss of the well and a relief well had to be drilled. In another instance offshore, the barite pill settled on the drillpipe as it was being pulled and the drillpipe parted, causing loss of the well.

UA
LA
Kill Line
Choke Line
Shear Range
Closed
A B
SR
VR
E F
UR
G H
LR
C D
Mud Line
30°
200 m
20°
500 m
5" Drill Pipe
1200 m
13 3/8"
2000 m
PLT. 13.0 ppg EM.W.
2050 m
9 5/8"
3000 m
PLT. 12.5 ppg EM.W.
6 1/2"
OPEN
HOLE
Bridge = Depth Unknown.
BHA
Depth = 4000 m

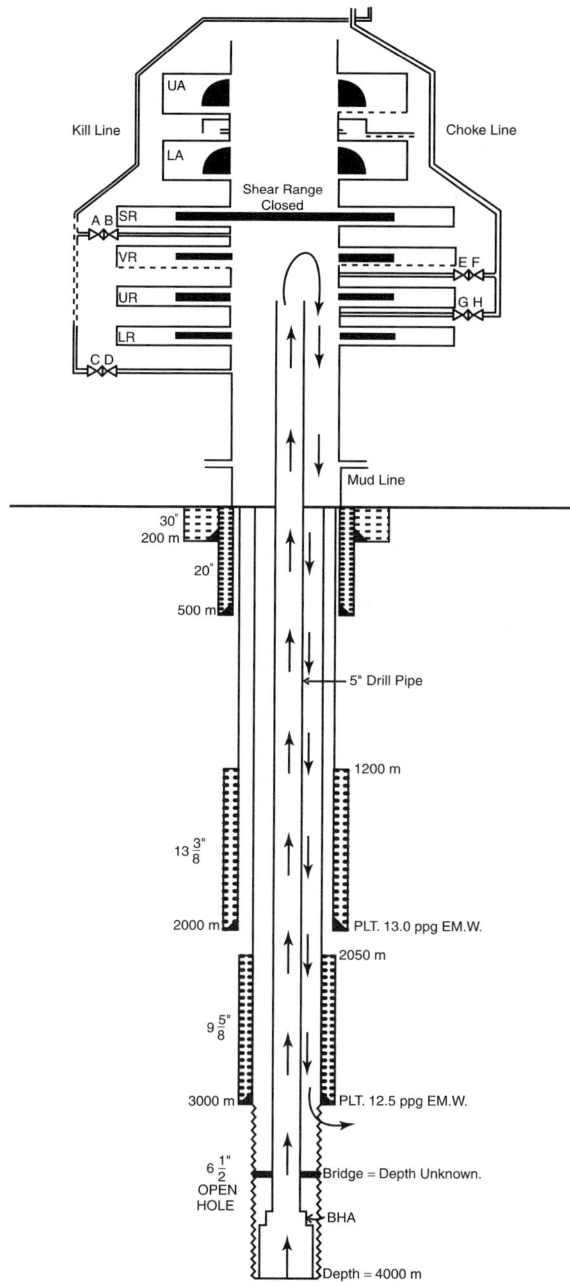

Figure 8.29

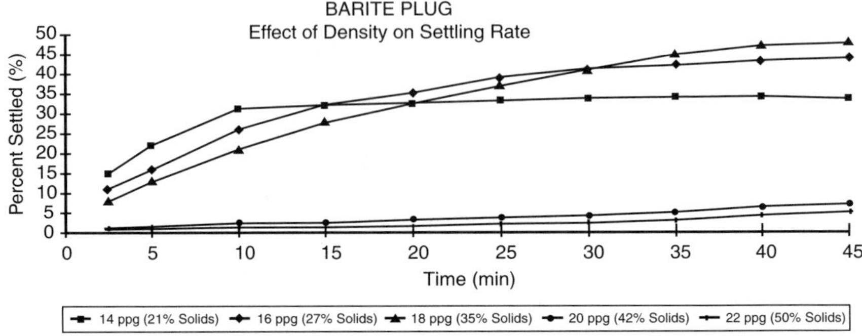

Figure 8.30

In both instances, millions of dollars were lost due to the unfortunate selection of barite and cement.

References

1.   Booth, Jake, "Use of Shallow Seismic Data in Relief Well Planning," *World Oil*, May 1990, page 39.

CHAPTER NINE

CASE STUDY: THE E. N. ROSS NO. 2

This case study of the blowout at the E. N. Ross No. 2 in Rankin County, Mississippi, offers a unique opportunity to understand many of the concepts discussed in this text from the perspective of the future and ultimate consequences. These analyses, evaluations, and commentary are intended to be instructive. Several differences of opinion are not presented. Interpretations of events are presented solely for the purpose of consideration and instruction. Nothing presented in this text is intended as a criticism of any operation conducted at the well. Certainly, those involved did not have the luxury of hindsight. The sole purpose of this chapter is to illustrate how the concepts and tools presented in this text can be used to assist in analyzing events and alternatives.

The E. N. Ross No. 2 was spudded on February 17, 1985. It was intended as a 19,600-foot Smackover development well in the Johns Field in Rankin County, Mississippi. Thirteen and three-eighths-inch casing was set at 5465 feet, $9\frac{5}{8}$-inch casing at 14,319 feet and a $7\frac{5}{8}$-inch liner at 18,245 feet with the top of the liner at 13,927 feet. With the exception of 1644 feet of 47 #/ft S-95 from 7335 feet to 8979 feet, the $9\frac{5}{8}$-inch casing consisted of 53.5 #/ft S-95. The 47 #/ft S-95 was run as a "weak link" so that in the event of excessive pressure on the wellbore, a casing failure would more likely occur below 7335 feet instead of at the surface. It was reasoned that, with the failure below 7335 feet, any deadly sour gas would most likely vent underground.

The Smackover was topped at approximately 18,750 feet and is well known to be an abnormally pressured, sour gas reservoir containing approximately 30 percent hydrogen sulfide and 4 percent carbon dioxide. A 17.4-ppg oil-base mud was used. A trip was made at 19,419 feet for a new bit. An insert bit was to follow two diamond bits. On June 10, 1985, on the trip in the hole with the new insert bit, the drill string became stuck with the bit at 19,410 feet.

Reported problems with the filter cake of the oil mud and the report that the pipe became stuck while sitting in the slips during a connection influenced the conclusion that the string was differentially stuck. Based upon the conclusion that the drill string was differentially stuck, it was decided to attempt to free the fish with a drill stem test (DST) tool. Accordingly, the string was backed off at 18,046 feet, which was 199 feet inside the $7\frac{5}{8}$-inch liner, leaving 1094 feet of spiral drill collars, stabilizers, and drillpipe. The wellbore schematic prior to commencing fishing operations is presented as Figure 9.1. Pursuant to the cement bond log, the top of the cement behind the $9\frac{5}{8}$-inch casing was at 850 feet.

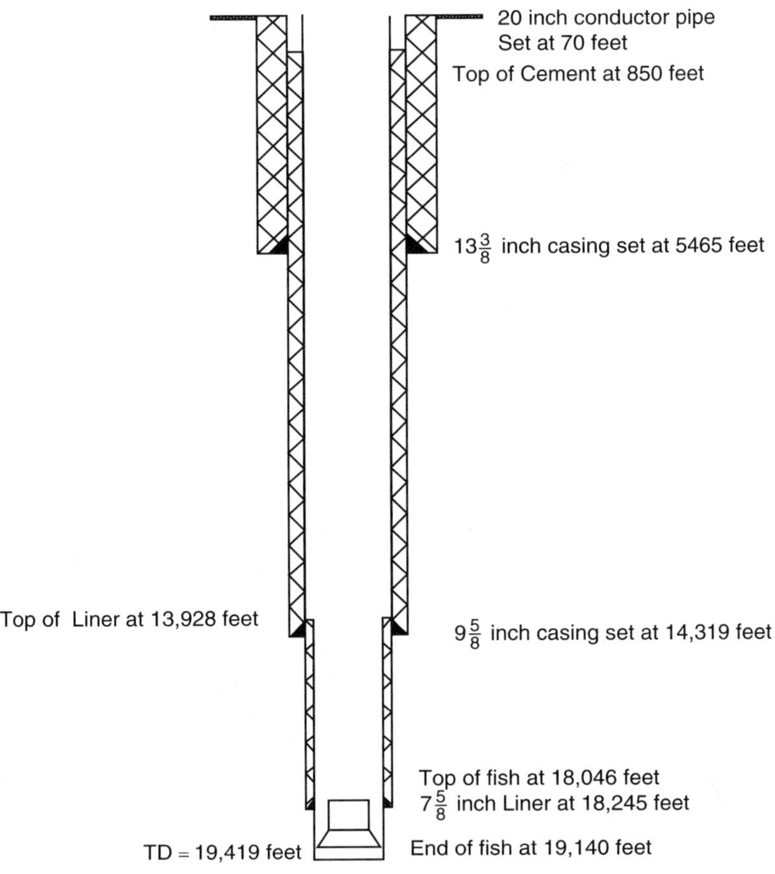

Figure 9.1 *Conditions Prior to Fishing.*

The first DST tool with an overshot on bottom failed to open. The mud inside the drillpipe was reversed. However, the overshot could not be released. After several hours of the overshot and packer assembly being worked, the string came free and was tripped out of the hole. When the DST tool was inspected on the rig floor, it was discovered that the bottom half of the packer, two subs, and the overshot were lost in the hole, which further complicated the fishing effort.

A taper tap was run without success. A mill was run and 2.5 feet were milled in three hours. On June 20, 1985, a second DST tool with a 2-inch spear was run. The drill string was filled to within 13 barrels of capacity with 100 barrels of 17.4-ppg oil-base mud followed by water, which brought the fluid level to approximately 800 feet from the surface. It was reported that the tool was opened, the well flowed for one minute, and an unrecorded quantity of fresh water was flowed to the surface. The fluid in the drillpipe was reverse-circulated to the surface. The hole was circulated conventionally. The density of the mud recovered from the bottom of the hole was gas cut from 17.4 ppg to 17.1 ppg. The DST tool was tripped out of the hole and the blind rams were closed. Approximately 9 inches of the spear were not recovered.

The DST tool was laid down and an overshot was picked up. The changing of the bottomhole assembly required approximately two hours. When the blind rams were opened, the well was flowing a small stream of mud.

With the well flowing, the bottomhole assembly consisting of the overshot, seven $4\frac{3}{4}$-inch drill collars, three stands of $3\frac{1}{2}$-inch drillpipe and one joint of 5-inch drillpipe were run into the hole and the well was shut in. An additional eight barrels of mud were reportedly gained during the time required to run the drill string, and the shut-in surface pressure was recorded at 500 psi. The end of the drill string was at 562 feet. The well was circulated through the choke with 18.4-ppg mud. After circulating the 18.4-ppg mud, the well was shut in and the shut-in surface pressure was 3000 psi. The gain increased by approximately 100 barrels.

As the shut-in pressure reached 3000 psi, the drillpipe began to move up through the derrick. The choke was opened and the bottom pipe rams were closed above a tool joint on the 5-inch drillpipe. The choke was once again closed and the shut-in surface pressure increased to 3250 psi. Attempts to close the safety valve on the kelly were unsuccessful due to the pressure at the surface. Therefore, the choke was opened a final time, the safety valve was closed, and the choke was closed. When the well was finally shut in on June 21, 1985, an estimated maximum total of 260 barrels had been gained and the shut-in surface pressure was 3700 psi.

After the well was shut in and before snubbing operations began, a detailed snubbing procedure was prepared. Basically, the snubbing procedure recommended that the $3\frac{1}{2}$-inch drillpipe be snubbed into the hole in 2000-foot increments. After each 2000-foot increment, the hole was to be circulated with 18.4-ppg mud using the classical Wait and Weight Method. That is, the drillpipe would be filled with the 18.4-ppg mud. The casing pressure would be held constant while the drillpipe pressure was being established. After the drillpipe pressure was established, it would be held constant while the 18.4-ppg mud was circulated to the surface. It was anticipated that this staging and circulating heavy-mud procedure would reduce the pressure on the $9\frac{5}{8}$-inch casing.

Further, some favored reverse circulating once the overshot was at the top of the fish. Accordingly, the two conventional back-pressure valves in the drill string had to be replaced with a single pump-out, back-pressure valve. The pump-out, back-pressure valve had a small ball and seat which could be pumped out by dropping a ball. Due to its small size, the pump-out, back-pressure valve was also very susceptible to erosion.

At the well site during the period from June 21 until July 13, preparations were made to snub the drill string into the hole. A snubbing unit was rigged up, a 400-foot flare line was laid, and a flow line was laid 1800 feet to the E. N. Ross No. 1. The surface pressure remained relatively constant at 3700 psi for 18 days from June 21 until July 9. From July 9 until July 13, the surface pressure declined to 2600 psi. By the end of the day on July 13, a total of 69 joints had been snubbed into the hole. The end of the work string was at approximately 2139 feet.

On July 14, an additional 60 joints were snubbed into the hole. The end of the work string was at approximately 4494 feet. The hole was circulated with 276 barrels of 18.4-ppg mud pumped in and 265 barrels of mud recovered pursuant to the snubbing program. The entire operation proceeded without incident and no gas was observed at the surface. After the 18.4-ppg mud was circulated, the surface pressure stabilized at 2240 psi.

On the morning of July 15, the surface pressure was recorded at 2350 psi. By the end of the day, the work string had been snubbed to

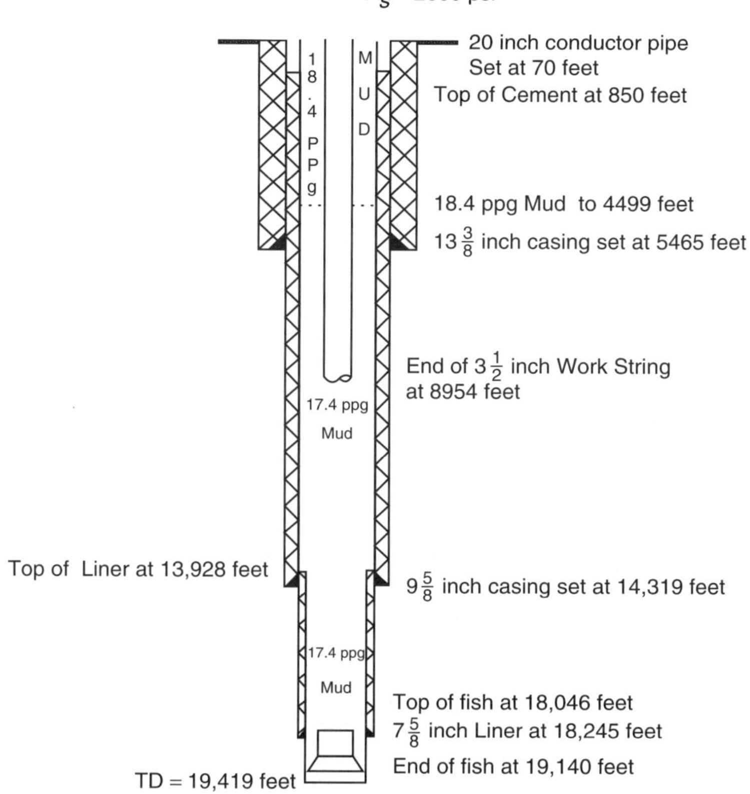

Figure 9.2   *July 15, 1985.*

8950 feet and the surface pressure had stabilized at 2000 psi. The wellbore schematic is illustrated as Figure 9.2. Two manually operated ball valves were installed in the top of the $3\frac{1}{2}$-inch drillpipe in the work basket. Chicksans and hammer unions were used to connect the pump trucks to the drillpipe as illustrated in Figure 9.3.

Although the snubbing procedure provided for 2000-foot increments, a total of approximately 4456 feet was run into the well. As provided in the snubbing procedure, circulation was initiated using the classic Wait and Weight Procedure. The casing pressure was held constant while the pump was brought to speed and the drillpipe pressure was established. With the annulus pressure constant at 2000 psi, the drillpipe pressure was recorded to be 2700 psi at 2 bpm and 2990 psi at 3 bpm. Once the drillpipe

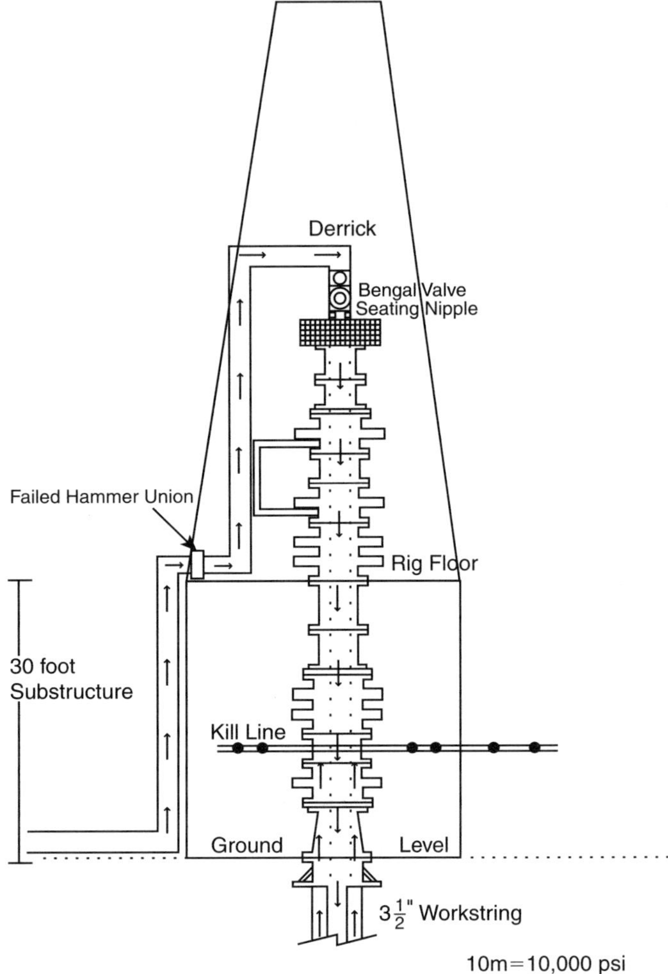

Figure 9.3   *Pump and Return Lines.*

pressure was established at 2990 psi, circulation commenced with the 18.4-ppg mud being pumped at 3 bpm.

The surface pressures, total barrels pumped, and total barrels recovered are recorded in Table 9.1 and illustrated in Figure 9.4. As shown in Table 9.1, pumping operations commenced at approximately 1643 hours. Thirty-five barrels were required to fill the drillpipe.

Table 9.1

Pumping History, E. N. Ross No. 2, July 15 1985

Time	Tubing Pressure	Casing Pressure	Barrels in	Net Barrels in	Barrels out	Average Rate in	Average Rate out	Gain	Remarks
16:43	0	2000	0	0	0	0		0	Tubing at 8958 feet
16:45	0	2000	10	0	0	5.0		0	2 bpm— 700 psi
16:48	0	2000	20	0	0	3.3		0	3 bpm— 2940 psi
16:52	0	2040	30	0	0	2.5		0	
16:59	2940	2060	40	5	0	1.4		0	
17:03	2880	2200	50	15	0	2.5		0	
17:08	2920	2125	60	25	24	2.0	4.8	−1	
17:12	2940	2150	70	35		2.5	2.5		
17:15	2880	2090	80	45	42	3.3	2.5	−3	
17:18	2880	2040	90	55	55	3.3	4.3	0	
17:22	2900	2010	100	65	65	2.5	2.5	0	18.6 ppg
17:25	2910	1990	110	75	76	3.3	3.6	1	
17:27	2880	1900	120	85	89	5.0	6.5	4	
17:32	2880	1900	130	95	100	2.0	2.2	5	
17:38	2880	1780	150	115	135	3.3	5.8	20	
17:41	2880	1700	160	125	160	3.3	8.3	35	
17:44	3000	1900	170	135	180	3.3	6.6	45	
17:47	3000	1900	180	145		3.3	6.0		
17:50	2920	1900	190	155		3.3	6.0		18.1 ppg
17:53	2940	1925	200	165		3.3	6.0		
17:57	3040	2100	210	175		2.5	6.0		17.8 ppg
18:00	2940	2200	220	185		3.3	6.0		

Continues

Table 9.1
Continued

Time	Tubing Pressure	Casing Pressure	Barrels in	Net Barrels in	Barrels out	Average Rate in	Average Rate out	Gain	Remarks
18:03	3020	2200	230	195	295	3.3	6.0	100	
18:06	3000	2300	240	205		3.3	4.0		
18:08			247	212	315	3.5	4.0	103	
18:09	3040	2400	250	215		3.0	5.5		
18:10			253	218	326	3.0	5.5	108	17.6 ppg
18:12	2980	2300	260	225		3.5	7.6		
18:13			263	228	349	3.0	7.6	121	
18:15	3040	2500	270	235		3.5	9.3		
18:18	2980	2500	280	245		3.3	9.3		
18:19			282	247	405	2.0	9.3	158	17.5 ppg
18:21	3080	2800	290	255		4.0	14.3		
18:22			294	259	448	4.0	14.3	189	17.4 ppg
18:25	2800	2740	300	265		2.0	10.5		
18:26			302	267	490	2.0	10.5	223	
18:28	2880	2740	310	275		4.0	8.6		
18:31	2980	2820	320	285		3.3	8.6		17.6 ppg
18:35	3040	3300	330	295		2.5	8.6		
18:37			336	301	585	3.0	8.6	284	
18:38	3040	3400	340	305		4.0	6.9		
18:42	3000	3480	350	315		2.5	6.9		
18:45	2960	3575	360	325		3.3	6.9		
18:47	3000	3700	370	335		5.0	6.9		18.2 ppg
18:52	3080	3900	380	345		2.0	6.9		
18:54	2840	3900	390	355		5.0	6.9		18.0 ppg

18:57	2940	3980	400	565		3.3	6.9		
19:01	2800	4100	410	375		2.5	6.9		
19:04	2980	3900	420	385		3.3	6.9		
19:08	2880	4310	430	395		2.5	6.9		
19:12	2900	4590	440	405		2.5	6.9		good flare
19:15	2980	4700	450	415		3.3	6.9		16.5 ppg
19:19	2900	4750	460	425		2.5	6.9		
19:23	2900	4700	470	435		2.5	6.9		flared
19:26	2900	4700	480	445	927	3.3	6.9	482	
19:28	3000	4500	490	455		5.0	7.9		
19:31	2990	4600	500	465		3.3	7.9		
19:39	4180	4800	530	495		3.7	7.9		
19:42		5100	540	505	1054	3.3	3.6	549	shut down
19:55	3780	5000				.7	3.6		start over
19:56	5200	5000	550	515		.7	3.6		
20:03	5200	5000	580	545		4.2	3.6		
20:07	5700	4990	600	565		50	3.6		
20:09	6000	4980				3.3	3.6		
20:10	5960	4900	610	575		3.3	3.6		
20:12	6060	4910	620	585		5.0	3.6		
20:13	6100	4940				5.0	3.6		
20:14	6100	4910	630	595		5.0	3.6		
20:15	6300	5000	640	605		10.0	3.6		
20:16	6240	5000				4.3	3.6		
20:17	6240	4960				4.3	3.6		
20:18	6460	4910	653	618		4.3	3.6		
20:19	6460	4890				8.3	3.6		
20:20	6500	4900				8.3	3.6		
20:21	6280	4990	678	643		8.3	3.6		
20:23	5940	5010	690	655		6.0	3.6		
20:25	5920	4950				3.7	3.6		
20:26	6100	5010				3.7	3.6		

Continues

Table 9.1
Continued

Time	Tubing Pressure	Casing Pressure	Barrels in	Net Barrels in	Barrels out	Average Rate in	Average Rate out	Gain	Remarks
20:27	6100	5000				3.7	3.6		
20:28	6100	4900				3.7	3.6		
20:29	6220	5100				3.7	3.6		
20:30	6220	4990				3.7	3.6		
20:31	6300	5050	720	685		3.7	3.6		
20:33	6360	5000	730	695		5.0	3.6		
20:34	6300	4900				5.0	3.6		
20:35	6400	5000	740	705		5.0	3.6		
20:36	6400	4950				5.0	3.6		
20:37	6560	5010	750	715		5.0	3.6		
20:38	6560	5000				5.0	3.6		
20:39	6650	5004	760	725		5.0	3.6		
20:41	6800	5010	770	735		5.0	3.6		
20:43	6750	5000				3.3	3.6		
20:44	6850		780	745		3.3	3.6		
20:45	6700	4700	790	755		10.0	3.6		
20:46	6750	4790				3.3	3.6		
20:47	6700	4710				3.3	3.6		
20:48	6650	4680	800	765		3.3	3.6		
20:49	6700	4700				5.0	3.6		
20:50	6720	4800	810	775		5.0	3.6		
20:52	6800	4750	820	785		5.0	3.6		
20:53	6700	4610				5.0	3.6		
20:54	6700	4625	830	795		5.0	3.6		
20:55	6740	4650				5.0	3.6		
20:57	6700	4580				5.0	3.6		
20:58	6750	4500	850	815		5.0	3.6		
21:00	6700	4650	860	825	1293	5.0	3.6	468	

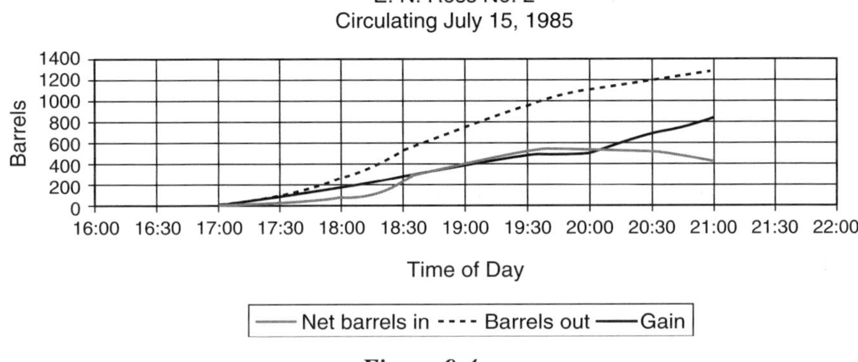

Figure 9.4

At 1727 hours with 120 barrels pumped into the drillpipe, the flow rate from the annulus began to exceed the pump rate, resulting in a net gain in the pits.

At 1854 hours, after 390 barrels were pumped and approximately 725 barrels were recovered, gas was reported at the surface. With the gas to the surface, the net gain was 369 barrels. At 1942 hours, the well was shut in with 5100 psi on the annulus and 4180 psi on the drillpipe. At that time, 540 barrels had been pumped, 1054 barrels had been released from the annulus for a net gain of 549 barrels and a good flare was reported. The back-pressure valve appeared to be holding.

After the well was observed for 13 minutes, pumping operations were resumed and the circulation rate was increased to six barrels per minute. The annulus pressure was held constant at 5000 psi and the drillpipe pressure was established at 5200 psi. Contrary to the snubbing procedure, the annulus pressure was held constant at 5000 psi until 2044 hours when the drillpipe pressure reached 6850 psi. The drillpipe pressure was then held constant at approximately 6700 psi. At 2110 hours, a leak developed in a hammer union on the rig floor and the pump-in line failed. The choke was closed. Flow from the drillpipe not only continued but increased, indicating that the back-pressure valve was not functioning. Subsequently, three men were carried to the snubbing basket in the crane's personnel basket. Their efforts to close the two ball valves were unsuccessful. The sour gas alarms sounded. The location was abandoned, the gas was ignited, and the rig was completely destroyed by the fire.

The well was subsequently capped and killed. Litigation continued for many years.

ANALYSIS OF THE BLOWOUT

Certainly, there were considerable opportunities for decisions during the course of this operation and each decision point was clouded with numerous alternatives.

THE DRILLING AND FISHING OPERATION

Utilizing a DST tool to free differentialy stuck pipe was common in shallow, low pressure or non-hydrocarbon-bearing formations. It was known to have been used on occasion in deep, high-pressure, hydrocarbon-bearing zones. For the operation to be successful, a differential into the wellbore must be created. There was no doubt that the Smackover had been penetrated and that it contained high-pressure sour gas. Therefore, if the operation was successful, an influx and well control problem could result.

Backing off inside the casing ultimately complicated the situation. Had the DST procedure been successful, any further question would have been moot. The considerations supporting backing off in the casing are valid. That is, it is much easier to engage the fish and set the DST packers in the casing as opposed to the open hole.

Further, as discussed in Chapter 4, the oil-base mud could mask the presence of an influx. No flow or improper filling was detected during the trip out of the hole. According to reports, the hole was properly filled during the trip out.

THE KICK

While out of the hole, the crew noticed that the well was flowing and immediately attempted to run pipe into the hole, a routine and common practice in 1985.

As outlined in Chapter 3, the best procedure is to shut in the well as soon as a kick is detected. Had the crew shut in the well when the flow was observed as the blind rams were opened, the gain and surface pressure would have been minimal.

By this time a very difficult well control problem was inevitable. With the top of the fish at 18,046, or almost 1400 feet from the bottom of the hole, it was not possible to trip the drill string to bottom and circulate out the influx.

As was common and routine in 1985, when the crew observed the kick, the drill collars and overshot were run along with a small amount of drillpipe. In light of later requirements, it would have been better if the drill string had not been run. The drill collars restrict the pump rate in the event that high kill rates are necessary. Further, with high surface pressures, snubbing the drill collars out of the hole is more difficult. If the drill collars were spiral drill collars, snubbing them out of the hole would have been even more difficult.

When the crew did succeed in running approximately 560 feet of drill string into the well and shutting in the well, the surface pressure was reported to be 500 psi and the additional gain was only eight barrels. Again, as was common and routine in 1985, the hole was circulated. However, there is insufficient change in density of 560 feet of mud to materially affect the bottomhole pressure of a 19,419-foot well.

Further, there is no good way to circulate the well when the bit is not on the bottom of the hole. As was discussed in Chapter 4, there is no classical procedure for circulating anywhere except on the bottom of the hole. The U-Tube Model is only applicable when the bit is on the bottom of the hole. With the bit anywhere other than the bottom of the hole, the U-Tube Model becomes the Y-Tube Model and it is impossible to understand exactly what is occurring in the wellbore. The Wait and Weight Method, the Driller's Method, and keeping the drillpipe pressure constant are meaningless when the bit is not on the bottom of the hole.

As illustrated in Chapter 4, the hole could be circulated using the Constant Pit Level Method. Another name for the Constant Pit Level Method is the Barrel In–Barrel Out Method. A more descriptive name is the Dead Well Method. These names all describe the manner in which a well is routinely circulated during any drilling operation when an influx is not in the hole.

As outlined in Chapter 4, all things must remain constant when this method is being utilized. Specifically, the pit level, choke size, drillpipe pressure, and annulus pressure must not change while the well is being circulated. In the event that any of these change, the well must be shut in.

In Chapter 2, it was suggested that a change in any of these parameters during a routine drilling operation was a signal that the well must be shut in. The same is true when a kick has been taken. Any change is a signal that the well must be shut in.

The pit level did not remain constant during the time that the well was being circulated. The gain increased from 8 barrels to approximately 100 barrels. When the well was finally shut in, the surface pressure increased from 500 psi to 3250 psi, and the wellhead force resulting from the shut-in surface pressure was sufficient to cause the drill string to be pushed out of the hole. The well had to be opened and allowed to flow while the tubulars could be secured. When finally shut in, the surface pressure was 3700 psi. The size of the total gain is disputed. For the purpose of this writing, the size of the total gain was determined from pressure analysis to be 260 barrels.

Finally, it will be remembered that a joint of 5-inch drillpipe was run in order that the rams could be closed, as there were no rams in the stack for the $3\frac{1}{2}$-inch drillpipe. As a result of this minor complication, the well had to be frozen in order to remove the 5-inch drillpipe and commence snubbing operations.

THE SNUBBING PROCEDURE

The snubbing procedure provided that the drill string was to be snubbed into the hole in 2000-foot increments with the well being completely circulated after each 2000-foot increment until reaching "bottom." Further, it was advised that the circulating pressure for each 2000-foot increment would be established by holding the casing pressure constant while bringing the pump to speed and establishing the drillpipe pressure, and thereafter, keeping the drillpipe pressure constant until the new mud was circulated to the surface. Finally, it was anticipated that reverse circulating at "bottom" might be the preferred procedure. Therefore, conventional snubbing wisdom, which dictated that two back-pressure valves be run in the drill string, was abandoned in favor of a single pump-out, back-pressure valve.

In 1985, as well as today, many involved in well control would have approached this problem in exactly the same manner. However, conceptually this snubbing procedure contained several theoretical difficulties.

As discussed in the previous section and demonstrated by the operations at this well, there is no good way to circulate with an influx when the bit is not on the bottom of the hole. The recommendation to use classical well control procedures to establish the drillpipe pressure and keep the drillpipe pressure constant while circulating is potentially problematic.

Classical circulating procedures can be successful if there is no influx migration. If the influx migrates, problems can develop. The rising influx can cause the surface pressures to increase. In response, the choke will be opened to keep the drillpipe pressure constant. In that event, additional influx will occur. Remember that, as just discussed, the condition of the well deteriorated when the rig crew circulated at 562 feet.

The techniques introduced in Chapter 4 can be used to illustrate that, theoretically, replacing the 17.4-ppg mud above the influx with 18.4-ppg mud would reduce maximum pressure at the surface. In addition, theoretically, the techniques can be used to demonstrate that circulating the influx to the surface in 2000-foot increments will reduce the maximum pressure at the surface.

With the $9\frac{5}{8}$-inch 47 #/ft "weak link" at 7335 feet, it is presumed that the surface pressure will not exceed approximately 5000 psi with the 17.4-ppg mud in the hole even though the $9\frac{5}{8}$-inch casing has been cemented to within 850 feet from the surface. Keeping in mind that the practical considerations render this approach very difficult and, in some instances impossible to accomplish, the theoretical aspects are instructive and interesting to consider. Consider Example 9.1:

Example 9.1

Given:

Figures 9.5 and 9.6

Specific gravity, $S_g = 0.785$

Gain $= 260$ bbls

Bottomhole pressure, $P_b = 16,907\,\text{psia}$

Surface pressure, $P_a = 3700\,\text{psia}$

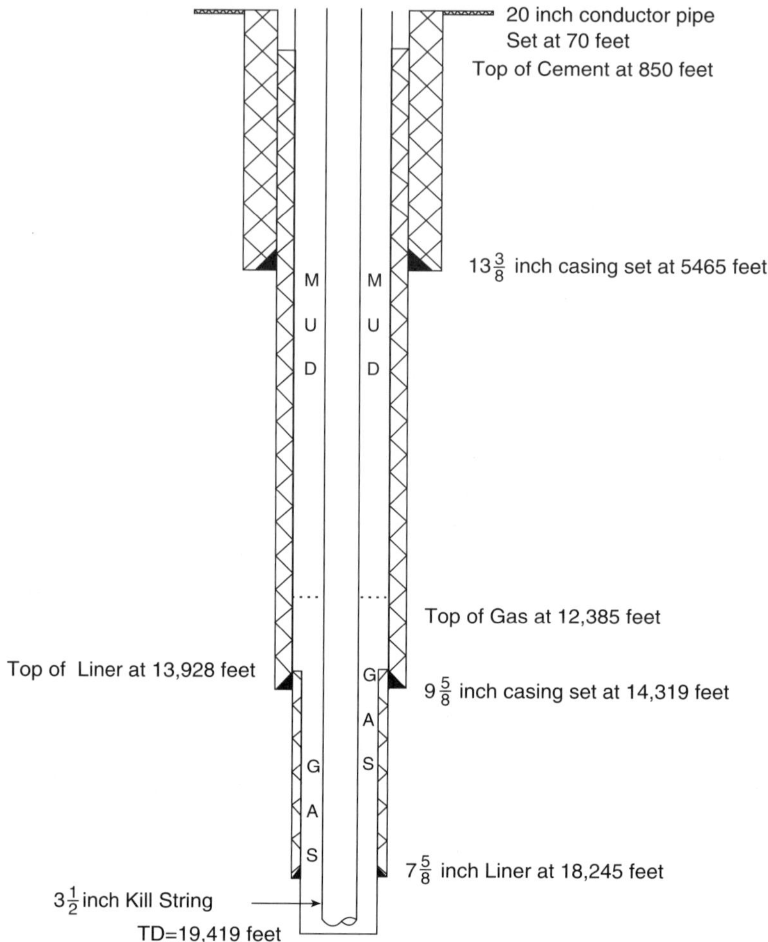

Figure 9.5 *Example 9.1 Wellbore Schematic.*

Bottomhole temperature, $T_b = 772°$ Rankine

Surface temperature, $T_s = 520°$ Rankine

Temperature gradient, $T_{grad} = 1.3°/100$ feet

Compressibility factor, $z_b = 1.988$

$$z_s = 1.024$$

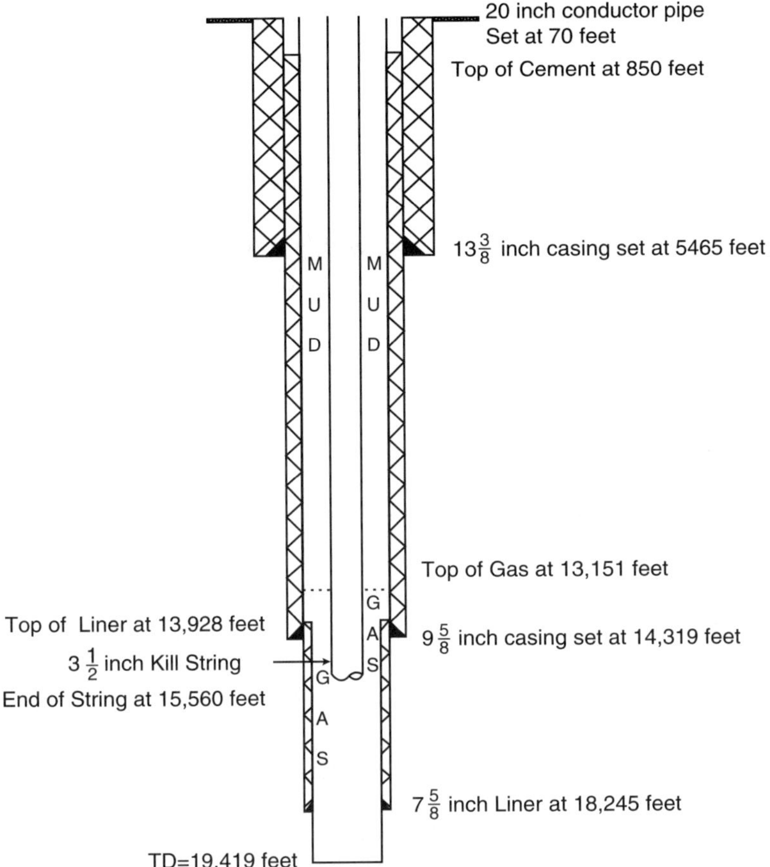

Figure 9.6   *Example 9.1 Part 5.*

Gas gradient, $\rho_f = 0.162\,\text{psi/ft}$

Casing annular capacity, $C_{dpca} = 0.0589\,\text{bbl/ft}$

Liner annular capacity, $C_{dpla} = 0.0308\,\text{bbl/ft}$

Top of gas, $TOG = 12{,}385$ feet

Required:
1.   The surface pressure with $3\frac{1}{2}$-inch-drillpipe on bottom and 17.4-ppg mud.

2.   The surface pressure with $3\frac{1}{2}$-inch-drillpipe on bottom and 18.4-ppg mud.

3.   The surface pressure with gas to surface in (1).

4.   The surface pressure with gas to surface in (2).

5.   The surface pressure with gas to the surface if only 2000 feet of influx are circulated out with 17.4-ppg mud.

6.   The surface pressure with gas to the surface if only 2000 feet of influx are circulated out with 18.4-ppg mud.

Solution:

1.   With pipe at 19,419 feet, the surface pressure with 17.4-ppg mud is

$$P_b = P_s + \rho_f h_b + \rho_m (D - h_b)$$

$$P_s = P_b - \rho_f h_b - \rho_m (D - h_b)$$

$$h_b = D - TOG$$

$$h_b = 19{,}419 - 12{,}385$$

$$h_b = \textbf{7304 feet}$$

$$P_s = 16{,}907 - 0.162(7034)$$

$$- (0.052)(17.4)(19{,}419 - 7034)$$

$$P_s = \textbf{4562 psi}$$

2.   With pipe at 19,419 feet, the surface pressure with 18.4-ppg mud is

$$P_s = 16{,}907 - 0.162(7034)$$

$$- (0.052)(18.4)(19{,}419 - 7034)$$

$$P_s = \textbf{3918 psi}$$

3. With pipe on bottom at 19,419 feet, the surface pressure with gas to the surface and 17.4-ppg mud is given by Equation 4.22:

$$P_{xdm} = \frac{B}{2} + \left[\frac{B^2}{4} + \frac{P_b \rho_m z_x T_x h_b A_b}{z_b T_b A_x} \right]^{\frac{1}{2}}$$

$$B = P_b - \rho_m (D - X) - P_f \frac{A_b}{A_x}$$

$$B = 16,907 - 0.9048(19,419 - 0) - (0.162)(7034)$$

$$\times \left(\frac{0.7806(0.0308) + 0.2194(0.0589)}{0.0589} \right)$$

$$B = -1378$$

$$P_{0dm} = \frac{-1378}{2} + \left[\frac{-1378^2}{4} \right.$$

$$\left. + \frac{16,907(0.9048)(1.024)(520)(7034)(0.0370)}{1.988(772)(0.0589)} \right]^{\frac{1}{2}}$$

$$P_{0dm} = \textbf{4202 psi}$$

4. With pipe on bottom at 19,419 feet, the surface pressure with gas to the surface and 18.4-ppg mud is given by Equation 4.21:

$$B = 16,907 - 0.9568(19,419 - 0)$$

$$- (0.162)(7034) \left(\frac{0.0370}{0.0589} \right)$$

$$B = -2389$$

$$P_{0dm} = \frac{-2389}{2} + \left[\frac{-2389^2}{4} \right.$$

$$\left. + \frac{16,907(0.9568)(1.024)(520)(7034)(0.0370)}{1.988(772)(0.0589)} \right]^{\frac{1}{2}}$$

$$P_{0dm} = \textbf{3926 psi}$$

5.   The surface pressure with gas to the surface, pipe not on bottom, 17.4-ppg mud, 2000-foot increment of influx, and the following:

234 bbls of gas in liner

26 bbls in intermediate $= 368$ feet

$TOG = 13,560$ feet

$TOL = 13,928$

Maximum depth of pipe $= 15,560$ feet

Total influx encountered:

$$L_{liner} = 15,560 - 13,928$$

$$L_{liner} = \mathbf{1632} \text{ feet in } 7\tfrac{5}{8}\text{-inch liner}$$

$$L_{casing} = 13,928 - 13,560$$

$$L_{casing} = \mathbf{368} \text{ feet in } 9\tfrac{5}{8}\text{-inch casing}$$

$$I_{tot} = 1632(0.0428) + 368(0.070)$$

$$I_{tot} = \mathbf{96 \text{ bbls}}$$

The new top of gas, *TOG*, is then given by

$$TOG = 13,928 - \left(\frac{96 - 0.0308(1632)}{0.0589} \right)$$

$$TOG = \mathbf{13,151 \text{ feet}}$$

The new height of the influx, h_b is

$$h_b = 15,560 - 13,151$$

$$h_b = \mathbf{2409 \text{ feet}}$$

The new area for the bottom of the influx, A_b, would be

$$A_b = \frac{777}{2409}(0.0589) + \frac{1632}{2409}(0.0308)$$

$$A_b = \mathbf{0.0399}$$

The bottomhole pressure is

$$P_b = 16{,}907 - 0.162(19{,}419 - 15{,}560)$$

$$p_b = \mathbf{16{,}282 \ psi}$$

The new surface pressure could then be calculated as

$$B = 16{,}282 - 0.9048(15{,}560)$$

$$-0.162(2409)\left(\frac{0.0399}{0.0589}\right)$$

$$B = \mathbf{1939}$$

$$P_{0dm} = \frac{1939}{2} + \left[\frac{1939^2}{4}\right.$$

$$+ \frac{16{,}282(0.9048)(1.024)(520)(2409)(0.0399)}{1.988(772)(0.0589)}\Bigg]^{\frac{1}{2}}$$

$$P_{0dm} = \mathbf{4016 \ psi}$$

6.   The surface pressure with pipe not on bottom with 18.4-ppg mud and the same conditions as (5):

$$B = 16{,}282 - 0.9568(15{,}560)$$

$$-0.162(2409)\left(\frac{0.0399}{0.0589}\right)$$

$$B = \mathbf{1130}$$

$$P_{0dm} = \frac{1130}{2} + \left[\frac{1130^2}{4} \right.$$

$$\left. + \frac{16{,}282(0.9568)(1.024)(520)(2409)(0.0399)}{1.988(772)(0.0589)} \right]^{\frac{1}{2}}$$

$$P_{0dm} = \textbf{3588 psi}$$

Example 9.1 assumes that there was no fish in the hole and that it was possible to snub the $3\frac{1}{2}$-inch drillpipe to the bottom of the hole at 19,419 feet. As shown in part 1, with the drill string on bottom and 17.4-ppg mud in the hole, the surface pressure would have increased from 3700 psi to 4562 psi because the influx was longer. As illustrated in part 3, which assumes that the entire influx is circulated to the surface from the bottom of the hole, the surface pressure with gas to the surface would have been only 4202 psi, or 360 psi less than when the influx was on bottom.

According to part 5, if a 2000-foot increment of influx was circulated to the surface with 17.4-ppg mud, the maximum pressure with gas to the surface would have been 4016 psi, which is approximately 200 psi less than that experienced when circulating the entire influx from bottom. As a matter of interest, the surface pressure prior to circulating with the bit at 15,560 feet in part 5 is calculated to be 3996 psi.

By comparison, with the 18.4-ppg mud and the bit on the bottom of the hole, the surface pressure would have been 3918 psi (part 2). If the entire influx was circulated to the surface as assumed in part 4, the maximum surface pressure with gas to the surface would be 3927 psi, or about the same as when circulation commenced. If a 2000-foot interval was circulated to the surface as assumed in part 6, the maximum pressure with gas to the surface would have been 3588 psi.

The unusual pressure behavior is due to the presence of the fish, the long $7\frac{5}{8}$-inch liner, and the size of the influx. The influx entirely fills the liner. Therefore, when circulation commences, the influx is immediately circulated into the $9\frac{5}{8}$-inch casing where it is considerably shorter. Consequently, the annulus pressure declines dramatically. As the influx is circulated up the $9\frac{5}{8}$-inch casing to the surface, the annulus pressure increases.

Analysis of Example 9.1 illustrates that the snubbing procedure was theoretically sound in that replacing the 17.4-ppg mud with 18.4-ppg mud would reduce the surface pressure. In addition, the snubbing procedure was theoretically sound in that circulating the influx to the surface in 2000-foot increments would further reduce the surface pressure. The maximum calculated difference would be 3600 psi compared to 4600 psi.

The advantage of the lower pressure must be considered from the operational perspective. That is, the worst-case-scenario surface pressure of 4600 psi is less than the maximum allowable of 5000 psi. Therefore, considering the time, difficulty, and mechanical problems associated with circulating at 2000-foot intervals, circulating at intervals is not the best theoretical alternative.

It must be remembered that there is a theoretical problem in Example 9.1. It was assumed that the 17.4-ppg mud could be replaced by the 18.4-ppg mud and that the influx could be circulated to the surface in 2000-foot increments. There is no theoretically correct procedure to accomplish that which was assumed and analyzed in this example.

The snubbing procedure discussed circulating at 2000-foot increments and circulating at "bottom." With the fish in the hole, "bottom" for the snubbing operation was 18,046 feet, which was 1400 feet from the bottom of the hole.

The theoretical implications of the concept of reverse circulating the influx through the drillpipe are interesting to consider. Certainly, the pressures in the annulus would be less if the influx was reversed. However, the pressure on the drillpipe must be considered. Consider Example 9.2:

Example 9.2

Given:

Assume 2000 feet of influx is to be reverse circulated to the surface given the conditions in Example 9.1 and the capacity of the drill string, C_{ds}, is 0.00658 bbl/ft. The wellbore schematic is presented as Figure 9.7.

Required:

The maximum pressure on the drillpipe.

Solution:

As determined in Example 9.1, when the first 2000 feet of influx was encountered, the top of the gas would have been at 15,560 feet.

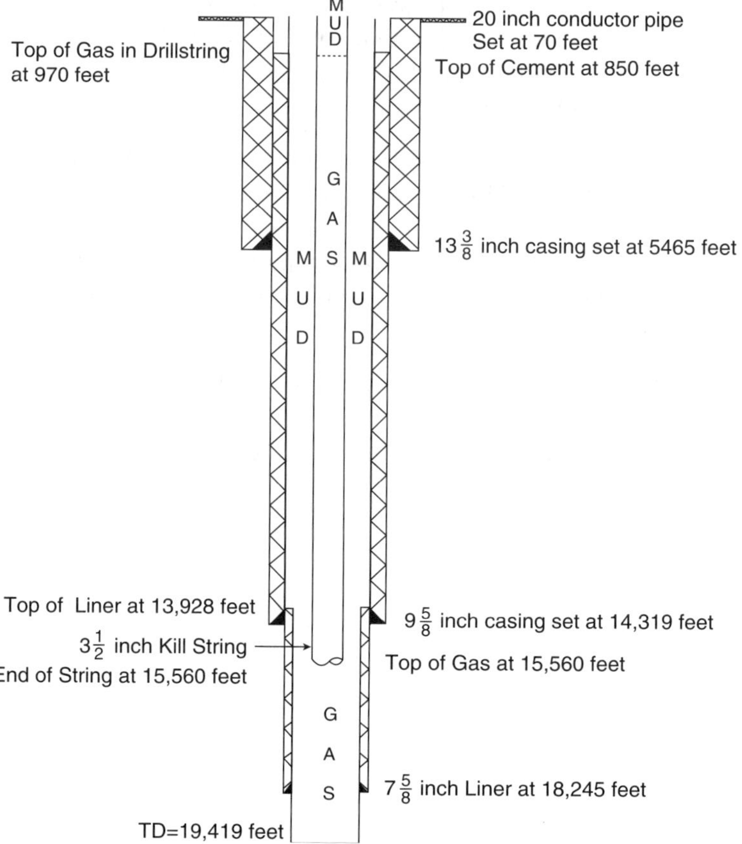

Figure 9.7   *Example 9.2.*

The volume of influx to be circulated to the surface would be 96 bbls.

Therefore, the length of the gas column inside the drillpipe would be

$$h = \frac{V}{C_{ds}}$$

$$h = \frac{96}{0.00658}$$

$$h = \mathbf{14{,}590 \ feet}$$

The drillpipe pressure would be

$$P_s = P_b - \rho_m(D - h_b) - \rho_f h_b$$

$$P_s = 16{,}282 - (0.9568)(15{,}560 - 14{,}590)$$

$$- 0.162(14{,}590)$$

$$P_s = \mathbf{12{,}990 \ psi}$$

As illustrated in Example 9.2, had the reverse circulation procedure been attempted, the drillpipe pressure could have been anticipated to reach a maximum of approximately 13,000 psi. It is unlikely that anyone would have been comfortable with pressures of this magnitude on the drillpipe and surface connections especially when that pressure is the result of gas containing 30 percent hydrogen sulfide.

THE SIGNIFICANCE OF THE SURFACE PRESSURES

The changes in surface pressure from the time of the initial influx on June 21 until July 15 are most interesting to analyze and study. During the first 18 days until July 9, the surface pressure remained relatively constant at 3700 psi. As discussed in Chapter 4, this constant surface pressure indicated that the influx did not migrate during this 18-day period.

After July 9, the surface pressure began to decline until July 15, the day of the surface failure and fire. The surface pressure had stabilized at approximately 2000 psi. As illustrated in Example 4.1, the decline in the surface pressure was the result of the migration of the influx. During the six days between July 9 and July 15, the influx had migrated to approximately 10,750 feet. The top of the influx on July 15 is substantiated by the analysis of the pumping operation prior to the fire, which is presented later in this text.

THE SNUBBING OPERATION TO JULY 14

The snubbing operation proceeded routinely to July 14. On July 14, the end of the drill string was at approximately 4494 feet. The surface pressure was allowed to increase to 3800 psi prior to circulating the 18.4-ppg

Table 9.2

Snubbing Record, E. N. Ross No. 2, July 14, 1985

Pipe Snubbed	Total Depth of Pipe	Displacement Volume	Volume Bled	Pressure
30.4	2161.37	0.36		3010
30.4	2191.77	0.36		3040
30.4	2222.17	0.36		3060
30.4	2252.57	0.36		3100
30.4	2282.97	0.36		3140
30.4	2313.37	0.36		3150
30.4	2343.77	0.36		3200
30.4	2374.17	0.36		3240
30.4	2404.57	0.36		3260
30.4	2434.97	0.36		3300
30.4	2465.37	0.36		3340
30.4	2495.77	0.36		3380
30.4	2526.17	0.36		3400
30.4	2556.57	0.36		3430
30.4	2586.97	0.36		3450
30.4	2617.37	0.36		3480
30.4	2647.77	0.36		3500
30.4	2678.17	(1.14)	1.5	3520
30.4	2708.57	0.36		3540
30.4	2738.97	0.36		3560
30.4	2769.37	0.36		3600
30.4	2799.77	0.36		3600
30.4	2830.17	0.36		3620
30.4	2860.57	0.36		3640
30.4	2890.97	0.36		3600
30.4	2921.37	0.36		3600
30.4	2951.77	0.36		3600
30.4	2982.17	0.36		3780
30.4	3012.57	0.36		3850

mud by not bleeding the quantity of mud displaced by the work string. Table 9.2 is a portion of the snubbing record for July 14. It is important to note that the 29 joints snubbed represent a displacement of 10.4 barrels. However, only 1.5 barrels were bled.

As discussed in Chapter 4, the displacement of each joint or stand should be bled as that joint is snubbed. Otherwise it is more difficult to understand the hole condition, determine the proper surface pressure, or analyze influx migration. However, with the 3000-psi surface pressure at the beginning of the day, the top of the influx was well below the end of the work string.

The snubbing procedure was followed to establish the drillpipe pressure pursuant to the classical procedure. That is, the annulus pressure was held constant at 3800 psi and the drillpipe pressure was established. The 18.4-ppg mud was then circulated to the surface with 276 barrels being pumped in and 265 barrels being recovered, the difference being the volume required to fill the drillpipe. After the pumping operation, the surface pressure stabilized at 2240 psi. No gas was circulated to the surface.

Since there was no gas in the drillpipe annulus and the influx did not migrate during the pumping operation, the displacement procedure proceeded without difficulty and was successful. For further discussion, refer to the section in Chapter 4 concerning circulating off bottom and the discussion previously presented in this section.

THE SNUBBING OPERATION, JULY 15

On the morning of July 15, the surface pressure was 2390 psi. The snubbing operation proceeded routinely. The snubbing procedure was altered to permit a bigger increment. Accordingly, the work string was snubbed to 8954 feet for a 4464-foot increment as opposed to the recommended 2000-foot interval. During the day the surface pressure declined until the end of the day when it stabilized at 2000 psi. According to the analysis in Example 4.1, the top of the influx was at 10,250 feet. By rigorous analysis, the top of the influx was at 10,750 feet. The conditions at the E. N. Ross No. 2 on the evening of July 15 are illustrated in Figure 9.8.

THE CIRCULATING PROCEDURE, JULY 15

As outlined in the snubbing procedure, the casing pressure was held constant at 2000 psi and the drillpipe pressure was established at 2900 psi at a rate of 3 bpm. Circulation commenced pursuant to Table 9.1 and Figure 9.4. Given the wellbore schematic as illustrated in Figure 9.8, the work string annular volume was 532 barrels. As previously noted,

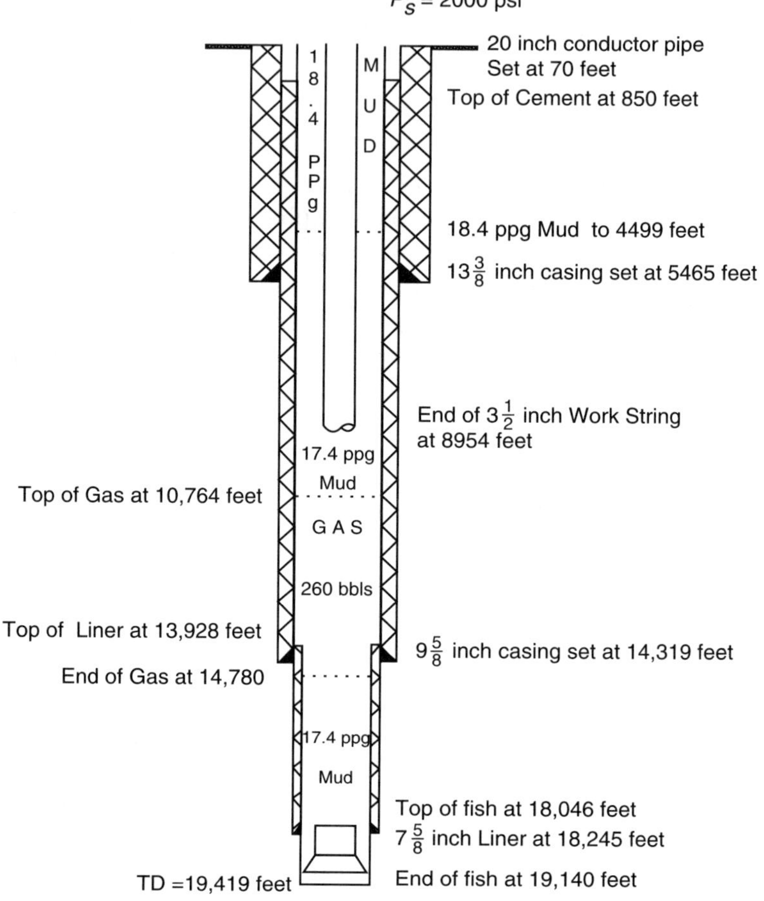

Figure 9.8   *July 15, 1985.*

the pit level began to increase after pumping approximately 120 barrels into the drill string. It is important to recognize that at that time, which was 1727 hours, the drillpipe pressure was constant at 2900 psi and the annulus pressure had declined to 1900 psi. Thereafter, the annulus pressure decreased to 1700 psi and then began to increase as the drillpipe pressure was held constant.

As illustrated in Table 9.1 during the 11-minute period of 1826 hours to 1837 hours, 95 barrels of mud were recovered from the well for an average rate out of 8.64 bpm. The average pumping rate during the same time was 3.09 bpm for a net gain of 5.55 bpm. Therefore, interpolating Table 9.1, the annular volume of 532 barrels was recovered at approximately 1831 hours.

As previously stated, gas did not reach the surface until approximately 1854 hours, or almost one-half hour after the recovery of the drill string annular volume and after a total of 725 barrels had been recovered. Therefore, it can be concluded that there was no gas present in the drill string annulus when the circulating operation commenced. It can also be concluded that the pit level gain, which commenced at 1727 hours after 120 barrels were pumped, was not the result of gas expansion in the drill string annulus.

It follows then that the gain and increase in annulus pressure was due to migration of the influx or an additional influx. Had the guidelines presented in Chapter 4 been followed, the operation would have been terminated when the pit level began to increase at 1727 hours. It could have been reasoned that the operation was not proceeding according to plan and procedure when the work string annulus volume had been recovered, no gas had been observed, and the annulus pressure was continuing to increase.

The top of the influx when circulation commenced can be substantiated by analyzing Table 9.1. Consider Example 9.3:

Example 9.3

Given:

Figure 9.8

Table 9.2

Drill string annular capacity, $C_{dpca} = 0.0589$ bbl/ft

$9\frac{5}{8}$-inch casing capacity, $C_{ci} = 0.0707$ bbl/ft

Gas to surface with 390 barrels in and 725 barrels out

Required:

Estimate the top of the influx when circulation commenced.

Solution:

The volume of the drill string annulus is

$$V_{dpca} = D_{dpca}(C_{dpca})$$

$$V_{dpca} = 8954(0.0589)$$

$$V_{dpca} = \textbf{527 bbls}$$

Volume pumped with 527 bbls out is approximately 320 bbls.

Volume in with gas to surface is 390 bbls.

Volume out with gas to surface is 725 bbls.

Volume pumped between annular volume out and gas to surface is

$$= 390 - 320$$

$$= \textbf{70 bbls}$$

Volume contributed by annulus is

$$= 725 - 527 - 70$$

$$= \textbf{128 bbls}$$

Top of influx when circulation commenced is

$$= \frac{128}{0.0707} + 8954$$

$$= \textbf{10,764 feet}$$

As illustrated in Example 9.3, 725 barrels of mud had been recovered when gas reached the surface (which is 198 barrels more than the drill string annular capacity). During the period in which the 198 barrels were being recovered, only 70 barrels were pumped into the well. Therefore, 128 barrels had to come from the hole below the end of the drill string. It follows that the top of the influx was at approximately 10,764 feet. A rigorous determination of the surface pressure and appropriate hydrostatics confirms this calculation to accuracy well within the limits of the data and accuracy of the pressure gauges being used.

The well was shut in and observed at 1942 hours. After 540 barrels of mud were pumped into the well and 1054 barrels were recovered, the annulus pressure had increased from 2000 psi to 5000 psi. It was decided to proceed, keeping the annulus pressure constant and pumping at 6 bpm. As reflected in Table 9.1, from that time until the pump-in line ruptured, the drillpipe pressure increased from 2900 psi to 6700 psi.

As previously stated, there is no correct model for circulating off bottom with gas in the drill string annulus. Therefore, circulating with the drillpipe pressure constant at 3 bpm is theoretically as proper as circulating at 6 bpm with the annulus pressure constant. However, as illustrated in Figure 9.4, the net gain decreased after increasing the pump rate to 6 barrels per minute, indicating that the condition of the well was improving.

The pump and return lines for July 15 are schematically presented as Figure 9.3. As illustrated, the hammer union failed on the rig floor. There were no check valves between the failure and the drill string, and the pump-out, back-pressure valve failed. With the annulus closed, the full force of the well was concentrated on the drillpipe and the flow from the well continued to increase. Efforts to close the two ball valves on the top of the drill string were unsuccessful although both valves were represented to be applicable for the conditions experienced.

As noted in Chapter 1, ball valves can be problematic in well control problems. In this particular instance, the uppermost ball valve was of the type commonly known as a low-torque valve and is considered efficient for use in difficult circumstances. (The lower valve was of the simple kelly-valve variety.) During controlled tests after the incident, at various rates and pressures, the valves could not be consistently operated using the associated closing bars. Laboratory conditions are usually much less traumatic than the actual conditions experienced at the wellsite. A hydraulic valve would probably have operated. Also, a gate valve can always be closed on flow; however, it could erode if the flow rate is high or closing requires considerable time.

With all of the problems encountered, control of the well was being regained when the hammer union on the pump-in line washed out.

Efforts to close the two valves were not successful under the conditions encountered even though they were represented to be applicable to those conditions. The well began to vent at the rig floor and the hydrogen sulfide alarms began to sound. The rig was soon inaccessible due to the toxic nature of the gas, and there was little alternative but to ignite the gas.

ALTERNATIVES

It is interesting to consider the alternatives available to control the well. With the bottom 1400 feet of the hole lost to the fish, it is difficult to remove the influx from that portion of the hole. Therefore, there is no good, easy, foolproof procedure to control this well.

One alternative would have been to attempt to bullhead the influx back into the formation. However, with 1100 feet of open hole it is unlikely that the influx would have been displaced into the Smackover. It is more likely that the fracture gradient at the shoe would have been exceeded and the influx above the shoe would have been displaced into the interval immediately below the shoe. The influx remaining in the bottom 1100 feet of open hole would have continued to be difficult.

The most intriguing aspect of the entire situation is that, for the first 18 days after the initial kick, the influx did not migrate and that, in the next six days, it only migrated less than 4000 feet to approximately 10,750 feet. One alternative well control procedure would have been to let the influx migrate to the surface. The procedures outlined in Chapter 4 could be used to design a migration schedule and anticipated annular pressure profile. The volumetric procedure is merely the Driller's Method with zero circulation rate. Consider Example 9.4:

Example 9.4

Given:

Examples 9.1, 9.2, and 9.3

Assume that the influx migrates to the surface.

The $3\frac{1}{2}$-inch drill string remains at 8954 feet.

The well contains only 17.4-ppg mud.

Assume the maximum surface pressure is 5000 psi.

Required:

1. The annulus pressure profile

2. The procedure for migration

3. The procedure for displacing the influx

Solution:

1. The procedure for determining the annulus pressure profile is the same as presented in Example 9.1. The surface pressure, under these conditions, with gas to the surface and a fish in the hole would be:

$$P_{0dm} = \textbf{3510 psi}$$

When first shut in, the top of the influx was at 13,546 feet and

$$P_a = \textbf{3700 psi}$$

The gas at 8954 feet, P_x, is given by Equation 4.21:

$$P_{xdm} = \frac{B}{2} + \left[\frac{B^2}{4} + \frac{P_b \rho_{m^z x} T_x h_b A_b}{z_b T_b A_x} \right]^{\frac{1}{2}}$$

$$B = P_b - \rho_m (D - X) - P_f \frac{A_b}{A_x}$$

$$B = 16{,}907 - 0.9048(19{,}419 - 8954)$$

$$-0.162(5873) \left(\frac{0.0426}{0.0707} \right)$$

$$B = \textbf{6892}$$

$$P_{8954dm} = \frac{6892}{2} + \left[\frac{6892^2}{4}\right.$$

$$+ \left.\frac{16,907(0.9048)(1.586)(636)(5873)(0.0426)}{1.988(772)(0.0707)}\right]^{\frac{1}{2}}$$

$$P_{8954dm} = \mathbf{10{,}314 \ psi}$$

$$P_a = P_x - \rho_m X$$

$$P_a = 10{,}314 - 0.9048(8954)$$

$$P_a = \mathbf{2212 \ psi}$$

The annulus pressure profile is presented in Figure 9.9.

2. The procedure for migration in this case is very simple. Assuming that the maximum casing pressure is 5000 psi, merely let the casing pressure increase to 5000 psi and bleed the mud necessary to keep the casing pressure constant at 5000 psi. The volume bled should not exceed the

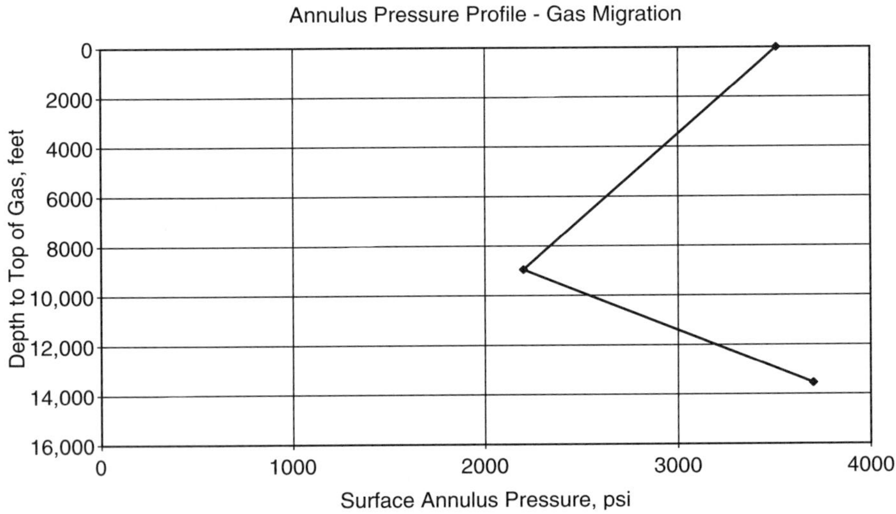

Figure 9.9

increased volume of the influx.

$$V_s = \frac{P_b z_s T_s}{P_s z_b T_b} V_b$$

$$V_s = \frac{16{,}907(0.775)(520)}{3510(1.988)(772)}(260)$$

$$V_s = \textbf{329 bbls}$$

Therefore, the additional gain is

$$\Delta V = 329 - 260$$

$$\Delta V = \textbf{69 bbls}$$

No more than 69 barrels of mud should be bled from the annulus as the influx migrates at 5000 psi.

3. Once the surface pressure stabilizes, the influx is at the surface and can be displaced using the Driller's Method, provided that the anticipated pressures and volumes are as predicted by the analysis. If not, the well must be shut in. Alternatively, a volumetric procedure could be used. The well could have been pressured to 5000 psi through the drill string while carefully measuring the mud pumped. After reaching 5000 psi, the annulus could have been bled to the stabilized surface pressure less the hydrostatic of the volume pumped.

For example, if P_a = 3510 psi and V_s = 329 bbls:

$$V_2 = \frac{P_1 V_1}{P_2}$$

$$V_2 = \frac{3510(329)}{5000}$$

$$V_2 = \textbf{231 bbls}$$

Or the volume of mud pumped to obtain 5000 psi would be

$$329 - 231 = \textbf{98 bbls}$$

The effective hydrostatic of the additional 58 bbls is

$$\Delta Hyd = \frac{V_m}{C_{dpca}}(\rho_m - \rho_f)$$

$$\Delta Hyd = \frac{98}{0.0589}(0.9048 - 0.162)$$

$$\Delta Hyd = \textbf{1236 psi}$$

Therefore, the surface pressure could be bled from 5000 psi to $3510 - 1236$ or

$$P_a = \textbf{2274 psi}$$

The procedure could be repeated until the influx is released. In this procedure, only **GAS** can be released.

As illustrated in Example 9.4, influx migration was a viable alternative. The maximum permissible surface pressure would not have been exceeded. The influx could have been displaced or released volumetrically.

Realizing that the influx had migrated to 10,750 feet would have offered an interesting alternative to consider. With the top of the influx at 10,750 feet, the bottom of the influx would have been at 14,780 feet. The work string could have been snubbed to the top of the fish and the influx circulated out using the Driller's Method. The anticipated pressures would be the same as determined in Example 9.1 with the exception that the influx would have begun above 10,000 feet as opposed to beginning from bottom. However, the risk is that, if there is additional influx below the end of the work string, the problems could be the same as occurred. Therefore, any displacement procedure must be carefully and completely analyzed. If there is any deviation from the analysis, the well must be shut in and evaluated.

Another alternative would have been to connect the well to the plant and flare lines and flow the well. Historically, the wells in the area produced only a few million cubic feet of gas per day. Therefore, the well might have been flowed for a short period until the flowing bottomhole pressure was drawn down. After that, it could have been killed utilizing the dynamic momentum concepts described in Chapter 5. This would have been a good contingency for the effort to circulate from below the influx as previously described. As a matter of interest, when the well was finally capped after the fire, it was controlled as just described. A heavy mud was pumped at a high rate with the well flowing. It died.

OBSERVATIONS AND CONCLUSIONS

There are very interesting and educational observations to be made. There were many instances in which the procedure was predicated on reaching "bottom." However, after the fish was left in the hole, it was no longer possible to reach total depth of 19,419 feet. With the top of the fish at 18,046 feet, "bottom" became 1373 feet less than total depth. Therefore, any theoretical procedure based on "bottom" was practically impossible. The concept of stripping to "bottom" and circulating out was theoretically impossible. With the drill string as deep as possible, 1373 feet of gas would remain below the end of the drill string.

In well control operations, circulating off bottom is rarely a good idea. In this case, the condition of the well deteriorated, in most instances when circulation was attempted.

CHAPTER TEN

CONTINGENCY PLANNING

Daily Drilling Report

Monday, January 12, 1988
We came out of the hole with the core last night and noticed that the well was flowing. So, we picked up the bottom hole assembly and started back in the hole. We got to 1500 feet and the well was flowing too hard to continue. We shut in and had 500 psi at the surface. We picked up the kelly and circulated for one hour keeping the drillpipe pressure constant. Then, we shut in again and had 4200 psi at the surface. What do you want to do now?

For as long as oil and gas wells have been drilled, there have been kicks, blowouts, well fires, and other well control problems. It is certain that these problems will continue. In fact, a recent statistical study concluded that there are as many problems today as there were in the 1960s—which is rather startling considering the emphasis on regulation and training.

Well control in any and all forms can be accomplished by anyone. That is, anyone can put out a well fire and anyone can kill a well. The Al-Awda Project in Kuwait proved this point beyond a shadow of doubt. In Kuwait, fires were extinguished and wells killed by professionals, engineers, tool pushers, roughnecks, snubbing hands, school teachers, bartenders, mud engineers, farmers, Americans, Iranians, Chinese, Romanians, Hungarians, Russians, as well as Kuwaitis. Interestingly, the average time required to control each well in Kuwait was essentially the same regardless of the composition or national origin of the team.

The consequences of a blowout can be staggering, as can the consequences of a bad decision during control attempts. There is the potential for loss of life, revenue, equipment, and vast quantities of natural resources. From the advent of well control as a profession in the early 20th century, until the early 1980s, only one life was lost—Myron Kinley's brother,

Floyd, was killed on a rig floor while fighting an oil well blowout and fire. This is remarkable considering the conditions under which well control efforts are conducted. Since the early 80s, several have been killed, including five (three professionals) at a blowout in Syria. Considering the present state of the industry, including the well control service providers, more are sure to be killed in the future.

Why has there been more loss of life? There are several reasons, in the opinion of this writer. One reason is that since the boom of the early 80s and since the Gulf War, there are more well control "professionals." Another reason is that since the bust of the 1980s, the industry has lost a wealth of practical field experience. As the worldwide drilling manager for Amoco, George Boykin, in an address to the SPE/IADC Drilling Conference in the spring of 1998, complained that the industry is experiencing a crew change.

Well control operations are generally expensive to very expensive. Several blowouts in recent years have been reported to cost in excess of $100 million. One was reported to cost in excess of $250 million. Recovery and control operations at the Apache Key 1-11, which blew out in 1981 and was referred to by many as the biggest blowout in the history of the state of Texas, were reported to have cost in excess of $50 million.

The value of the forever-lost natural resources can exceed the recovery costs. At one international operation, more than two-thirds of one of the largest gas fields in the world have been lost to an underground blowout. As with many industries, environmental considerations can be a major expense. A burning gas well has very little environmental impact. However, with a spewing oil well, the cost of restoring the environment can exceed the recovery costs.

Costs and problems do not always end when the well is killed. The litigious nature of today's world results in extensive litigation, which often exceeds the life of the well and costs more than the control effort. For example, after the blowout at the Apache Key, the litigation included hundreds of litigants, cost hundreds of millions of dollars, and threatened the very existence of the company. The final legal issues were resovled in 1998, 17 years after the blowout.

The indirect costs and consequences to the energy industry can be equally staggering. The blowout in the Santa Barbara Channel significantly impacted the regulatory and environmental problems faced by the domestic industry. In Canada, a large sour gas blowout in 1985 forever changed the regulatory requirements, increased government intervention, and added to the cost of exploration. In the North Sea, the incident at the Piper Alpha platform and at the Saga blowout had similar consequences.

The operations group is faced with substantial responsibility with regard to well control and is challenged to minimize the impact of any well control event. As with any operation, management, planning, and execution are fundamental. In well control, a contingency plan is vital to minimizing the impact of the problem.

In the opinion of this writer, one of the most important functions of the contingency plan is to determine when additional expertise is required in any situation, under any circumstance. Specifically, at what point and under what circumstances does the driller involve the tool pusher and the tool pusher involve the drill site supervisor and the drill site supervisor involve the office, etc.? The basic answer to this series of questions is that the next level of supervision must become active when the expertise of the first level is exceeded, a critical decision must be made, or when a crisis may be looming. Perhaps more important, when does the operator involve outside help in the form of a well control consultant? Again, the answer depends on the level of expertise of the operator. All team members should be encouraged to seek the next level of assistance. It is always better (and cheaper) to have help and not need it, than to need help and not have it.

The contingency plan must first and foremost emphasize the safety of the personnel at the well site. Any act or action that will expose well site personnel to the risk of personal harm must be discouraged. It is the responsibility of management to define the point at which the well and perimeter will be secured and the well site abandoned. It can never be forgotten that the principal components required for a fire are fuel, oxygen, and an ignition source. Anytime these three are present, all personnel in the immediate vicinity are in grave danger. The situation is even more serious if the wellbore fluids include toxic components such as hydrogen sulfide. Equipment can be replaced.

It is not possible to globalize a contingency plan to the extent that the only requirement is to fill in the blanks. A fat plan with endless flow charts and irrelevant information usually gathers dust on a shelf and is worse than no plan. It must be sufficiently site-specific to identify the personnel, equipment, and services required to address a potential well control scenario. For example, an operator in south Texas need not identify the location and capacity of various cargo planes, whereas an operator in the jungles of Africa needs that information.

For outside well control expertise, the operator faces a challenge. Historically, a well control "expert" was primarily a fire fighter. Since only about one surface blowout in a hundred catches fire, fire fighting expertise is only one consideration. A well may blowout at the surface or it may blowout underground. The blowout may require surface or subsurface intervention or a relief well may be necessary. Choosing the appropriate form of expertise is a challenge.

The operator must evaluate the qualifications of the individuals to be utilized by the potential service providers. Bright coveralls and an arrogant attitude do not qualify someone to be a well control expert. Experience and qualifications may be difficult to determine. Well control operations are often cloaked in secrecy. The operator may be embarrassed by the loss or may fear litigation. As a result, follow-up investigations may not be sincere efforts. The service providers are not usually interested in divulging "trade secrets" such as their methodology, philosophy, or well control techniques for fear of losing a real or perceived competitive edge. Egos are huge. Claims on any job may be many. In some instances, a job has been brilliantly conducted. In other instances, the well control efforts have turned a serious well control problem into a major disaster. Unfortunately, both may be proclaimed as a triumph. Ignorance and arrogance are a dangerous combination.

The best contingency plan will include a well control consultant with nothing to sell but his time. Some operators identify and depend on individual experts for various functions regardless of their company affiliation. It has become popular for some large service providers to offer total services. However, that approach is not always economical. The service provider may be more interested in providing unnecessary pumping services, if that is the primary service line, at the expense of the project.

CHAPTER ELEVEN

THE AL-AWDA PROJECT:
THE OIL FIRES OF KUWAIT

No text on advanced pressure control would be complete without a brief history and overview of this historical project. I am proud to have served the Kuwait Oil Company and the Kuwaiti people in their effort. I consider my involvement one of the greatest honors of my career.

No picture can capture and no language can verbalize the majesty of the project. It was indeed beyond description by those present—and beyond complete appreciation by those not present. A typical scene is shown in Figure 11.1 which was photographed during the day. The smoke turned day to night.

With these extensive oil fires, the rape of Kuwait was complete. The retreating Iraqi troops had savagely destroyed everything in the oil fields. There was nothing left to work with. There were no hand tools, no pump trucks, no cars, no pickups, no housing—nothing. Everything necessary to accomplish the goal of extinguishing the fires and capping the wells had to be imported.

The world owes the valiant Kuwaitis a great debt. To extinguish almost 700 oil well fires in eight months is an incredible accomplishment, especially in light of the fact that Kuwait is a very small country of only 1.5 million people that had been completely and ruthlessly pillaged. No one worked harder or longer days than the Kuwaitis. Many did not see their families for months and worked day after day from early morning until long after dark for days, weeks, and months in their tireless effort to save their country.

OVERVIEW OF THE PROJECT

As understood by the author, the basic organization chart, effective after August 1, 1991, as pertained to fire fighting and well control,

418

Figure 11.1

is presented as Figure 11.2. The oil fields of Kuwait are shown in Figure 11.3. Greater Burgan Field is the largest oil field in Kuwait. Larry Flak was the coordinator for the fields outside Burgan, which were Minigish and Raudatain. Texaco was responsible for the wells in the neutral zone and the British Consortium was responsible for Sabriyah. The Kuwaiti Wild Well Killers, or KWWK, were responsible for the wells in Umm Gudair. Larry Jones, a former Santa Fe employee, was charged with contracts and logistics. Abdoulla Baroun, an employee of Kuwait Oil Company, liaisoned between Kuwait Oil Company and the multinational teams.

By August 1, eight to twelve teams from four companies had controlled 257 wells with most of the wells being in the Ahmadi and Magwa Fields, which are nearest to Kuwait City. After Magwa and Ahmadi, the primary emphasis was on the Burgan Field. However, as additional teams arrived, the original teams were moved to the fields outside Burgan. By the end of the project in early November, there were 27 fire fighting teams deployed in Kuwait as shown in the organization structure of Figure 11.2.

Thousands were involved in these critical operations and all deserve mention. Almost all of the support was provided by Bechtel under

AL-AWDA PROJECT
Kuwait Petroleum Company
Kuwait Oil Company

Saud Al-Nashmi
Manager of Drilling and Well Control

SUPPORT SERVICES
BECHTEL
Tom Heischman
Manager

WELL CONTROL
Neal V. Parry
Senior Coordinator

Communications
Transportation
Water Supply
Accounting
Purchasing
Reporting
Logistics
Housing

Third Party Services
Mine Clearing
Construction
Demolition
Equipment
Support
Supply

Robert D. Grace
Burgan Field
Coordinator

Larry Jones
Contracts
Coordinator

Larry Flak
Minigish and Roudatain
Coordinator

Abdullah Baroun
International
Coordinator

Red Adair Company   3 Teams
Cudd Pressure Control   2 Teams
Abel Engineering   1 Team
Alert   2 Teams
Kuwait Wild Well Killers 1 Team
USSR Fire Fighters   1 Team
China Fire Fighters   1 Team
Rumanian Fire Fighters   1 Team
Hungarian Fire Fighters 1 Team
Red Flame   1 Team
Iranian Fire Fighters   1 Team

Wild well control, Inc. 3 Teams
Safety Boss   3 Teams
Boots and Coots   3 Teams
French Fire Fighters   2 Teams

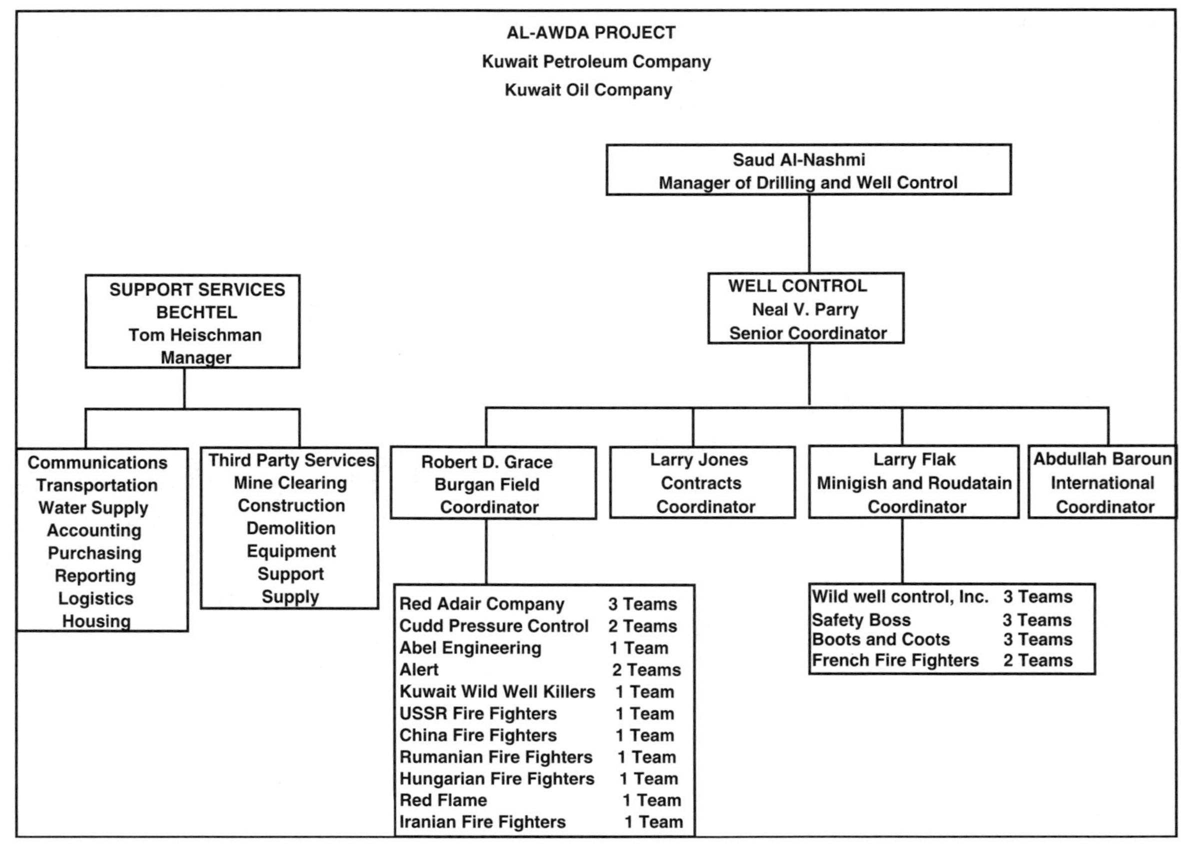

Figure 11.2

Figure 11.3

the very capable management of Tom Heischman. Texaco furnished support in the Neutral Zone, and the British Consortium furnished most of the support in Sabriyah. A substantial contribution was made by the management and employees of Santa Fe Drilling Company, many of whom were among the first to return to Kuwait after the war. One of the many contributions made by Santa Fe Drilling Company was the supply of heavy equipment operators who worked side by side with fire fighters to clear the debris and extinguish the fires.

An early report of the status of the fields and wells in Kuwait is presented as Table 11.1. In most of the fields, it was easy to determine the status of each individual well. Such was not the case in Burgan Field. In Burgan, the well density was very high. The smoke reduced visibility to a few feet, and access to some parts of the field was impossible until the very end of the project. Even in the last few weeks, there was disagreement concerning the status of individual wells. However, the totals were very accurate considering the circumstances. A typical day in Burgan Field is shown in Figure 11.1.

The majority of the wells in Kuwait are older and shallow (less than 5000 feet) with surface pressures less than 1000 psi. Typically, they were completed with $3\frac{1}{2}$-inch tubing inside 7-inch casing and produced through both the casing and the tubing. The older wells had the old-style Gray Compact Head that houses all of the casing hangers in one body in progressively larger mandrels. A Gray Compact Head is pictured in Figure 11.4. The newer and deeper wells had higher pressures and more conventional wellheads.

The Iraqi troops packed plastic explosives around the bottom master valve on the tree as well as on the wing valves on the "B" section. Sand bags were then packed on top of the explosives to force the explosion into the tree. The force of the explosion was tremendous and the damage indescribable.

In fortunate instances, such as pictured in Figure 11.5, the tree was blown off at the bottom master valve and there was no other damage. In that case, the fire burned straight up and the oil was almost totally consumed. However, the vast majority of cases were not so simple. In most instances the destruction of the tree was not complete and, as a result, oil flowed out of multiple cracks and breaks. As a consequence, the combustion process was incomplete. The unburned oil collected around the wellhead in lakes that were often several feet deep (Figure 11.6). The ground throughout the Burgan Field was covered with several inches of oil. Ground fires covering hundreds of acres were everywhere (Figure 11.7). In addition, some of the escaping unburned oil was cooked at the wellhead and formed giant mounds of coke (Figure 11.8).

While the coke mounds were an immediate problem for the fire fighters, perhaps they were a benefit to Kuwait and the world. The coke accumulations served to choke the flow from the affected wells. In all

Table 11.1

The Al-Awda Project: Oil Wells Survey Data by Field

Field	Total Number of Wells					Remarks
	Drilled	On Fire	Oil Spray	Damaged	Intact	
Magwa	148	99	5	21	15	
Ahmadi	89	60	3	17	6	
Burgan	423	291	24	27	67	
Raudatain	84	62	3	5	3	
Sabriyah	71	39	4	9	2	
Ratqa	114	0	0	0	8	Figure excludes shallow wells
Bahra	9	3	2	*	*	*Denotes uncertain about figure
Minagish	39	27	0	7	1	
Umm Gudair	44	26	2	10	2	
Dharif	4	0	0	0	3	
Abduliyah	5	0	0	0	4	
Khashman	7	0	0	1	1	
Total	1037	607	43	97	112	

Figure 11.4

Figure 11.5

Figure 11.6

Figure 11.7

Figure 11.8

probability, the reduction in flow rate more than offset the additional time required to cap the wells.

THE PROBLEMS

THE WIND

The wind in Kuwait was a severe problem. Normally, it was strong from the north to northwest. This strong, consistent wind was an asset to the fire fighting operation. However, during the summer months, the hot wind was extremely unpleasant. In addition, the sand carried by the wind severely irritated the eyes. The only good protection for the eyes was to wear ordinary ski goggles. It was expected that the summer "schmals" or wind storms would significantly delay the operations. But such was not the case. The oil spilled onto the desert served to hold the sand and minimize the intensity of the wind storms. As a result, the operation suffered few delays due to sand storms.

The wind was most problematic on those occasions when there was no wind. During these periods, it was not uncommon for the wind direction to change 180 degrees within fifteen minutes. In addition, the wind direction would continue to change. Any equipment near the well might be caught and destroyed by the wind change. In any event, all of the equipment would be covered with oil, and the operation would be delayed until it could be cleaned sufficiently to continue. These conditions could persist for several days before the wind once again shifted to the traditional northerly direction.

The humidity in the desert was normally very low. However, when the wind shifted and brought the moist Gulf air inland, the humidity would increase to nearly 100 percent. When that happened, the road would become very slick and dangerous. On several occasions there were serious accidents. In one case, a man was paralyzed as a result of an automobile accident caused by the slick road.

LOGISTICS

The first problem was to get to the location. Access was provided to the fire fighters by the EOD, who cleared the area of explosives remaining from the war, and by Bechtel, who was responsible for furnishing the location and supplying water for fire fighting. Everyone involved in this aspect (and there were many) did an incredible job. Close cooperation was vital in order to maintain efficiency in strategy planning. The goal was to keep all teams working to the very end of the project.

During the height of the activity in September and October, it was not unusual to haul 1500 dump truck loads of road and location building material each day and several hundred loads during the night.

It was not possible to survey the locations in the oil lakes for munitions. Therefore, access was safely gained by backing the truck to the end of the road and dumping a few cubic yards of material into the lake, hoping that the dirt would cover any ordinance. A dozer would then spread the material and the process would be repeated until the location was reached.

WATER

The water system was constructed from the old oil gathering lines (illustrated in Figure 11.9). Most of the lines had been in the ground for

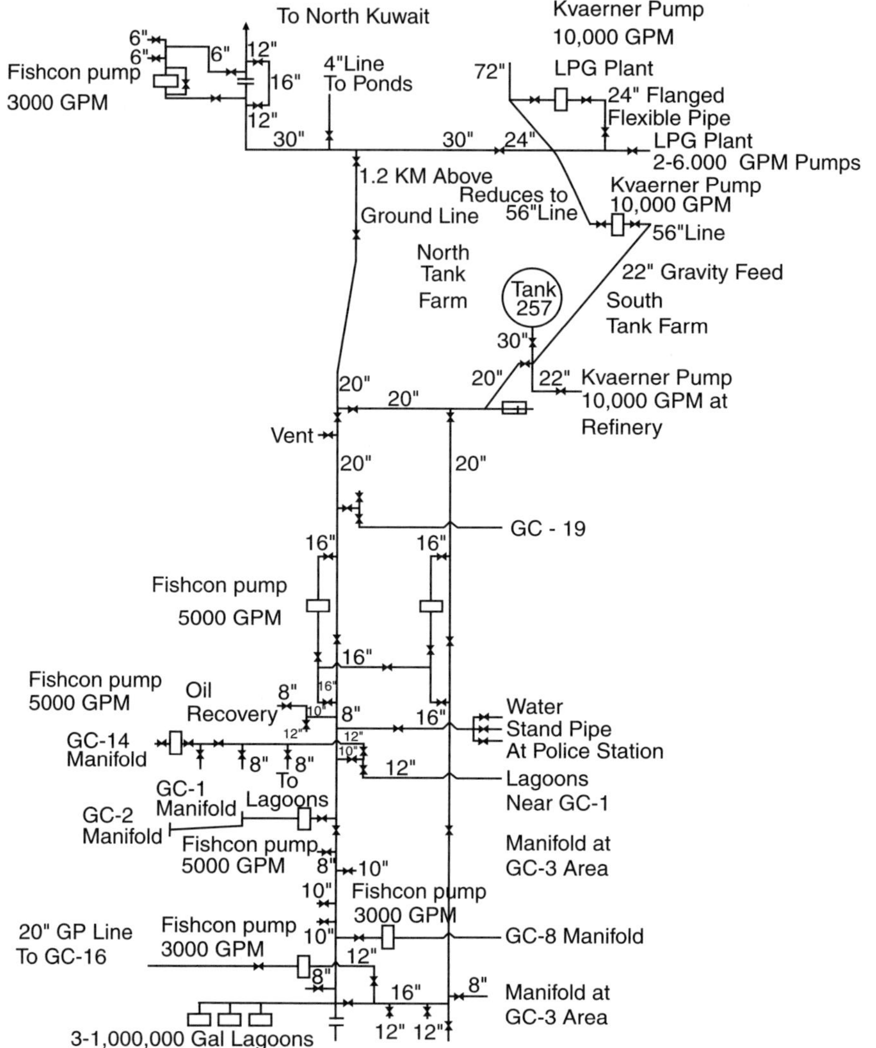

Figure 11.9   *The Water Supply System.*

many years and all had been subjected to munitions. Therefore, it was not unusual for a water line laid across the desert to look like a sprinkler system. In spite of everything, 25 million gallons of water were moved each day. Lagoons were dug and lined at each location, when the gathering system was capable of supporting the lagoon with water. The capacity of the lagoons was approximately 25,000 barrels.

Sufficient water was always a problem. Water was continually pumped into the lagoon during the fire fighting phase of the operation. Generally, the lagoon would fill overnight. During the last days of the project, two lagoons were constructed in the Burgan Field at locations predicted to be difficult. Because the team concentration increased near the end, the demand on the isolated working area was very great. In spite of all obstacles, in these last days, water was not a problem for the first time during the project. In areas where water could not be transported by pipe line, trac tanks were used and water was trucked from nearby loading points.

GROUND FIRES

Once the location was reached, the fire fighters took over and spread the material to the well. In the process, ground fires had to be controlled and were a major problem. The ground fires often covered tens of acres. In many instances, we were not able to identify the well. The ground fires were fed by the unburned oil flowing from the coke mounds or from the wells themselves after the fire had been extinguished. Most of the time, the wild well fighters worked on a live well with a ground fire burning less than 100 yards downwind.

The worst ground fires were in the heart of Burgan. In anticipation of the problems, together with Safety Boss personnel, a unit was specially designed to fight the ground fires. It consisted of a 250-bbl tank mounted on Athey Wagon tracks. A fire monitor was mounted on top of the tank and a fire pump was mounted on the rear of the tank. It was pulled by a D-9 Cat and followed by a D-8 for safety. It had a built-in inductor tube to disperse various foaming agents commonly used in fire fighting. The crew routinely worked between the blowing well and the ground fire and would be covered with oil every day. "Foamy One," pictured in Figure 11.10, and her crew made a tremendous difference and contribution to the effort.

The Rumanian and Russian fire fighters were particularly good at fighting and controlling ground fires. They both brought fire trucks capable of spreading powder and chemicals to smother the ground fires. Their efforts contributed significantly to the success of the project. In all cases, the hot spots had to be covered with dirt to prevent re-ignition. Some of the hot spots continued to burn for months. They often were not visible during

Figure 11.10

the day. However, at night the desert fire was clearly visible. One of the last projects was to cover these persistent devils.

OIL LAKES

In addition to the ground fires, unburned oil had gathered into huge lakes. Often the lake surrounded the well. In one instance, fire came out one side of the coke pile and a river of oil flowed down the other side of the coke pile. The lakes could be several feet deep. Often the lakes caught fire and burned with unimaginable intensity, producing tremendous volumes of smoke.

Working in the lakes was very dangerous because the access roads could become bounded by fire, trapping the workers. After the oil weathered for several days, it was less likely to burn. Therefore, most lake fires could be extinguished by eliminating the source of fresh oil. In the latter days, the problem was solved by surrounding the burning well in the lake with a road approximately 50 feet in width. The road was then crossed by fire jumps in strategic locations that isolated the fire from the fresh oil in the lake. The approach was successful. Had some of the bigger lakes caught fire, they would have burned for weeks (see Figure 11.6).

THE COKE PILES

Once the well was reached, the wellhead had to be accessed. Typically, each fire fighting team had fire pumps, Athey Wagons, monitor sheds, cranes, and two backhoes—a long reach and a Caterpillar 235. The first step was to spray water on the fire from the monitor sheds in order to get close enough to work. The average fire pump was capable of pumping approximately 100 barrels of water per minute and two were usually rigged up on each well.

Although the lagoons contained approximately 25,000 barrels of water, they could be depleted very rapidly at 200 barrels per minute. Using the water cover, the monitor sheds could be moved to within 50 feet of the wellhead. The long reach backhoe could then be used to dig away the coke pile and expose the well head. This operation is illustrated in Figure 11.11.

As previously mentioned, some of the unburned oil cooked around the wellhead to form a giant coke pile. The coke pile formed like a pancake 100 feet in diameter. At the wellhead itself, the coke pile might be as large as 30 to 40 feet high and 50 to 70 feet in diameter. It had the appearance of butter on top of a pancake.

In some instances the coke piles were very hard and difficult to dig. In other instances, the coke was porous and easily removed. It was not unusual for fire to burn out one side of the coke pile while oil flowed out another side.

In the northern fields, berms had been constructed around the wellheads. These quickly filled up with coke. Digging the coke resulted in a pot of boiling, burning oil.

CONTROL PROCEDURES

After exposing the wellhead, the damage was assessed to determine the kill procedure. Eighty-one percent of the wells in Kuwait were

Figure 11.11

Table 11.2
Summary of Kill Techniques

Stinger	225
Kill Spool	239
Capping Stack	94
Packer	11
Other	121
Total	690

controlled in one of three procedures. The exact proportions are presented in Table 11.2.

THE STINGER

If the well flowed straight up through something reasonably round in shape, it would be controlled using a stinger. As shown in Table 11.2, a total of 225 wells were controlled using this technique. A stinger was simply a tapered sub that was forced into the opening while the well flowed and sometimes while it was still on fire. The stinger was attached to the end of a crane or Athey Wagon. The kill mud was then pumped through the stinger and into the well. If the opening was irregular, materials of irregular shapes and sizes were pumped to seal around the stinger.

Due to the low-flowing surface pressures of most of the wells, the stinger operations were successful. However, stingers were not normally successful on openings larger than seven inches or on higher pressured wells. A typical stinger operation is schematically illustrated in Figure 11.12.

THE CAPPING SPOOL

Another popular alternative was to strip the wellhead to the first usable flange. Using a crane or an Athey Wagon, a spool with a large ball valve on top was then snubbed onto the usable flange remaining on the tree. The valve was then closed and the well killed through a side outlet on the spool below the ball valve. This procedure was referred to as a capping spool operation and is illustrated in Figures 11.13 and 11.14. As shown in Table 11.2, 239 wells were controlled using this technique. This operation was performed after the fire had been extinguished.

THE CAPPING STACK

Failing the aforementioned alternatives, the wellhead was completely stripped from the casing. This operation was accomplished with and without extinguishing the fire. In some instances, the wellhead was pulled off with the Athey Wagon. In other instances, it was blown off with explosives. In the early days of the effort, it was cut off using swabbing units and wirelines. In the final four months, the procedure often involved the use of high-pressure water jet cutters. (Cutting will be discussed later.) After the wellhead was removed, the casing strings were stripped off using mechanical cutters, commonly known as port-a-lathes, leaving approximately 4 feet of the string to receive the capping stack.

The wells were normally capped with a capping stack on the 7-inch production casing. Fortunately, all of the casing strings in the wells in Kuwait were cemented to the surface. Therefore, when the wellheads were severed from the casing, they did not drop or self-destruct as would

Pump-In Line
Stinger
Athey Wagon
Boom
Pump
In Line
Athey Wagon Boom

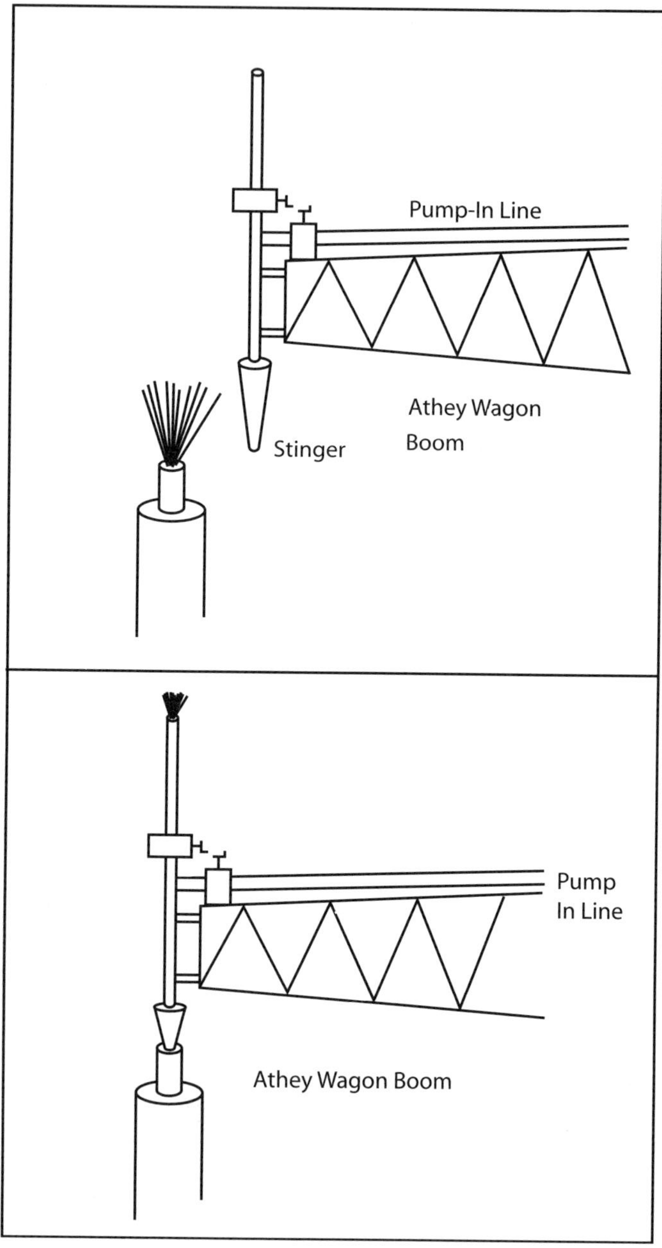

Figure 11.12

Figure 11.13

commonly occur in most wells. The capping stack consisted of three sets of blowout preventers (Figure 11.15). The first set were slip rams designed to resist the upward force caused by the shut-in well. The second set were pipe rams turned upside down in order to seal on the exposed 7-inch casing. A spool with side outlets separated the slip rams from the uppermost blind rams.

After the wellhead was removed, the fire was extinguished. The casing was stripped, leaving approximately 4 feet of 7-inch casing exposed. Normally, a crane would be used to place the capping stack on the exposed casing. Once the capping stack was in place, the pipe rams were closed, followed by the slip rams. The well would now flow through the capping stack.

When the pump trucks were connected and all was ready, the blind rams were closed. The pump trucks then pumped into the well. This operation has been performed elsewhere on burning wells, but was performed in Kuwait only after the fires were extinguished. The capping stack operation was performed on 94 wells in Kuwait. BG-376 was the last blowout in

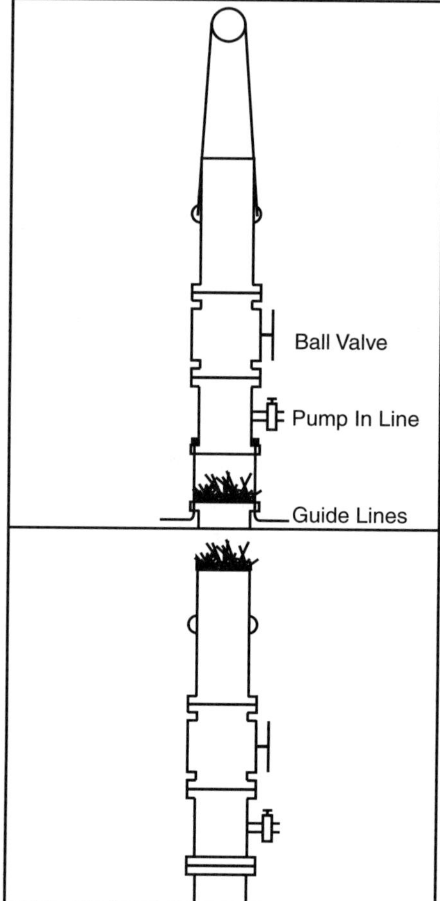

Figure 11.14   *Kill Spool.*

the Burgan Field and was controlled on November 2, 1991. Figure 11.16 depicts the capping stack operation at BG-376.

EXTINGUISHING THE FIRES

WATER

Extinguishing the fires in Kuwait was the easiest part of the entire operation. Three basic procedures were employed, though no official

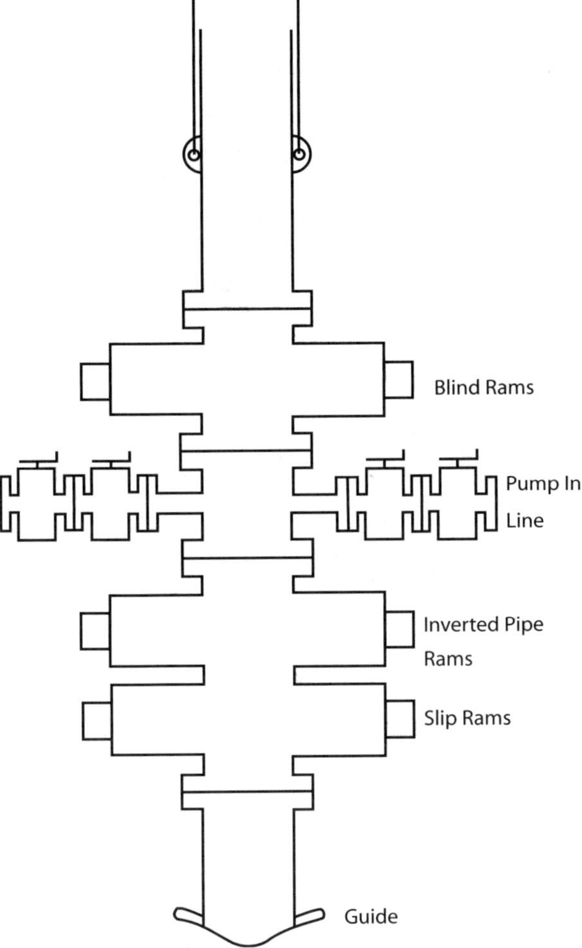

Figure 11.15	*Capping Stack.*

records were kept. However, the vast majority of the fires were extinguished using the water monitors. The oil contained asphalt and was of low gravity; therefore, it was less volatile than most crudes. Usually three to five monitors were moved close to the fire, and the flow was intensified at the base of the fire. As the area cooled, the fire began to be interrupted. One monitor sprayed up along the plume to further cool the fire. Very large well fires were extinguished in just minutes. Figure 11.17 illustrates the use of the fire monitors to extinguish a fire.

Figure 11.16

Figure 11.17

Some of the teams used fire suppressant materials and chemicals to effectively extinguish the fires and minimize their water requirements. A much broader use of these materials would have more effectively conserved the precious water supplies.

NITROGEN

Prior to the availability of the required volumes of water, many fires were extinguished using nitrogen. A 40-foot chimney attached to the end of an Athey Wagon was placed over the fire, causing the flow to be directed up through the chimney. The fire burned only out of the top of the chimney. Nitrogen was then injected into the chimney through an inlet at the base. A typical chimney is shown in Figure 11.17.

EXPLOSIVES

Of course, some of the fires were extinguished using explosives. Explosives effectively rob the fire of the oxygen necessary to support combustion. The fire monitors were used to cool the fire and the area around the fire in order to prevent re-ignition. Charge size ranged from five pounds of C_4 to 400 pounds of dynamite. The charges were packed into a drum attached to the end of an Athey Wagon boom. Some included fire suppressing materials along with the explosives. The drum was wrapped with insulating material to insure that the explosives did not merely burn up in the fire. The explosives were then positioned at the base of the plume and detonated. Figure 11.18 illustrates a charge being positioned for detonation.

NOVEL TECHNIQUES

One technique captured considerable publicity and interest. The Eastern Bloc countries—Russia, Hungary, and Rumania—used jet engines to extinguish the fire. The Hungarian fire fighters were the most interesting. Their "Big Wind" consisted of two MIG engines mounted on a 1950s vintage Russian tank. Water and fire suppressants were injected through nozzles and into the vortex by remote control. The tank was positioned approximately 75 feet from the fire and then the water lines were connected.

Figure 11.18

The engines were turned on at low speed and the water started to protect the machine as it approached the fire. The tank was then backed toward the fire. Once positioned, the speed of the engines was increased and the fire was literally blown out as one would blow out a match. The Hungarian "Big Wind" is illustrated in Figure 11.19.

CUTTING

In the early days of the fire fighting operation, a steel line between two swab units was used to saw casing strings and wellheads. This technique proved to be too slow. By early August, pneumatic jet cutting had completely replaced the swab lines. Water jet cutting is not new technology even to the well control business; however, techniques were improved in Kuwait.

Two systems were primarily used in Kuwait. The most widely used was the 36,000 psi high pressure jet system using garnet sand. A small jet was used with water at 3 to 4 gallons per minute. Most often, the cutter

Figure 11.19

ran on a track around the object to be cut. In other instances, a handheld gun was used, which proved to be very effective and useful. More than 400 wells employed this technique.

One limitation was that the fire had to be extinguished prior to the cut, and the cellar prepared for men to work safely at the wellhead. Another aspect had to be evaluated before widespread use was recommended. In the dark of the smoky skies or late in the evening, the garnet sand caused sparks as it impacted the object being cut. It was not known if under certain circumstances these sparks would have been sufficient to ignite the flow. However, re-ignition was always of concern.

Another water jet system used a 3/16 jet on a trac or yoke attached to the end of an Athey Wagon boom. The trac permitted cutting from one side with one jet while the yoke involved two jets and cut the object from both sides. The system operated at pressures ranging from 7500 psi to 12,500 psi. Gelled water with sand concentrations between 1 and 2 ppg was used to cut. The system was effective and did not require the men to be near the wellhead. In addition, the system could be used on burning wells, provided the object to be cut could be seen through the fire. This technology was used on 48 wells.

Conventional cutting torches were used by some. A chimney was used to elevate the fire and the workers would cut around the wellhead. Magnesium rods were also used because they offered the advantage of being in 10-foot lengths that could be telescoped together.

STATISTICS

The best authorities predicted that the operation would require five years. It required 229 days. The project's progress is illustrated in Figures 11.20 and 11.21. Originally, there were only four companies involved in the fire fighting effort. At the beginning of August, additional teams were added, forming a total of 27 teams from all over the world. The companies that participated and the number of wells controlled by each company are shown in Table 11.3. The number of wells controlled by each company listed in Table 11.3 is not significant because some companies had more crews and were in Kuwait for a longer period of time. What is significant is that the most difficult wells were controlled after August 1. As shown in Figure 11.21, the number of team days per well was essentially constant at approximately four team days per well. That is not to say that some wells were not more difficult and that some teams were not better.

As also illustrated in Figure 11.21, the high was reached in the week ending October 12, when a record 54 wells were controlled in

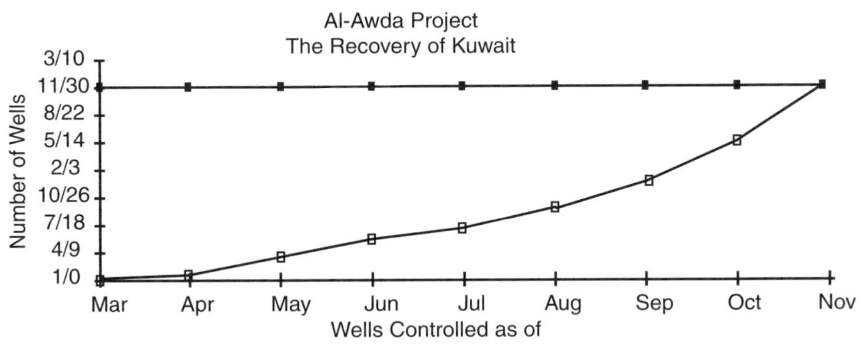

Figure 11.20

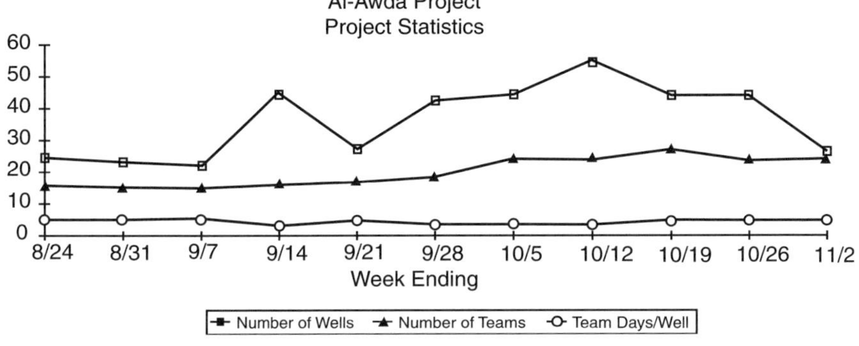

Figure 11.21

Table 11.3

The Al-Awda Project: Oil Wells Secured and Capped

Contractor's Name	Total
Red Adair (American)	111
Boots & Coots (American)	126
Wild Well Control (American)	120
Safety Boss (Canadian)	176
Cudd Pressure Control (American)	23
NIOC (Iran)	20
China Petroleum	10
Kuwait Oil Company (KOC)	41
Alert Disaster (Canadian)	11
Hungarian	9
Abel Engineering (American)–KOC	8
–WAFRA	31
Rumanian (Romania)	6
Red Flame (Canadian)–KOC	2
–WAFRA	5
Horwell (French)	9
Russian	4
British	6
Production Maintenance	9
Total	727

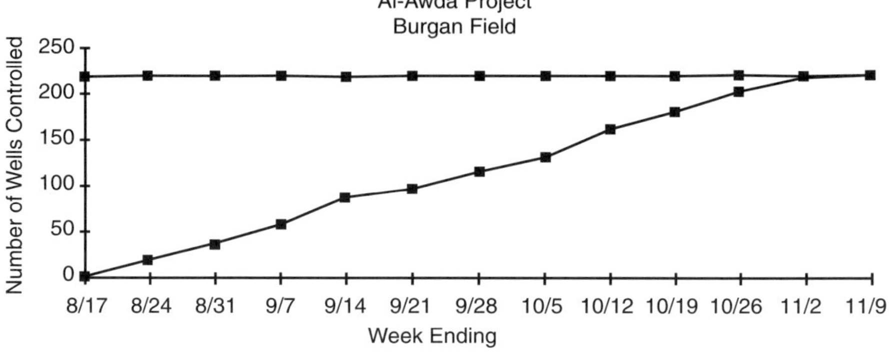

Figure 11.22

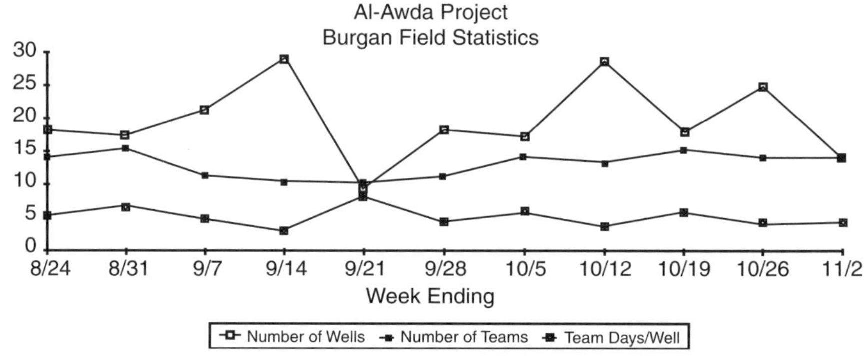

Figure 11.23

seven days. For the month of October, 195 wells were controlled for an average of 6.3 wells per day. The record number of wells controlled in a single day was 13.

Figures 11.22 and 11.23 illustrate the statistics for the Burgan Field. A high of 15 teams were working in Burgan in late August. Beginning in early September, the teams were moved to the fields in the north and west. They were replaced with teams from around the world. Figures 11.21 and 11.23 show that the number of team days per well actually decreased between August and November from an average of five team days per well to an average of four team days per well.

As illustrated in Figure 11.23, the progress in Burgan was consistent. As can be noted in Figure 11.23, the best progress was recorded during the week ending September 14 when 29 wells were controlled by 10 teams at just over two team days per well. The tougher wells in Burgan were controlled in the latter days of the project.

SAFETY

There were 11 fatalities associated with the Kuwait fires as of the official end of the project on November 6, 1991.

Every fire fighting team was assigned an ambulance and a medic. These men were lifesavers in this harsh environment. They were always present with water and other drinks for the parched workers. In addition, they were able to treat minor illnesses and provide support for medical helicopters. It was routine to have an injured man in the hospital within 10 to 15 minutes of an injury being reported.

The British Royal Ordnance did a marvelous job of clearing the well sites of mines and other munitions. A path and area for the fire fighters were routinely checked. Any explosives too near the flames would have been consumed by the fire. Therefore, the fire fighters were safe from these problems. However, by the end of the project, EOD had lost two men to land mine clearance. One was killed while clearing a beach. The other was killed while working a mine field in the Umm Gudair Field when the mine inexplicably detonated.

The worst accident occurred very early in the project. Smoke from a burning well along the main road between Ahmadi and Burgan routinely drifted across the road. Just as routinely, the crews became accustomed to the situation and regularly drove through the smoke. On this particular day, there was fire in the smoke and in the ditches beside the road. Three service company men in one vehicle and two journalists in another vehicle apparently became disoriented in the smoke and drove into the fire. All five perished. In other instances, one man was killed in a pipeline accident and another in a road accident involving heavy equipment.

Of the fire fighters, the Chinese team had one man severely burned, but his life was never in danger. In addition, at the end of the project, two members of the Rumanian team were badly burned late one evening when the wind died and gas fumes gathered. The fumes were ignited by a hot spot from a previously extinguished ground fire. Both were airlifted to Europe but later died.

CONCLUSION

Pursuant to documents filed with the United Nations Security Council in 1996, the total damage inflicted by Iraq, excluding the value of the lost reserves, was $951,630,871, of which the Al-Awda Project cost was $708,112,779, or over $3 million per day. It was estimated that in the beginning five to seven million barrels of oil per day were being consumed by the fires. More than 10,000 non-military personnel from 40 countries and all continents except Antarctica supported the fire fighters.

It is estimated that the wells emitted approximately 5000 tons of smoke daily in a plume 800 miles long. More than 3.5 million meals were served during the Al-Awda Project.

At the official end of the Al-Awda Project on November 6, 1991, there were 696 wells on the report. The first well was secured on March 22 and the last fire was extinguished on November 6 for a total of 229 days. Remarkable!

EPILOGUE

The Kuwait story began, for me, in June 1990. At that time, I was teaching a Drilling Practices Seminar in London. I had six participants from Kuwait, two of which would later become part of the Kuwaiti Wild Well Killers (KWWK—Kuwait's own fire fighting team during the Al-Awda Project). These were wonderful fellows. We endlessly discussed the situation in the Middle East. It was obvious they were not fans of Saddam Hussein and Iraq.

Shortly after our seminar ended, the Iraqi army invaded their homeland. As Iraqi journals would later reveal, within days of the initial invasion, the Iraqis began to deliberately and systematically prepare to destroy the Kuwaiti oilfields. Explosives were placed around the wellheads and wired to common points in order that one man could destroy many wells with the push of a button. Kuwaitis familiar with the oil fields were forced to help.

One of the major contributors to the planning and preparation for the Al-Awda project was Adel Sheshtawy, a native Egyptian and U.S. citizen. Adel taught Petroleum Engineering at the University of Oklahoma from 1973 to 1978, roughly. In that capacity, he had many Kuwaiti engineers in his classes and those engineers had moved into responsible positions within Kuwait Oil Company (KOC).

In addition, Adel had offered short courses in various petroleum engineering subjects. In one of those short courses, he had the opportunity to meet Saud Al-Nashmi. Ultimately, Saud became responsible for the Al-Awda Project, battling the oil fires of Kuwait.

Soon after the invasion, Saud called Adel and the two met in Houston. Everything was chaotic. There was a presidential directive that no one was to deal with a Kuwaiti or Iraqi company. However, Saud had received reports from drilling operations that were continuing in Kuwait;

explosives were being placed in some of the wells and he was deeply concerned.

Mike Miller, president of Safety Boss, the Canadian well control group, recalled that Adel contacted him in late September or early October 1990 to begin planning for the return to Kuwait. Mike recalls that at that time the estimated maximum number of fires was 50.

For a short period of time, the Kuwaitis made their headquarters at a hotel. KOC had a long association with the Houston-based company O'Brien Goins Engineering (OGE). W.C. "Dub" Goins and T.B. O'Brien were, even at that time, legends in the petroleum industry. Adel and several Kuwaitis subleased office space next to OGE and began planning for the return to Kuwait. One of OGE's employees, Larry Flak, would later become the first coordinator of the Al-Awda Project.

According to Mike Miller, there were at least four meetings in Houston during the fall of 1990, numerous telephone calls, and hundreds of hours of planning. The contracts were in place with the Kuwaiti government in exile in Tief, Saudi Arabia, but were not signed until the ground war broke out. Each group was responsible for bringing everything they needed to sustain a full year of operation. All of the contractors had a substantial degree of independence. The well control company had complete control in preparing the well for the kill operation.

In the fall of 1990, I was managing a relief well for the Mil-Vid #3 blowout near Vidor, Texas, for Amerada Hess. On the evening of November 30, I was visited by Adel Sheshtawy and Fahed Al-Ajmi. Al-Ajmi is now a deputy director of KOC. Adel and Fahed were in my trailer house all evening, and we primarily discussed relief well options. Adel did most of the talking. We discussed state-of-the-art relief well technology. They seemed particularly interested in high angle/horizontal relief wells.

Adel was of the opinion that Iraq primarily wanted two things. The Kuwaitis had been drilling in the southern tip of the Rumaila Field, which is one of the biggest fields in Iraq with only a small portion extending into Kuwait. Iraq felt that the Rumaila Field was Iraqi and resented Kuwait's trespass.

Further, he said that Iraq's goal was to have access to a deep-water port in the Persian Gulf and access had been offered via the Shatt al

Arab—the confluence of the Tigris with the Gulf. For those reasons, Adel believed there would be no war.

Adel went on to say that, although it was generally accepted that there would be no war, a contingency plan had been prepared. It was well known that Iraq had blown up oil wells during its war with Iran, and it was believed that similar operations were planned for the Kuwaiti oilfields. At that time, he did not let on that he suspected the wells were being wired for destruction.

The contingency plan, as Adel described it to me that evening, envisioned Iraq blowing up 120 wells, maximum. Kuwaiti officials in exile had contracted with four well control companies to furnish two teams each. Each team was to consist of a minimum of four men including a team leader. The companies were to furnish all the equipment needed to sustain a year-long fire fighting effort. The four companies were Red Adair, Boots & Coots, and Wild Well Control—(all from Houston)—and Safety Boss, the Canadian company.

From the beginning, and even during the planning phases, there was considerable resentment on the part of the Houston-based groups toward the Canadians. Although Safety Boss had been in business longer than any of the Houston companies, the Houston companies were much better known worldwide, thanks largely to the showmanship of Red Adair.

The Houston groups were all from the same gene pool. The Canadians had different ideas and presented a different perspective. As we shall see, the Canadians were unencumbered by the conventional Houston wisdom concerning fire fighting and, as a result, were much more mobile and efficient in the Al-Awda Project. This mobility and efficiency were the primary reasons that the Canadians handled almost 40 percent more fires than any of the Houston companies.

On January 15, 1991, Adel's prophecies were disproved when Iraq failed to withdraw and the war began. By the end of the war on February 23, the withdrawing Iraqi forces had blown up more than 700 oil wells, far more than the contingency plan had envisioned.

The Kuwaiti government in exile immediately set about implementing their contingency plan and working frantically to expand it to include anyone and everyone that could help.

I received numerous clandestine calls in the middle of the night from someone claiming to be a member of the royal family of Kuwait or a special friend of the royal family with the authority to gain an audience and a contract. Everyone wanted to be my agent or special contact for Kuwait. Some even wanted money upfront to make the right connections. It was crazy.

The problem was basically this. The well control industry had never been large for the simple reason that there were never many blowouts and oil well fires. Myron Kinley dominated the business single-handedly until the late 1950s, and Red Adair continued that monopoly until the late 1970s.

The boom of the late '70s and early '80s created a demand for well control services that exceeded the supply. This excess demand resulted in a window of opportunity. In the late 1970s Adair had a dispute with his two right-hand men, Boots Hansen and Coots Mathews, and they quit to form Boots & Coots. About the same time, Joe Bowden started Wild Well Control. In this same general timeframe, Bob Cudd left Otis and formed Cudd Pressure Control.

The new companies were generally staffed with men that had worked for one of the older companies. One exception was Bob Cudd and Cudd Pressure Control. Although best known for his expertise in snubbing, Bob had considerable previous experience in oil well fire fighting and capping operations.

Safety Boss was a Canadian firm established in 1954, which operated worldwide and was owned and operated by my old friend Mike Miller. However, within the industry at large, little was known about their operations. Safety Boss was started by Mike's father and was first active in lease fires. The company used more conventional fire fighting techniques, including specially designed fire trucks with an ordinary appearance.

Each of the four firms selected as a part of the contingency plan had a small number of employees capable of managing a blowout. Pursuant to the contingency plan, each firm was to supply two (later three) full-time teams, which, in reality, meant four teams (later six) working 28-day rotations. Each team was to be composed of a minimum of four experienced men. That meant that each firm was to supply 16 (later 24) experienced fire fighters. The branch of the industry that provided emergency fire fighting services was not that large and few personnel were available.

As a result, most of the fire fighting companies were scrambling for warm bodies to fill the team requirements. The mystic veil surrounding oil well firefighting had been removed. There were roughnecks, engineers, snubbing hands, work-over hands, school teachers, and bartenders, to name a few, working as fire fighters. The bottom line was that most of the well fires in Kuwait were fairly simple and almost anyone could do it.

OGE was to provide the well control engineering. That meant that each fire fighting team was assigned an engineer from OGE. The fire fighting team was to extinguish the fire and cap the well. Then, the well control engineer was to kill the well and secure it for later recovery.

Another key organization was Santa Fe Drilling Company headquartered in Santa Fe Springs, California. Kuwait Oil Company had purchased Santa Fe Drilling Company in 1981 and had confidence in its management. Santa Fe Drilling was to provide support services for the fire fighting effort. Specifically, each fire fighting team was assigned a team of four or more heavy equipment operators and roustabouts. Ultimately, all teams would have backhoes, forklifts, and dozers at their disposal. The Santa Fe men would operate the equipment.

Bechtel, the large construction firm, had overall responsibility for providing support for the fire fighting effort. Specifically, Bechtel was to provide housing, heavy equipment, cars, etc. From my perspective, two of the biggest areas of responsibility for Bechtel were water for fire fighting and civil works for roads and road-building material.

Overall coordination and responsibility remained with officials of KOC. The primary burden rested upon Saud Al-Nashmi, who was a relatively young but very capable engineer.

When the war ended and the extent of the damage was revealed, the Kuwaitis were anxious to implement their contingency plan and to further expand the plan to include anyone else capable of contributing to the fire fighting effort. The aforementioned infrastructure was not then well known.

After scores of unsuccessful leads and hundreds of frustrating telephone calls, I finally learned that Ahmed Al-Khatib, a Palestinian national I believe, was the proper contact person within Santa Fe Drilling Company's office in Santa Fe Springs, California.

I contacted him, and it was clear that I was on the right trail. He told me that the Kuwaitis were currently looking for others to work in Kuwait on the well fires. He said there was a committee that screened everything and that the committee was moving to London. He went on to say that those interested were to send their information to him and he would pass it on with comments. After that, if the applicant was judged qualified, he would be invited for an interview.

I informed my old friend Bob Cudd, and we sent our information. Very shortly, an interview was arranged and Bob and I went to California. On Tuesday, March 12, 1991, we met with Ahmed Al-Khatib and John Alford at Santa Fe Drilling Company's office.

We were informed that the situation in Kuwait was much worse than expected and that KOC was anxious to identify everyone capable of contributing. They were particularly interested in identifying additional well control engineering services as well as oil well fire fighting capabilities.

It was in this timeframe that I first heard about Neal Parry. Neal had begun his career with Santa Fe Drilling Company as a driller and had retired in 1988 as executive vice president, after 36 years of service. In his various capacities, he had spent considerable time in Kuwait and had gained the Kuwaitis' confidence. I would later get to know, work for, and greatly admire Neal. In a recent interview, Neal commented that his job in Kuwait was the highlight of his career. That comment from a man who ascended from driller to executive vice president of a large company speaks volumes about the Al-Awda Project.

Mr. Al-Khatib told us that we had just missed Neal. He had left California to join the committee in London. That committee was to consider new technology as well as additional services.

On March 28, Neal called from London. He told me that Bob Cudd and I were to go to Kuwait and survey the situation. Since the airport had been bombed out and there were no commercial flights to Kuwait, we were to travel to Dubai and then go by charter to Kuwait City. We would spend a few days in Kuwait and then return to London to meet with the committee. The committee shared an office with Santa Fe Drilling Company near Victoria Station in London.

Bob and I left for London on April 3. We went to Santa Fe's London office and met with the committee, which consisted of Ahmed Al-Awadi,

Farouk Kandil, Mustafa Adsani, and Neal Parry. Jim Dunlap assisted Neal. We got our first insight into the conditions in Kuwait but nothing could have prepared us for what we were about to see.

On Tuesday, April 9, Bob and I traveled to Dubai and spent the night there. The next morning, we boarded an Evergreen Charter for Kuwait City. As we approached Kuwait City, the smoke from the fires became clearly visible. I had seen many oil well fires but no one had seen 700-plus oil well fires packed into an area roughly 40 miles long and 40 miles wide.

I'll never forget landing at the airport in Kuwait City and seeing the devastation. There was a big bomb crater in the center of the runway. There was smoke and fire everywhere. Stan Petree, drilling superintendent with Santa Fe, met us and escorted us to our quarters. For the next several days, we toured Kuwait.

During those days, I took hundreds of photographs and hours of video. Even with all the photos and all the words in the English language, I am unable to describe what I saw. The worst was in greater Burgan Field, where the well density is the highest (Figure 1).

Figure 1

Figure 2

Greater Burgan is the oldest and largest field in Kuwait and one of the largest oil fields in the world. It is the crown jewel of Kuwait. The small city of Ahmadi, on the north end of Burgan, is the center of the oil field community. The Burgan highway runs south from Ahmadi to GC-1 (Gathering Center One) and the Emir's Desert Palace.

In the middle of the Burgan field immediately adjacent to GC-1, the Emir's family had built a beautiful, luxurious palace complete with gardens, wild animals, and some of the finest horses in the world.

This symbol of Kuwaiti sovereignty at the end of the Burgan highway now lay in ruins among the fires (Figure 2). My first trip down the Burgan highway was sobering. The sun could not be seen. It was black as midnight. Oil rained down on our vehicle, reducing visibility to a few feet. There were oil wells on fire and ground fires everywhere I looked.

For most of the way, oil was halfway up our tires. I don't know why the heat from our SUV's exhaust didn't ignite the oil beneath us. I would later learn that the oil in Kuwait contains lower-end hydrocarbons and has a high ignition point—a fact that served us well in the fire fighting effort.

Figure 3

All along the highway from Ahmadi to Burgan, fires raged on both sides of the road. The smoke was so thick in places that I couldn't see the front of the car and couldn't tell we were on the road. I don't mind admitting it was a little scary. We turned around at the gathering center, or what was left of it. All of the gathering centers were completely leveled (Figure 3).

Later I quizzed members of our military about the destruction of the gathering centers, and they were just as puzzled. One officer commented that evidence indicated intense heat. He was not sure whether the damage had been done by the allies or the Iraqis, but it looked almost nuclear to him.

The Iraqi destruction was senseless. They had sacked and destroyed the KOC headquarters in Ahmadi. They had stolen or destroyed every tool and vehicle in the Kuwaiti oil fields. Other than the equipment recently brought in by Bechtel or the fire fighters, I did not see and could not find a piece of equipment that was operational. I did not even see any hand tools—no hammers, screwdrivers, pliers, nothing.

On the lighter side, the camps housing the fire fighters were fun. Those guys had gathered enough munitions to supply a small army!

There were AK-47s, antitank guns, land mines, and every kind of bomb imaginable. I thought it a miracle that they didn't blow themselves up.

I went to one meeting conducted by our army. The topic for discussion was the various land mines we could expect. As the military ordnance expert showed pictures of the various mines and explained how dangerous they were, I could hear someone in the background telling his friends how he had two or three of those in his room.

In one instance, a crew decided to try out an Iraqi antitank weapon. The door was open on their pickup and the brave soul holding the weapon was standing in front of the open door aiming at an abandoned tank. He fired and his aim was true. However, the backfire from the weapon opened a neat hole in the door of his vehicle. The frightened crew waited until dark, returned to Ahmadi, got a backhoe, and buried the vehicle! The team leader later found out about the incident and reimbursed KOC for the vehicle.

On this trip, I met on numerous occasions with Miles Shelton, the local manager for Santa Fe Drilling Company. At that time, I got the distinct impression that Miles, and perhaps others, envisioned Cudd Pressure Control and the other newcomers working in the north of Kuwait, independent from the operation in Ahmadi. Santa Fe would supply support for the fire fighters as Bechtel was doing in the south and GSM, my company, would provide the well control engineering. Certainly, that was consistent with the impressions I got in California. I heard that scenario on many occasions from different reliable sources, but, personally, I don't think it was ever anything the Kuwaitis seriously considered.

On April 16, Bob and I returned to London. We met with the committee the following day. We made our presentation and were told that it was just a matter of time before we would be working in Kuwait. On April 20, we returned home to wait.

On May 9, I went to Trinidad to fix a blowout there and finished the job on May 17. Just as that job was finishing, I got a call from Bob Cudd that we had been summoned to London to negotiate the contracts. Consequently, I went directly from Port of Spain to London.

Around June 1, we returned to the U.S. to wait. It seemed as though everything was moving in slow motion during that time. Bechtel could not or would not supply the teams working in Kuwait. Therefore, there was no need to bring in more teams.

In the meantime, I got a call from Ludwig Pietzsch in Germany. His group was interested in contributing in the new technology arena. At the end of June I went to Frankfurt to meet with Pietzsch and his group. They had very interesting ideas involving laser-guided cranes, advanced jet cutting, and optics which could see through the smoke and fire for a better look at the well heads. I continued to work with Pietzsch and his group. Unfortunately for them, the industry, and all of us, the project had been completed before any of their ideas could be tested. Undoubtedly, some would have made significant contributions to fire fighting technology.

On July 29 things began to happen. I was notified that Neal Parry had officially been named the new coordinator under Saud Al-Nashmi. Bechtel was behind, but the new teams were on their way. In addition, the international teams would be added. Bechtel was to supply the commercial teams and the international teams would supply themselves. The British were to be in charge of all the work in Sabriyah, including the fire fighting, and the French were to be working in Raudatain. Of course, neither the British nor the French companies had any fire fighting capabilities. Consequently, they hired their hands off the streets of Houston or wherever they could. Nevertheless, they did some good work.

On August 5, Neal Parry called and asked me to come to Kuwait to work for him on his management team. He said that he didn't claim to know anything about fire fighting, but he did know how to get a job done. He said that I would be in charge of the field operations, and he would get me whatever I needed. He went on to say that the bad blowouts were ahead of us. He wanted the fire fighters to cap and kill wells as fast as possible. He did not want any of the fire fighting teams to bog down on one well. If a well proved too difficult, it was to be passed and left for later.

By that time, the wells around Kuwait City had been capped. The greatest challenge was Burgan Field. In Burgan, the wells were very close together, the smoke was the thickest, oil lakes were everywhere, and ground fires raged. Burgan was to be my responsibility.

It was recognized that the vast majority of wells would be classically extinguished, capped, and killed. However, it was generally believed at that time that there would be many wells that would require much more time, energy, and effort. Some, it was thought, would require relief wells. In the event that wells of that nature were identified, it was my job to plan the well control strategy and ultimately conquer the last wells.

I asked Neal how long the project would last. He told me to plan on staying five years and taking a field break every 90 days. On Saturday, August 10, 1991, I kissed my family goodbye and left for Kuwait. I arrived in Kuwait City on August 12.

My room was in the Ahmadi House. It had been used since March by the fire fighters, and the floor and rug were black from oil, mud, and soot. My first impression was that the project was in chaos. There was no coordination or organization. Everyone was running around like a bunch of loose cannons. In general, the fire fighters simply worked on the wells they wanted to work on, when and where it pleased them.

There were several different fire fighting strategies. The disciples of Red Adair operated in essentially the same manner. They would set at least two large, skid-mounted, diesel-powered centrifugal pumps on the edge of the 25,000 barrel lagoon. An eight to twelve-inch flanged line was run to a staging manifold. The fire monitors were shielded using reflective tin. All of the plumbing was screwed or bolted together. It was very time-consuming to move in on a fire. Some, like Cudd Pressure Control, used irrigation piping that was lighter and more easily assembled. Everything had to be loaded on trucks to be moved very far.

Safety Boss, the Canadian Company, used specially designed fire trucks. Their water was hauled by transport and stored in 500-barrel frac tanks. The Canadians wore fire-retardant coveralls and bunker suits when necessary. As a result, they routinely worked closer to the fire and used less water. They used conventional fire hoses to connect their pumps to their monitor houses. As a result, they were very mobile and flexible. They could drive to a fire, roll out their hoses, and be working in a manner of minutes. At the end of the day, they would roll up their hoses and drive the truck to Ahmadi. In addition, they used more fire-retardant chemicals than the Houston-based fire fighters.

There were advantages to their mobility and flexibility. When the wind shifted, the exposed equipment would be soaked with oil. As a result, it could not be operated until the steam cleaners had removed the oil. Since the fire trucks were driven to town each evening and washed if necessary, they were ready to go to work the next morning.

Their mobility was an advantage in some instances. In one instance, using the existing pipelines, I simply couldn't get sufficient water to fill the lagoon at a well on the far west side of Burgan. I asked Mike Miller and his Safety Boss crew to fix it. It was fixed in less time than it would have taken the conventional fire fighters to unload their trucks. In my opinion, these advantages were primarily responsible for the fact that, although the Canadians started several weeks later than everyone else, they accounted for 40 percent more wells than any of the other original four. The Canadians, as a result of their effort, and Cudd Pressure Control, as a result of the dedication of Bob Cudd, were the only companies retained after the Al-Awda Project for the post-capping operations.

The basic strategy for most of the teams was the same. They would work to get the fire burning straight up and then extinguish the fire. After the fire was extinguished, the well would be capped.

The international teams used a variety of techniques. The Eastern Bloc countries such as Romania, Hungry, and Russia used fire trucks, powder trucks, and jet engines. Typically, one or more jet engines were mounted on a truck or tank and would be used to literally blow the fire out. The Hungarian "Big Wind" was the most spectacular (Figure 4).

Figure 4

I routinely kept the newcomers on the edge of Burgan and out of the way until they gained experience and demonstrated their abilities.

When I got there, the Houston-based fire fighters were working primarily in the areas around Ahmadi. The Canadians were working in the south of Burgan. As the prevailing wind was from the north, the Canadians were always working in the thick, black smoke while the Houston group was working in the sunshine. The resentment and discrimination only served to inspire the Canadians.

As I got organized, it was abundantly clear that the biggest problem was the water supply. Bechtel was a zoo. They had notebooks of organization charts. On one of the first days, I went through the organization charts trying to find out who was responsible for the water supply. I even went to Bechtel's headquarters and went from office to office. No one seemed to know who was in charge of anything. Most were sitting around reading a newspaper. On that day, I was unable to determine who was responsible for the water supply. It was frustrating.

A little later, I was in the field and met a very pleasant New Zealander named Randy Cross. Randy was very capable and seemed to have field responsibility for supplying water and locations. I wasn't sure what his job was but I soon learned he could get a lot done. He did a remarkable job.

The decision had been made to use the existing pipeline system to pipe water from the sea to a lagoon at each well. That was not the best idea. With all of the bombs that had been dropped, the pipelines were in poor condition. When we turned the water on, it looked like a giant sprinkler system. Randy maintained that it would have been more efficient to build a new system.

I learned that Randy had a big and vital job. He was responsible for restoring the roads, preparing the location, digging and lining the lagoons, and filling them with water. His job was to stay ahead of the fire fighters. My challenge and motto became "Catch the Kiwi!"

As we worked into Burgan, Randy was having difficulty keeping ahead of us. He came to me one day with a problem. He needed to open a road into the heart of Burgan. He needed the road in order to build a material pit for roads and locations. He also needed it for access to the pipelines.

Since I needed someone with mobility, I turned to my old friend Mike Miller with Safety Boss. We surveyed the road Randy wanted opened. There were oil well fires, ground fires, oil, and smoke everywhere. Many of the oil well fires were immediately adjacent to the road Randy wanted opened. Accepting the challenge would mean Mike and his men would have to work in the raining oil with fires all around them and in thick smoke. Mike never hesitated for a second. He said that if it had to be done, he would do it.

Mike and his men went to work. They worked on some tough fires under the most adverse conditions. Their equipment was uniquely suited for this type of operation. They could drive their fire trucks to the location, set up, and be spraying water on the fire in about an hour. At the end of the day, they could roll up their hoses and drive their equipment to Ahmadi. A minimal amount of equipment was left unattended and exposed. The road was opened in what seemed like record time. I visited their operation daily. It was a miracle to me that they could continue—but they did.

If there were any heroes in Kuwait, Randy Cross would have to be at the top of the list. The recovery effort in Burgan was organized such that the fire fighters had the wind to their backs. That meant that Randy and his men had to work in the smoke and ground fires rebuilding roads, preparing locations, digging pits, and restoring pipelines.

The smoke was so thick that visibility was zero every day. The oil rained down so badly that I had to have my car washed almost daily. Ordnance had to be cleared. Randy did his work using Filipino labor. They hauled 1500 loads of road material each day and sometimes as many as 300 loads a night.

On a list of heroes, the men clearing ordnance would also have to be at the top. The British Royal Ordnance cleared in Burgan. There was only one way to do it. They had a seat on the front fender of several Land Rovers. They would mark off grids and then drive the grids. Those poor guys would show up at the mess hall for lunch, and their vehicles would be black with oil. They would be soaked from head to foot. I don't know how they survived. Some of them didn't. Some of them were killed by land mines.

Royal Ordnance would clear and mark a road and location. It was best to stay within the perimeter as marked. There were unexploded bombs and mines everywhere in the desert. That more were not injured or killed is a testimony to the good work of these guys. I know that often when the ground fires broke out in a new area around a well, it would sound like the 4th of July.

When an ammunition dump was located, ordnance would call on one of the explosives experts with the fire fighters to light it up. They would come around and advise that there was going to be an event at a particular time and everyone in the vicinity was to take cover five minutes prior to the scheduled event. One of those poor guys had to remain within 100 yards of the scheduled event to make sure no one inadvertently wandered into the area. When one of those dumps went up, it was spectacular. I don't know how the watchman survived. It would shake the ground for miles (Figure 5).

After about one month, Bechtel opened up the Latifa Towers, and all of the fire fighters moved into them. The Latifa Towers were two 17-story apartment buildings overlooking the Gulf. Bob Cudd and I shared

Figure 5

a three-bedroom suite on the 16th floor of one of the towers. We had news channels and movies. A large mess hall was built across the road and we ate there.

Shortly after arriving, I felt we would be finished by Easter, 1992. A little later, I told my wife I'd be finished by Christmas. After a few more days, I thought we would be finished by Thanksgiving, 1991. It was simply a matter of getting organized and going to work.

The next few months were a blur. We got up every morning at about 0400, had breakfast, and went to work. We finished about 1700, went to the Towers, showered, had dinner, went to bed, and did it all over the next day. Pretty soon, I didn't know what day it was or even what month it was.

Neal was true to his word. After he got there, we got everything we needed. He truly knew how to manage a project. Oil well fire fighters are well known for their egos and independent ways. I don't think anyone else could have commanded their respect and kept them moving together in the right direction.

By the middle of October it was apparent that the project was progressing more rapidly than anyone had imagined. At this point, I was fearful that someone would become over-confident or careless and make a fatal mistake. To that point, there had been no serious injuries and we were determined to maintain the high safety standards.

One of the most dangerous times in the fire fighting was when the wind changed direction. The vast majority of the time the wind blew strongly from the north. Occasionally, the wind direction would shift from the north. On those occasions, the wind just seemed to swirl around without direction. Consequently, the oil would swirl, soaking the ground and all of the exposed equipment.

Late one afternoon, the north wind died. We all knew what that meant. The Rumanian team had been working on ground fires. When the wind died, they were in the process of shutting down for the night when suddenly a ground fire backfired and engulfed one of their crews. Two of their men were badly burned.

Our safety net was excellent. We were able to transport anyone who was injured to the hospital in Ahmadi in less than 15 minutes. I recall that when I visited the injured Rumanians, their faces were badly burned

Figure 6

and swollen. They were stabilized and air-lifted to Europe. Unfortunately, both men later died as a result of their injuries.

Things did continue remarkably well. I reported to Neal on November 2, 1991. "It is my pleasure to inform you that there are no fires burning in the Burgan Field!" We finished the project and scheduled the closing ceremony for November 6, 1991. Together with the Kuwaiti fire fighters, BG 118 was equipped to be the "last" fire in Kuwait. The Emir would extinguish it himself at the closing ceremony. A platform was built about 100 yards away from the burning well and a special control panel was installed for the Neal told me that when the Emir flipped the switch, that fire had better go out or my butt was toast!

The closing ceremony was a sight to behold. Huge tents (Figure 6) were set up in the desert to house the many dignitaries who came from everywhere. There was an abundance of food, music, and dancing. The Kuwaitis were dressed in their finest traditional attire (Figure 7)—an interesting sight to us Westerners. There were numerous speeches and many ceremonies. Finally, the Emir made his appearance. All of us in the fire

Figure 7

fighting got to pass by for review and shake the Emir's hand as a token of his gratitude.

All the while BG 118, burning in the background, was a focus of attention. Then, the final moment came and the Emir stepped to the platform. I held my breath as he reached for and flipped the switch. The last fire in Kuwait went out without a whimper. Thunderous applause accompanied loud cheering. The Al-Awda Project was finished.

Fortunately, everything had gone better than planned. I returned home just before Thanksgiving, 1991. Hopefully, it was a once in a lifetime experience—perhaps even a once in the history of the world event. It was a wonderful example of what men from many different cultures can overcome when they work diligently together. We did a good job and I'm proud to have been a part of it! I wouldn't have missed it.

INDEX